Chemistry and Chemical Reactivity

SEVENTH EDITION

John C. Kotz
State University of New York, Oneonta

Paul M. Treichel
University of Wisconsin, Madison

John Townsend
West Chester University of Pennsylvania

Prepared by

Roxana V. Allen
St. John's School, Upper School, Cypress Texas

BROOKS/COLE
CENGAGE Learning

Australia • Brazil • Japan • Korea • Mexico • Singapore • Spain • United Kingdom • United States

ISBN-13: 978-0-495-55739-5
ISBN-10: 0-495-55739-0

Brooks/Cole
10 Davis Drive
Belmont, CA 94002-3098
USA

Cengage Learning is a leading provider of customized learning solutions with office locations around the globe, including Singapore, the United Kingdom, Australia, Mexico, Brazil, and Japan. Locate your local office at: **www.cengage.com/global**

Cengage Learning products are represented in Canada by Nelson Education, Ltd.

For your course and learning solutions, visit **www.cengage.com**

Purchase any of our products at your local college store or at our preferred online store **www.ichapters.com**

READ IMPORTANT LICENSE INFORMATION

Printed in the United States of America
1 2 3 4 5 6 7 12 11 10 09

To Mike, Christian and Russell

Table of Contents

Preface

This AP® Instructor's Guide was written and compiled from various *Chemistry and Chemical Reactivity, 7th Edition* (by Kotz/Treichel/Townsend's) supplemental materials and information obtained at the College Board website (www.apcentral.collegeboard.com) to help AP® instructors prepare their students for the AP® Chemistry exam. This book is not intended to replace the curricular materials available on the College Board website. It is intended to enable teachers to better correlate the AP® course expectations with the text.

Each chapter includes the following elements:

- A detailed chapter outline

- Chapter objectives

- A lesson plan, which provides:

    o A correlation to the AP® Chemistry syllabus

    o Suggested time for each chapter

    o Pointers to the answers to exercises, as well as solutions to the conceptual questions, general questions, and practicing skills in the Instructor's Resource Manual (IRM) and Student Solutions Manual (SSM).

    o A pointer to the test bank for that chapter

- A chapter glossary

- A media menu which lists the multimedia resources available to instructors using *Chemistry and Chemical Reactivity, 7th Edition*

    o Exercises

    o Tutorials

    o Active Figures

    o Additional Resources

- Diagnostics quiz questions mapped to the chapter goals

- A test bank with answers

Before the chapter-by-chapter materials are introductory essays which will guide AP® instructors on making AP® accessible, preparing an effective curriculum map, preparing students for the

AP® exam, and utilizing library resources. Additionally, there is information on submitting an AP® syllabus to College Board.

Making AP® Accessible

Though its Equity and Access Initiative, the College Board believes that all students should be able to receive an education that is rigorous and challenging. As a result, it advocates an "open-door policy" for AP® enrollment. This policy encourages schools to end certain practices for AP® eligibility, such as GPA requirements, "honors" or "gifted" student status, and teacher nomination.

A roadmap to a strong curriculum
Although the College Board's initiative intends to open doors to AP® enrollment, it is important that students enrolling are prepared to walk through. For this preparation, students must be introduced to the rigors and personal requirements of an AP® course. The AP® Chemistry Course Description is updated biennially and will reflect current test concepts and percentages. It is anticipated that the test will undergo major revisions in the 2010 school year which will be presented on the 2011 AP® Chemistry exam.

Curriculum mapping
Many school districts have curriculum map requirements. A district-wide curriculum map for K-12 is an excellent tool for identifying where, when, and how content and skills are taught. If your district's curriculum is limited to your state's standards and assessment requirements and does not include AP®, you can work with other teachers within your school and district to enhance your existing maps or to create ones from scratch. Once you have captured the skills and content for your curriculum map, you can determine which areas of instruction need to be modified or enhanced to provide the prerequisites for AP®.

The AP® Course Audit
The AP® Course Audit was created at the request of secondary school and college university members of the College Board who sought a means to provide teachers and administrators with clear guidelines on the curricular and resource requirements that must be in place for AP® courses. The AP® Course Audit also helps colleges and universities better interpret secondary school courses marked "AP®" on students' transcripts. To receive authorization from the College Board to label a course "AP®," schools must demonstrate how their courses meet or exceed these requirements, which colleges and universities expect to see within a college-level class.

The AP® program unequivocally supports the principle that individual schools must develop their own curriculum for courses labeled "AP®." Rather than mandating any one curriculum for AP® courses, the AP® Course Audit instead provides each AP® teacher with a set of expectations that college and secondary school faculty nationwide have established for college-level courses. AP® teachers are encouraged to develop or maintain their own curriculum that either includes or exceeds each of these expectations; such courses will be authorized to use the AP® designation. (From the 2008 College Board Chemistry AP Course Description)

Introduction

AP Vertical Teams®
AP Vertical Teams® are teams of teachers at different grade levels (usually high school and feeder middle schools) who create a scope and sequence of instruction that incrementally prepares students for AP® courses. The teams align the curriculum to the AP® standards and provide preparation for the requirements of AP®. Students learning the curriculum are also given instruction in and demonstration of necessary time managements and study skills.
College Board AP Vertical Teams® are also supported by Pre-AP®. This program provides professional-development resources and support for teachers preparing students for college and AP®. More information about this program and AP Vertical Teams® can be found at the College Board's website; www.apcentral.collegeboard.com.

The AP® Exam
One of the challenges you face as an AP® instructor is that of getting all your students to take the AP® exam. For various reasons, even after a successful performance in an AP® course, some students are reluctant to take the AP® course exam. Listed below are common reasons students give for not taking the exam and strategies you can use to overcome them.

"I'd like to but we can't afford it."
Reassure the student that there are ways to reduce or even eliminate the fees associated with the AP® exam. The College Board and many state and federal grants provide financial aid for students who cannot afford to take an AP® exam. Because ability to pay is a personal issue for many students, you should be prepared to find out if your school, district, or state participates in a grant program and advise students about these programs, without calling attention to their situation.

"I don't think I'd get anything out of it."
Direct students to this College Board website: www.collegeboard.com/apstudents/. On this page is a "College Finder" with a number of informative features, including a list with hyperlinks, of every institution that offers college credit for specific AP® scores. Although these schools provide credit for AP® exam scores, it is unlikely that they will provide credit for taking the course alone.

Encourage your students to check the schools that interest them and see how much credit they may be able to obtain. Also inform them that college credit awarded as a result of AP® exam scores, saves time and money and provides an early start for their college career.

Even students who are not going to college immediately can benefit from the enhancement that high AP® scores give to their high school transcripts.

"It's too hard."
Many students who do perfectly well in class are intimidated by the AP® exam. Provide all students with plenty of classroom test exposure and practice. Work with other AP® instructors to offer additional tutoring/test preparation sessions after school or on Saturdays. Above all, provide encouragement to all students and fortify them with test-taking strategies.

"I'd rather take the course again in college."
This is certainly the student's (or parents') prerogative. However, tell students that a good AP®
exam score could provide cost-saving benefits and access to courses that otherwise would have
to wait until students fulfill prerequisites. Finally, many degree programs today take longer to
complete than the traditional four years. Completing the AP® exam can be a step toward this
goal.

Other special needs
Testing accommodations are available for students with special needs. Information can be found
at www.apcentral.collegeboard.com. Talk to your school's administration to determine your
facility's level of compliance.

What is a Curriculum Map?

A curriculum map is essentially a scope and sequence of what students should know, be able to
do, or be taught at each grade level for a specific curricular discipline. Maps can provide an
expanded view of your entire school district at all grade levels, or they can focus on benchmark
grade levels (elementary, middle, or high school) for a particular subject area.

Curriculum maps are usually structured so that you can identify what is needed for student
scholastic achievement. A good curriculum map should;

- Include essential content and skills that are to be taught at each grade level or in
  courses for your school or district.
- Show how content and skills are taught in all classrooms across grade levels for a
  specific discipline.
- Indicate how the students' achievement or acquisition of knowledge and skills is
  assessed, including formal and informal assessment activities.
- Be in an easily accessible visual format that can be used to monitor the curriculum
  taught in your school or district.

Introduction

Sample Curriculum Map

Subject____

School____

Grade Level(s)____

Content/Knowledge	Skills	Instruction	Assessment
• What students should know	• What students should be able to do	• Multimedia presentation • Assigned reading • Investigations	• Formal Assessment -Unit Tests -Chapter Tests • Informal Assessment -Group presentations -Investigations

Overall, a curriculum map should provide a complete view of what students will be learning throughout a specified portion of their academic career. The map enables you to structure your instruction so that concepts and skills taught at previous grade levels are not repeated in subsequent grade, rather they're reinforced or expanded upon. Valuable time can then be committed to providing instruction for the next level of necessary skills and concepts.

How do I know which concepts and skills should be included in a curriculum map?
Almost every state has established curriculum standards to assure greater accountability and scholastic achievement. These standards are the guiding principles for the concepts and skills that all students must learn. In addition, many school districts prescribe curriculum standards that align to state objectives.

In response to No Child Left Behind (NCLB), states have also created standards for assessment. Because these standards are usually essential for student promotion and/or graduation, they too should be included in curriculum maps.

Mapping Advanced Placement (AP®)
Like state and district requirement, the College Board also has student expectations for each of their course offerings. These expectations can be found in the content outline for the course(s) you teach. AP® Course Descriptions are available for free downloading at www.apcentral.collegeboard.com.

To help ensure that students are prepared for the rigors of AP® courses and exams, an alignment of AP® course outlines to a curriculum map can allow you to see if students' previous course work provides sufficient instruction and preparation. An alignment can be created by comparing commonalities between your state and/or district curriculum and AP® course outlines.

Using teaching materials to build curriculum maps

The chapter lesson plans found in your AP® Instructor's Guide are quick, one-stop sources for building curriculum maps for your AP® course. Each plan is grouped by chapter and aligned appropriate expectations of your AP® course outline. You are informed of essential instruction and instruction that enriches learning after students have completed the AP® exam. The plans list all the ancillary resources and activities needed to ensure thorough, sound instruction. References for formal and informal assessment options can also be found.   To assist you with curriculum mapping, each lesson contains:

Content/Knowledge	Skills	Instruction	Assessment
• Chapters aligned to AP course outlines • Objectives listing the concepts and skills to be taught in each chapter • Chapter glossary	• Chapters aligned to AP course outlines • Objectives listing the concepts and skills to be taught in each chapter	• Suggested time for each chapter (for teachers on tradition A/B block schedules) • References to the group activities for that chapter	• References to conceptual questions, problem sets, and quizzes • References to the test bank

When you use your teaching resources to build your curriculum map, it becomes more than a device that captures the curriculum taught in your school or district; your map is now a tool to be used for everyday classroom instruction.

Preparing for the AP® Exam

Your goal as a teacher is to ensure that your students acquire the knowledge and skills they will need in future academic careers and life in general. As an AP® instructor, in addition to striving to meet this goal, you must prepare your students for your course's AP® exam.

Introduce the AP® curriculum and exam to your students

During your first classroom session with your students, review the AP® course outline provided by the College Board, your syllabus for the year (or quarter or semester), and the format of the AP® exam.

Introduce to students the types of questions they will encounter on the exam. Let them know that in addition fo multiple-choice questions, they will have to answer free-response questions.

Create AP® exam conditions

Some students may have never taken an exam quite like an AP® exam. It could be the longest test they have taken up to this point in their academic career. Because students may have never had to write timed problem-solving responses, they may be fearful of the exam. Your task as an

Introduction

AP® instructor is to create an environment that prepares and provides students with the self-confidence to take a test as rigorous as an AP® exam.

Practice Quizzes and Tests

Whenever possible, create tests that conform to the AP® model:
- Five-response multiple-choice items
- Multi-part free-response questions

With exposure to practice quizzes and tests similar to those on the AP® exam, students can see the types of questions they are likely to encounter and can use the thinking skills needed to answer each question.

Free-Response

Because the writing ability of students varies, free-response questions often cause them the most anxiety. Early in the course, students can benefit from class discussions and activities that let them apply problem-solving skills and other strategies to their writing.

Throughout the year, assign problems from unit quizzes and tests and group-based problem/solution projects.

When grading students' writing assignments, use rubrics that model those of the AP® exam review panel. Provide thorough feedback and praise good work. You may also want to use AP® scale when grading, instead of a traditional scale. If you decide to use the AP® scale, suggest additional information that would garner, for example, a *4* instead of a *3* on the real AP® exam.

Practice Exams

So that students can truly experience the exam, you should administer a full, mock AP® exam at least once before students take the real exam in May. You can create your exam using questions from your test bank or from one of the released exams on the College Board website: www.apcentral.collegeboard.com.

Depending on the length of your class periods, you may have to administer the mock exam outside of school hours. You can also divide the exam into sections over more than one day. When scoring the exams, use the rubrics provided by the College Board and give students feedback as soon as possible, especially if this is their last chance before the real exam is administered.

Strategies for taking the AP® exam

Many students become very nervous and anxious about taking an AP® exam. They look to you, as their instructor and coach, for help. Before test day, review with students the sound test-taking strategies listed below. You can also find "Exam Tips" on the CB website.

Before Test Day…

- Study for the exam. This may seem like common sense, but this is something that you need to do throughout the year. The week or night before the exam is not the time to cram an entire course of information in your head.
- Make sure that your calculator is one that is acceptable for use during the AP® exam. Also, know how to use the functions on your calculator.
- Familiarize yourself with the constants, conversions, formulas, and equations that you may need for the AP® exam. You will not need to memorize these values because they are provided during the free-response portion of the exam, but you should understand how to use them.
- Get a good night's sleep the night before and eat a good breakfast on the day of the exam

Taking the Test…

- Read directions carefully. Understand how and where you should respond to the question. Make sure that you are providing what the question asks for. For example, if a question asks you to justify your response, you will need to provide evidence to support your answer.
- Read questions carefully, making note of key words and terms such as "NOT," "EXCEPT," "TRUE," and "FALSE." Overlooking these terms can result in zero credit for an otherwise well-written answer.
- Use estimation strategies to check the reasonableness of your answers. This will allow you to eliminate answer choices when answering multiple-choice questions and double check your calculations for free-response questions.
- When answering free-response questions, show ALL your work. This way, if you are unable to completely answer the question, you can receive partial credit for what you do get correct. If you continue your answer onto another page, provide clear information for the rest of the answer's location.
- If you get "stuck" on a question, skip it and answer the questions you know. If there is time, you can come back to the question later.
- For multiple-choice questions, be sure to fill in your answers fully. Make sure that stray marks and changed answers are completely erased and that only one correct answer is visible.
- If you finish a section of the exam early, use the time to double-check multiple-choice answer, proofread your free responses, or check you equations and work.
- Relax and think positively. Remember, you have been preparing for this exam all year!

Utilizing Library Resources

Advanced Placement courses attract motivated high school students, most of who are bound for college. These students need to develop critical-thinking and research skills to prepare them for their college courses. Exposing students to resources that are used in college libraries helps prepares students for their future courses. Students should become familiar with standard reference works, with professional organizations that publish authoritative magazines, and with

Introduction

journals that maintain Website containing reliable articles written by professionals. Library materials can provide the most up-to-date information on the latest scientific research. Library materials also can allow students to probe information in depth and search for information on countless topics. Students need to learn how to evaluate the library resources they use. They should determine who wrote the material, when the material was created, what organization published the material, and whether or not the writer advocates a particular viewpoint that may bias the information.

The following bibliography should be helpful to both teachers and students as they navigate through the AP® Chemistry curriculum. In addition to familiarizing students with specific resources, using these library resources should help students gain an understanding of the different types of library materials and the kinds of information each type of material contains. Reference books contain general information that helps students gain an understanding of chemistry topics. Chemistry magazines and journals are an important resource for students, as they contain articles and updates from professional meetings and conferences and often have the most current information available. Students also will gain knowledge about accessing these publications, many of which are available in full-text in numerous on-line databases. Many publications are also available through professional organizations such as the American Institute of Chemists (AIC) and the American Chemical Society (ACS). These organizations maintain websites that contain a wealth of information helpful to students in AP® Chemistry classes, including lists of additional print materials and links to reliable online information.

Bibliography of Sources for AP® Chemistry Students

Reference Books

ACS Style Guide: Effective Communication of Scientific Information, 3rd Edition
A grammar and style guide for professionals and students in writing chemistry papers and lab work. Published by the American Chemical Society, 2006.

CRC Handbook of Chemistry and Physics
Classic reference for authoritative and up-to-date data on physics and chemistry. Contains glossary of terms and charts and tables of chemical and physical data. 89th edition published in 2009. Also available in an online version.

Encyclopedia of Chemistry
Includes descriptive entries of chemistry topics, essays, recent discoveries and biographies. Additional material includes lists of websites, resources, a bibliography, and cited references. Published in 2005 by Facts on File.

McGraw-Hill Concise Encyclopedia of Chemistry
Based on the McGraw-Hill Encyclopedia of Science and Technology, an informative and concise volume of up-to-date chemistry information, articles, and supplemental reading. 2004

McGraw-Hill Encyclopedia of Science and Technology
Comprehensive, authoritative 20-volume references set containing articles written by experts in all major disciplines of science and technology. The newest edition, updated in 2002, covers

recent discoveries and developments. Contains illustrations, study guides and an extensive index. Also available in an online version.

Lange's Handbook of Chemistry
Extensive collection of facts, data, and experimental findings in every area of chemistry. This reference book has been made for professionals, educators, and students alike for 70 years. Currently available in a 70th Anniversary Edition. 2004.

World of Chemistry
Subject encyclopedia containing entries on theories, discoveries, pioneers, and issues. Includes current scientific information, an extensive index, and a chronology of developments in chemistry. Published in 2000 by Thomson/Gale.

<u>Journals and Magazines</u>

Journal of Chemical Education
Contains editorials for educators, as well as articles about chemistry topics and learning materials. Published by the Division of Chemical Education of the American Chemical Society.

Journal of the American Chemical Society
The official publication of the American Chemical Society. It has professional articles, editorials, and news about the latest in chemical research and industry.

Chemical Reviews; Accounts of Chemical Research
Both journals contain in-depth articles about specific chemistry topics, including biochemistry and technology and the most recent research advances in the chemical world. Published by the American Chemical Society.

<u>Websites</u>

American Institute of Chemists (AIC) (http://www.theaic.org)
Contains easily readable summaries of journal articles and professional papers that explain the latest developments in chemistry research. Published by AIC.

American Chemical Society (ACS) (www.acs.org)
Contains current news, highlights, article summaries, and more about advances in chemistry and topics that are pertinent to the ACS. The "Educators and Students" button has a detailed list of useful articles, education site, and resources for educators and students alike for all grade levels up to post-college level.

Chemie.DE: The Chemistry Information Service (http://www.Chemie.ED/news/e/)
Provides information, news, and articles about the international chemistry world.

Eric Weisstein's World of Chemistry (http://scienceworld.worfram.com/chemistry/)
An online encyclopedia that contains information about chemical reactions, organic/inorganic concepts and quantum chemistry. Published by Wolfram Research.

Acknowledgements

This book is dedicated to my family: Mike, Christian and Russell, who have put up with the countless hours of "mommy is on the computer" when I might have been doing something with them. It has been an enormous yet surprisingly satisfying endeavor for me to embark upon and I am grateful to have had this opportunity. I want to thank Jack Kotz for recommending me to Ashley Summers and Lisa Lockwood of Brooks/Cole, a division of Cengage Learning. I appreciate the patience with which Ashley has worked with me on this project. I want to encourage any chemistry teacher to take the time to get to know your textbook and ancillary materials as intimately as I have through this process. There are tremendous resources at your disposal where someone else has done all the background work. Your students will benefit from the increased time you'll have to actually teach by using these resources. I am blessed to be teaching amazing students at an incredible school where teaching AP Chemistry has become a daily inspiration due to the students' commitment and dedication and the professional and supportive faculty and administration.

Roxie Allen
St. John's School
Houston, Texas

About the Editor

Roxie Allen has taught honors chemistry, organic chemistry and Advanced Placement Chemistry at St. John's School in Houston, Texas since 1990. Prior to that, she taught in the Houston Independent School District and the Midland Independent School District. Mrs. Allen has been very active in both local and state chemistry teacher organizations, serving on the Executive Board of the Associated Chemistry Teachers of Texas for ten years between 1993 and 2008. Mrs. Allen has conducted workshops for teachers at the local, state and national levels, for ACT_2, ChemEd, the American Chemical Society Southwest Regional meetings, and NSTA.   Recently Mrs. Allen was invited to speak to the National Academies of Science at the Chemical Sciences Roundtable, concerning the status of Chemical Education in Secondary Schools in the United States. In 2002, Mrs. Allen was appointed to a term as a mentor to the American Chemical Society United States Chemistry Olympiad team where she served until 2006. In 2008 Mrs. Allen was awarded the American Chemical Society Southwest Regional High School Chemistry Teacher award. Mrs. Allen holds a Bachelor of Science Degree in Biochemistry from Texas A&M University and a Master of Science Degree in Chemistry from the University of Houston.

Chapter 1
BASIC CONCEPTS OF CHEMISTRY

CHAPTER OUTLINE

CHAPTER OBJECTIVES

- Understand the nature of hypothesis, laws, and theories (1.1)
- Apply the kinetic-molecular theory to the properties of matter (1.2)
- Classify matter (1.2)
- Recognize elements, atoms, compounds and molecules (1.3-1.4)

- Identify physical and chemical properties and changes (1.5-1.6)

LESSON PLAN

Suggested Time:
One traditional class or ½ block. This chapter should be entirely review and students should be advised to read through the chapter and bring any questions to class.

ASSESSMENT

Exercises:
    1.1   Elements p. 13
    1.2   Physical Properties (p. 15)
    1.3   Chemical Reactions and Physical Changes (p. 19)
Answers to Exercises p. A40
Practicing Skills pp. 20-21, Answers IRM pp. 2-3
General Questions pp. 21-23, Answers IRM pp. 306

GLOSSARY

atom(s) The smallest particle of an element that retains the characteristic chemical properties of that element

chemical bond An interaction between two or more atoms that holds them together by reducing the potential energy of their electrons

chemical change(s) A change that involves the transformation of one or more substances into one or more different substances.

chemical compound Matter that is composed of two or more different kinds of atoms chemically combined in definite proportions.

chemical equation(s) A written representation of a chemical reaction, showing the reactants and products, their physical states, and the direction in which the reaction proceeds

chemical formula Symbols which represent the composition of any pure substance

chemical property An indication of whether and how readily a material undergoes a chemical change

density The ratio of the mass of an object to its volume

element(s) Matter that is composed of only one kind of atom

extensive properties Properties that depend on the amount of matter present

CHAPTER 1: Basic Concepts of Chemistry

heterogeneous A mixture in which the properties of one region or sample are different from those in another region or sample

homogeneous A mixture in which the properties are the same throughout, regardless of the optical resolution used to examine it

hypothesis A tentative explanation or prediction based on the experimental observations

intensive properties Physical properties that do not depend on the amount o matter present

ion(s) An atom or group of atoms that has lost or gained one or more electrons so that is no longer electrically neutral

kinetic energy The energy of a moving object, dependent on its mass and velocity

kinetic molecular theory A theory of the behavior of matter at the molecular level

law A concise verbal or mathematical statement of a relation that is always the same under the same condition

macroscopic Processes and properties on a scale large enough to be observed directly

molecules The smallest unit of a compound that retains the composition and properties of that compound

particulate Representations of chemical phenomena in terms of atoms and molecules (also called submicroscopic)

periodic table Arrangement of the elements which includes the symbol and other information

physical change(s) A change that only involves physical properties

physical properties Properties of a substance that can be observed and measured without changing the composition of the substance

product(s) A substance formed in a chemical reaction

purified Separation of a mixture into its pure components

qualitative Nonnumerical experimental observations, such as descriptive or comparative data

quantitative Numerical experimental data, such as measurements of changes in mass and volume

reactant(s) A starting substance in a chemical reaction

solution(s) A homogeneous mixture

state Physical condition in which the substance exists

submicroscopic Representations of chemical phenomena in terms of atoms and molecules (also called particulate)

temperature A physical property that determine the direction of heat flow in an object on contact with another object

theory A unifying principle that explains a body of facts and the laws based on them

MEDIA MENU

The Media Menu section of this guide lists the interactive exercises, tutorials, active figures (animated, interactive figures) and additional resources available online for Chemistry Now, a unique web-based, assessment-centered personalized learning system for chemistry students. Go to www.cengagelearning.com/login.

Exercises

- Screen 1.5 Mixtures and pure substances
- Screen 1.12 Identifying physical and chemical changes

Self-study module

- 1.7 Elements and Atoms

Active Figures

- 1.7 States of Matter – Solid, Liquid, and Gas
- 1.8 Levels of Matter
- 1.9 Classifying Matter

Additional Resources

- Screen 1.6   Heterogeneous mixtures video
- Screen 1.13   Chemical Changes on the Molecular Scale video and animation
- Podcast Module 1   Elements and atoms

CHAPTER 1: Basic Concepts of Chemistry

DIAGNOSTIC QUIZ QUESTIONS

Goal #1 Nature of hypothesis, laws, and theories

1. A regularity observed in nature which can be stated as a mathematical relationship is best
    identified as

    a) a theory b) a hypothesis c) a law
    d) an observation e) a conclusion

    Answer: c

Goal #2 Apply the kinetic-molecular theory to the properties of matter

2. Which of the following statements concerning the kinetic-molecular theory are correct?

    1. Gas particles move faster when they are heated.
    2. Particles in a liquid are closely spaced, but are not confined to specific positions.
    3. Particles in a solid are closely spaced and are confined to specific positions.

    a) 1 only b) 2 only c) 3 only
    d) 1 and 2 e) 1, 2, and 3

    Answer: e

Goal #3 Classify Matter

3. Which of the following properties are consistent with liquids?
    1. The volume of a liquid is determined by the size of its container.
    2. A liquid has a rigid shape and a fixed volume.
    3. A liquid has a fixed volume that varies little with temperature or pressure changes.

    a) 1 only b) 2 only c) 3 only
    d) 2 and 3 e) 1, 2, and 3

    Answer: c

4. A pure substance composed of two or more elements is

    a) a heterogeneous mixture b) a homogeneous mixture c) an element
    d) a pure substance e) a chemical compound

    Answer: e

5.   A tablespoon of sugar will completely dissolve in a glass of pure water to give

   a) a chemical compound   b) a homogeneous mixture   c) a heterogeneous mixture
   d) an element   e) a pure substance

   Answer: b

6.   Concrete is composed of sand, gravel, and calcium oxide. Concrete is best described as

   a) a heterogeneous mixture   b) a homogeneous mixture   c) a pure substance
   d) a chemical compound   e) an element

   Answer: a

Goal # 4 Recognize elements, atoms, compounds and molecules

7.   Nitrogen, N_2, is a(n) __________ that is composed of two nitrogen __________.

   a) atom, molecules   b) compound, molecules   c) element, atoms
   d) atom, elements   e) atom, compounds

   Answer: c

8.   What is the name of the element having the symbol S?

   a) silicon   b) sulfur   c) selenium
   d) scandium   e) seaborgium

   Answer: b

9.   The symbol for cesium is __________.

   a) C   b) Ce   c) Cs
   d) Cm   e) Ca

   Answer: c

Goal #5 Identify physical and chemical properties and changes.

10.   Which of the following are chemical properties of iodine?

1. Iodine is a purple solid at $25^{\circ}C$.
2. Iodine reacts with sodium metal to form sodium iodide.
3. The density of iodine is $4.93 \ g/cm^3$.

a) 1 only   b) 2 only   c) 3 only
d) 1 and 3   e) 2 and 3

Answer: b

11.   Which of the following are physical properties of potassium?

1. Potassium reacts with water, producing hydrogen gas and aqueous potassium
   hydroxide.
2. Potassium conducts electricity.
3. Potassium is malleable at room temperature.

a) 1 only   b) 2 only   c) 3 only
d) 2 and 3   e) 1, 2 and 3

Answer: d

12.   You can identify a metal by carefully determining its density. A 23.1 g piece of an
       unknown metal is 1.23 cm long, 2.11 cm wide, and 1.00 cm thick. What is the identity of
       the element?

a) nickel, $8.90 \ g/cm^3$ b) aluminum, $2.70 \ g/cm^3$ c) zirconium, $6.51 \ g/cm^3$
d) chromium, $7.20 \ g/cm^3$ e) niobium, $8.57 \ g/cm^3$

Answer: a

13.   The density of liquid mercury is $13.5 \ g/cm^3$. What mass of mercury (in kg) is required to
       fill a hollow cylinder having an inner diameter of 2.00 cm to a height of 25.0 cm?

a) 1.06 kg   b) 4.24 kg   c) 0.171 kg
d) 1.71×10^{-4} kg   e) 2.12 kg

Answer: a

14.   Which of the following contains only extensive properties?

    a) melting point, density and color
    b) electrical conductivity and mass
    c) density, boiling point and volume
    d) volume and mass
    e) electrical conductivity and boiling point

Answer: d

TESTBANK

<u>1.1 Chemistry and Its Methods</u>

1. A hypothesis is a

    a) mathematical formula that models a pattern of behavior.
    b) concise statement of a behavior that is always the same under the same conditions.
    c) set of experiments designed to test a theory.
    d) well-tested, unifying principle that explains a body of facts.
    e) tentative explanation or prediction based upon experimental observations.

    Answer: e

2. A theory is a

    a) concise statement of a behavior that is always the same under the same conditions.
    b) tentative explanation or prediction based upon experimental observations.
    c) mathematical formula that models a pattern of behavior.
    d) well-tested, unifying principle that explains a body of facts.
    e) set of quantitative numerical data.

    Answer: d

<u>1.2 Classifying Matter</u>

3. According to the kinetic-molecular theory of matter, particles in a gas

    a) are far apart from each other, but they are confined to specific positions.
    b) are packed closely together like solids, but they are not confined to specific positions.
    c) move faster as their temperature decreases.
    d) vibrate back and forth about an average position.
    e) fly about randomly, colliding with themselves and the walls of their container.

    Answer: e

4. According to the kinetic-molecular theory of matter, particles in a solid

    a) vibrate back and forth about an average position.
    b) move slower as the temperature increases.
    c) fly about randomly, colliding with themselves and the walls of their container.
    d) expand to fill their container.
    e) are packed closely together, but are not confined to specific positions.

    Answer: a

5. Which one of the following statements is correct?

    a) Pure substances may be separated by filtration or distillation into at least two components.
    b) A heterogeneous mixture is also known as a solution.
    c) A heterogeneous mixture is composed of two or more substances in the same phase.
    d) The composition is uniform throughout a homogeneous mixture.
    e) The combination of two or more liquids always results in a homogeneous mixture.

    Answer: d

6. Which one of the following is most likely to be a homogeneous mixture?
    a) blood
    b) ground beef
    c) the air trapped inside an inflated balloon
    d) chocolate chip cookies
    e) mortar (a mixture of calcium carbonate and sand)

    Answer: c

7. Which one of the following is most likely to be a heterogeneous mixture?

    a) gasoline
    b) blood
    c) antifreeze (a mixture of water and ethylene glycol)
    d) sodium chloride (table salt) dissolved in water
    e) the air trapped inside an inflated balloon

    Answer: b

<u>1.3 Elements and Atoms</u>

8. Which of the following statements are true?

    1. Atoms are the smallest particles of an element that retain the element's chemical properties.
    2. Substances composed of only one type of atom are classified as elements.
    3. Of the 116 known elements, only about 90 are found in nature.

    a) 1 only　　　　b) 2 only　　　　c) 3 only　　　　d) 1 and 2　　　　e) 1, 2, and 3

    Answer: e

9. What is the correct symbol for silicon?

    a) S　　　　b) Sc　　　　c) Si　　　　d) Sm　　　　e) Sn

    Answer: c

10. What is the correct symbol for magnesium?

    a) M　　　　b) Ma　　　　c) Mg　　　　d) Mn　　　　e) Md

    Answer: c

11. What is the name of the element with the symbol Br?

    a) barium　　　　b) beryllium　　　　c) bismuth　　　　d) boron　　　　e) bromine

    Answer: e

12. What is the name of the element with the symbol Ca?

    a) cobalt　　　　b) carbon　　　　c) calcium　　　　d) cadmium　　　　e) californium

    Answer: c

13. Which one of the following substances is classified as an element?

    a) P_4　　　　b) NO　　　　c) KCl　　　　d) $C_6H_{12}O_6$　　　　e) NO_2

    Answer: a

<u>1.4 Compounds and Molecules</u>

14. An electrically charged atom or group of atoms is a(n)

    a) element.
    b) ion.
    c) chemical compound.
    d) heterogeneous mixture.
    e) homogeneous mixture.

    Answer: b

15. A pure substance composed of two or more different elements is

    a) an ion.
    b) a homogeneous mixture.
    c) a chemical compound.
    d) a solid.
    e) an atom.

    Answer: c

16. A(n) _________ is a pure substance that is composed of only one type of atom.

    a) ion
    b) homogeneous mixture
    c) element
    d) chemical bond
    e) chemical compound

    Answer: c

17. Which one of the following substances is classified as a chemical compound?

    a) He   b) H_2 c) Na   d) NaCl   e) Cl_2

    Answer: d

18. Which term best describes hydrogen peroxide, H_2O_2?

    a) homogeneous mixture
    b) heterogeneous mixture
    c) element
    d) chemical compound
    e) atom

Answer: d

19. Which of the following statements concerning water, H_2O, is/are true?

    1) H_2O is a heterogeneous mixture of hydrogen and oxygen.
    2) H_2O is a chemical compound.
    3) The percentage of hydrogen in H_2O is independent of where the water is obtained.

    a) 1 only   b) 2 only   c) 3 only   d) 1 and 2   e) 2 and 3

Answer: e

<u>1.5 Physical Properties</u>

20. Which one of the following statements is not a comparison of physical properties?

    a) Potassium reacts with water more quickly than calcium reacts with water.
    b) The electrical conductivity of aluminum is greater than copper.
    c) The density of ice is less than liquid water.
    d) Ethylene glycol (antifreeze) is more viscous than water.
    e) Gold is more malleable than titanium.

Answer: a

21. All of the following are physical properties EXCEPT

    a) the viscosity of motor oil.
    b) the rusting of iron.
    c) the boiling point of water.
    d) the melting point of gold.
    e) the conductivity of copper wire.

Answer: b

22. Which one of the following statements is not a comparison of physical properties?

    a) Mercury and gallium are both liquids at 50 °C.
    b) Carbon dioxide is more soluble in water than oxygen.
    c) Copper and silver are malleable metals.
    d) Oxygen and nitrogen are gases at 25 °C.
    e) Water may be split by electrolysis into hydrogen and oxygen.

Answer: e

23. An intensive property of a substance is

    a) independent of the amount present.
    b) dependent on its volume, but not its mass.
    c) not affected by its temperature.
    d) dependent only on its temperature.
    e) dependent only on its mass and volume.

Answer: a

24. Which of the following are extensive properties: mass, volume, and/or density?

    a) mass only
    b) volume only
    c) density only
    d) mass and volume
    e) volume and density

Answer: d

25. All of the following are examples of intensive properties EXCEPT

    a) melting point.
    b) color.
    c) mass.
    d) density.
    e) boiling point.

Answer: c

26.   The density of ice at 0 °C is 0.917 g/mL. What is the volume of 225 g of ice?

    a) 0.00407 mL   b) 18.7 mL   c) 206 mL   d) 226 mL   e) 245 mL

Answer: e

27. The density of liquid mercury is 13.5 g/cm^3. What mass of mercury will fill a 12.0 ounce
soda can? (1.00 oz = 29.6 mL, 1.00 g = 1.00 cm^3)

a) 0.0380 g   b) 26.3 g   c) 162 g   d) 369 g   e) 4.80 × 10^3 g

Answer: e

28. A piece of zinc metal displaces 8.22 cm^3 of water when submerged. If the mass of zinc is
58.6 g, what is its density?

a) 0.00208 g/cm^3 b) 0.140 g/cm^3 c) 7.13 g/cm^3 d) 50.4 g/cm^3 e) 482 g/cm^3

Answer: c

29. You can identify a metal by carefully determining its density. A 33.39 g sample of an
unknown metal is 1.50 cm long, 2.50 cm wide, and 1.00 cm thick. What is a possible identity
of the element?

a) nickel, 8.90 g/cm^3
b) aluminum, 2.70 g/cm^3
c) silver, 10.5 g/cm^3
d) iron, 7.87 g/cm^3
e) chromium, 7.20 g/cm^3

Answer: a

30. A cube of zirconium has a mass of 343 g. If each side of the cube has dimensions of 3.75 cm,
what is the density of zirconium?

a) 0.0410 g/cm^3 b) 0.154 g/cm^3 c) 6.50 g/cm^3 d) 24.4 g/cm^3 e) 18.1 g/cm^3

Answer: c

31. You can identify a metal by carefully determining its density. A 20.05 g cylinder of an
unknown metal is 2.00 cm long and has a diameter of 0.755 cm. What is a possible identity
of the element? (Volume $= \pi r^2 h$)

a) silver, 10.5 g/cm^3
b) iridium, 22.4 g/cm^3
c) gold, 19.3 g/cm^3
d) lead, 11.4 g/cm^3
e) nickel, 8.90 g/cm^3

Answer: b

32. At 25 °C, a bottle contains 2.00 L of water in its liquid state. What is the volume of the water after it freezes (at 0 °C)? The densities of liquid water and ice are 0.997 g/mL and 0.917 g/mL, respectively.

a) 1.09 L   b) 1.84 L   c) 2.00 L   d) 2.17 L   e) 2.19 L

Answer: d

1.6 Physical and Chemical Change

33. All of the following are examples of chemical change EXCEPT

a) the condensation of steam.
b) the rusting of iron.
c) the combustion of gasoline.
d) the tarnishing of silver.
e) the decomposition of cinnabar (HgS) to mercury metal upon heating.

Answer: a

34. Which of the following observations is/are examples of chemical change?

1. Sodium chloride melts at 801 °C.
2. The density of water decreases when it changes from a liquid to a solid.
3. The combustion of propane gas yields carbon dioxide and water.

a) 1 only   b) 2 only   c) 3 only   d) 1 and 2   e) 2 and 3

Answer: c

35. Which of the following observations is/are examples of physical change?

1. Hydrogen gas is evolved when zinc reacts with hydrochloric acid.
2. Aluminum melts when heated above 660 °C.
3. The density of water decreases when it solidifies.
a) 1 only   b) 2 only   c) 3 only   d) 1 and 2   e) 2 and 3

Answer: e

Short Answer Questions

36. Substances like hydrogen (H_2) and oxygen (O_2) that are composed of only one type of atom are classified as _________ .

Answer: elements

37. Properties, such as color and density, which can be observed or measured without changing the composition of a substance are called __________ properties.

Answer: physical

38. A mass of 10 g of table salt dissolves in water to form a __________ mixture (i.e., a mixture that is uniform throughout).

Answer: homogeneous

39. A(n) __________ is the smallest particle of an element that retains the characteristic chemical properties of that element.

Answer: atom

40. The density of a substance is defined as its mass per unit __________.

Answer: volume

Let's Review
THE TOOLS OF QUANTITATIVE CHEMISTRY

CHAPTER OUTLINE

LET'S REVIEW OBJECTIVES

- Use metric units correctly (1)
- Understand and use the mathematics of chemistry (2-3)

LESSON PLAN

Suggested Time:
One traditional class or ½ block. This material should be entirely review and students should be advised to read through the chapter and bring any questions to class. Math skills are critical to success on the AP so students should work as many problems as possible.

ASSESSMENT

Exercises:
1 Temperature Scales (p. 27)
2 Using Units of Length (p. 29)
3 Volume (p. 29)
4 Mass (p. 30)
5 Accuracy, Error, and Standard Deviation (p. 32)
6 Using Significant Figures (p. 38)
7 Using Dimensional Analysis (p. 39)
8 Graphing (p. 41)
9 Problem Solving (p. 43)

Answers to Exercies p. A40-41
Practicing Skills pp. 43-45, Answers IRM pp. 8-11
General Questions pp. 45-48, Answers IRM pp. 11-16
In the Laboratory pp. 48-49, Answers IRM pp. 16-18

GLOSSARY

absolute zero The lowest possible temperature, equivalent to -273.15°C, used as the zero point of the Kelvin scale

accuracy The agreement between the measured quantity and the accepted value

Celsium scale A temperature scale defined by the freezing and boiling points of pure water, defined as 0 $^\circ$C and 100 $^\circ$C.

conversion factor(s) A multiplier that relates the desired unit to the starting unit

dimensional analysis A general problem solving approach that uses the dimensions or units of each value to guide you through the calculations (also called factor label method)

18

Let's Review: The Tools of Quantitative Chemistry

exponential notation A way of presenting very large or very small numbers in a compact and consistent form that simplifies calculations (also known as scientific notation)

factor-label method A general problem solving approach that uses the dimensions or units of each value to guide you through the calculations (also known as dimensional analysis)

fixed notation Presenting numbers in their actual form.

International System of Units (SI) Uniform system of measurement units in which a single base unit is used for each measured physical quantify

Kelvin scale A temperature scale in which the unit is the same size as the Celsium degree but the zero point is the lowest possible temperature

liter(L) A unit of volume convenient for laboratory use

mass A measure of the quantify of matter in a body

metric system A decimal system for recording and reporting scientific measurements, in =which all units are expressed as powers of 10 times some base unit.

milliliter(mL) A unit of volume equivalent to one thousandth of a liter

percent error the difference between the measured quantity and the accepted value, expressed as a percentage of the accepted value

precision The agreement of repeated measurements with one another

scientific notation notation A way of presenting very large or very small numbers in a compact and consistent form that simplifies calculations (also known as exponential notation)

significant figure(s) The digits in a measured quantity that are known exactly, puls one digit that is inexact to the extent of +/- 1

standard deviation A measure of precision, calculated as the square root of the sum of the squares of the deviations for each measurement divided by the number of measurements

MEDIA MENU

The Media Menu section of this guide lists the interactive exercises, tutorials, active figures (animated, interactive figures) and additional resources available online for Chemistry Now, a unique web-based, assessment-centered personalized learning system for chemistry students. Go to www.cengagelearning.com/login.

Self-study module

- 1.17   Using Numerical Information
- 1.17   Dimensional Analysis

Active Figures

- 1 Comparison of Fahrenheit, Celsius, and Kelvin scales
- 1.8 Levels of Matter
- 1.9 Classifying Matter

DIAGNOSTIC QUIZ QUESTIONS

Goal #1 Use metric units correctly

1.   Thermostats are often set to 22°C. What is this temperature in Kelvin?

  a) 251 K   b) 284 K   c) 295 K
  d) 321 K   e) 322 K

  Answer: c

2.   Helium boils at 4.3 K. What is this temperature in Celsius?

  a) 268.9°C   b) 277.4 °C   c) -277.4 °C
  d) -268.9°C   e) -295.7 °C

  Answer: d

3.   The radius of a carbon atom is 7.7×10^{-11} m. What is the radius in picometers?

  a) 7.7 pm   b) 77 pm   c) 7.7×10^2 pm
  d) 7.7×10^3 pm   e) 7.7×10^{-1} pm

  Answer: b

4.   A typical volumetric flask holds a volume of 0.250 L. What is this volume in cubic centimeters?

  a) 0.25 cm^3 b) 2.50 cm^3 c) 2.50×10^2 cm^3
  d) 2.50×10^3 cm^3 e) 25.0 cm^3

  Answer: c

Goal # 2 Understand and use the mathematics of chemistry

5.   A student determines the density of a bar of silver by measuring its dimensions (2.00 cm by 1.5 cm by 1.00 cm) and determining its mass (25.3 g). It the true density of silver is 10.5 g/cm^3, what is the percent error in the student's measurement?

a) 1 %   b) 3%   c) 5%
d) 10%   e) 15%

Answer: c

6.   You and your lab partner are asked to determine the mass of a bar of silver. You use an analytical balance that measures mass to four decimal places (Method A). Your partner uses a top-loader balance that measures mass to two decimal places (Method B). The results are tabulated below:

	Method A (g)	Method B (g)
Measurement #1	3.3682	3.41
Measurement #2	3.3684	3.71
Measurement #3	3.3682	3.35
Measurement #4	3.3681	3.92
Average Mass	3.3682	3.60
Percent error	6.495%	0.1%

The actual mass of the silver bar is 3.6022 g. Which statements best describe the results?

a) Method A has good precision and poor accuracy. Method B has poor precision and good accuracy.
b) Method A has poor precision and good accuracy. Method B has good precision and poor accuracy.
c) Method A has poor precision and poor accuracy. Method B has good precision and good accuracy.
d) Method A has good precision and good accuracy. Method B has good precision and poor accuracy.
e) Method A has good precision and good accuracy. Method B has poor precision and poor accuracy.

Answer: a

TESTBANK

<u>1- Units of Measurement</u>

1. The SI unit of mass is the __________.

    a) Newton   b) liter   c) kilogram   d) mole   e) ounce

    Answer: c

2   The SI unit of temperature is the __________.

    a) kelvin   b) calorie   c) fahrenheit   d) absolute zero scale
    e) kilocalorie

    Answer: a

3. Which one of the following is not an SI base unit?

    a) liter   b) kelvin   c) calorie   d) meter   e) second

    Answer: c

4. Which method is correct for converting kelvin to Celsius?

    a) $T(^\circ C) = \left(\dfrac{5\ ^\circ C}{9\ K}\right) T(K) + 32$

    b) $T(^\circ C) = \left(\dfrac{9\ ^\circ C}{5\ K}\right) T(K) + 273.15$

    c) $T(^\circ C) = \dfrac{5\ ^\circ C}{9\ K} (T(K) - 273.15)$

    d) $T(^\circ C) = \dfrac{1\ ^\circ C}{1\ K} (T(K) - 273.15)$

    e) $T(^\circ C) = \dfrac{1\ ^\circ C}{1\ K} (T(K) + 273.15)$

    Answer: d

5. Ethanol boils at 351.7 K. What is this temperature in Celsius?

    a) 1.29 °C   b) 53.5 °C   c) 78.5 °C   d) 227.4 °C   e) 624.9 °C

    Answer: c

6. The boiling point of liquid helium is -269 °C. What is this temperature in kelvin?

    a) 0.985 K   b) 1.01 K   c) 4 K   d) 29 K   e) 542 K

    Answer: c

7. The temperature required to melt silver is 962 °C. What is this temperature in kelvin?

    a) 664 K   b) 689 K   c) 962 K   d) 1235 K   e) 1260 K

    Answer: d

8. Which of the following is the greatest length?

    a) 5.0×10^8 pm   b) 5.0×10^6 nm   c) 5.0×10^1 μm   d) 5.0×10^{-1} mm   e) 5.0×10^{-5} m

    Answer: b

9. Which of the following is the smallest volume?

    a) 5.0×10^1 cm^3 b)   0.50 dL   c) 0.50 mL   d) 5.0×10^3 uL   e) 5.0×10^{-2} L

    Answer: c

10. The radius of an aluminum atom is 0.143 nm. What is the radius in picometers?

    a) 1.43×10^{-12} pm
    b) 1.43×10^{-9} pm
    c) 1.43×10^{-2} pm
    d) 1.43×10^{-1} pm
    e) 1.43×10^2 pm

    Answer: e

11. The volume of space occupied by an oxygen atom is 1.2×10^{-30} m^3. What is this volume in units of pm^3?

    a) 1.2×10^{-66} pm   b) 1.2×10^{-18} pm   c) 1.2×10^{-6} pm   d) 1.2×10^{-3} pm   e) 1.2×10^6 pm

    Answer: e

12. The volume of space occupied by a carbon atom is 1.9×10^{-3} nm^3. What is the volume of a carbon atom in cubic meters?

a) 1.9×10^{-30} m^3 b) 1.9×10^{-21} m^3 c) 1.9×10^{-12} m^3 d) 1.9×10^{6} m^3 e) 1.9×10^{24} m^3

Answer: a

13. Which of the following volumes are equivalent to 125 cm^3?

1) 1.25 dL
2) 125 mL
3) 0.0125 L

a) 1 only   b) 2 only   c) 3 only   d) 1 and 2   e) 1, 2, and 3

Answer: d

<u>2- Making Measurements: Precision, Accuracy, Experimental Error, Standard Deviation</u>

14. Two electronic balances are tested using a standard weight. The true mass of the standard is 2.5000 g. The results of 5 individual measurements on each balance are recorded below.

	Balance A	Balance B
	2.4999 g	2.7022 g
	2.5003 g	2.5113 g
	2.5000 g	2.4940 g
	2.4998 g	2.2781 g
	2.5002 g	2.5145 g
average mass =	2.5000 g	2.5000 g

Which statement best describes the results?

a) A: poor precision, good accuracy. B: good precision, good accuracy.
b) A: good precision, good accuracy. B: poor precision, good accuracy.
c) A: good precision, good accuracy. B: good precision, poor accuracy.
d) A: poor precision, good accuracy. B: good precision, poor accuracy.
e) A: good precision, poor accuracy. B: poor precision, good accuracy.

Answer: b

15. A student uses a balance to determine the mass of an object as 12.28 grams. The correct mass of the object is 12.45 grams. What is the percent error in the student's mass determination?

a) 0.17%   b) 0.99%   c) 1.4%   d) 17%   e) 99%

Answer: c

16. Two students independently determine the volume of water delivered by a 10.00-mL pipet. Each student takes 8 measurements, then computes the average volume delivered and the standard deviation. The results are tabulated below.

	Average	St. Dev.
Student A	10.277 mL	± 0.003 mL
Student B	10.0 mL	± 0.9 mL

Which statement best describes the results?

a) A: good precision, good accuracy. B: good precision, good accuracy.
b) A: good precision, poor accuracy. B: poor precision, good accuracy.
c) A: poor precision, good accuracy. B: good precision, poor accuracy.
d) A: poor precision, poor accuracy. B: good precision, good accuracy.
e) A: poor precision, poor accuracy. B: poor precision, poor accuracy.

Answer: b

17. Assume that an average and a standard deviation are calculated using a large number of measurements. A single data point has a(n) _____ percent chance of falling within plus and minus one standard deviation of the calculated average.

a) 50%   b) 68%   c) 83%   d) 95%   e) 100%

Answer: b

3- Mathematics in Chemistry

18. Express 0.08010 in exponential notation.

a) 8.01×10^{-2} b) 8.01×10^{2} c) 8.010×10^{-2} d) 8×10^{-2} e) 8.010×10^{2}

Answer: c

19. Express 2.030×10^{-3} in standard notation.

a) 0.002030   b) 0.00203   c) 203000   d) 0.0203   e) 0.002

Answer: a

20. Convert 4.500×10^{-5} meters to micrometers and express the answer in standard notation using the correct number of significant figures.

a) 45 μm   b) 45.0 μm   c) 45.00 μm   d) 4500 μm   e) 4.500 μm

Answer: c

21. Convert 0.490 ng to grams and express the answer in scientific notation using the correct number of significant figures.

a) 4.90×10^{-8} g   b) 4.90×10^{-10} g   c) 4.9×10^{-10} g   d) 0.490×10^{-8} g   e) 0.490×10^{-10} g

Answer: b

22. An electronic balance is used to determine that a sample has a mass of 5.8234 g. If the balance's precision is ±0.1 mg, what is the correct number of significant figures for this measurement?

a) 2   b) 3   c) 4   d) 5   e) 6

Answer: d

23. What is the correct number of significant figures in 0.003040?

a) 2   b) 3   c) 4   d) 6   e) 7

Answer: c

24. Round 0.000680483 to 4 significant figures.

a) 0.000   b) 0.0007   c) 0.0006805   d) 0.00068048   e) 0.000680483

Answer: c

25. What is the correct answer to the following expression: $(86.11 - 85.49) \div 28.26$? Carry out the subtraction operation first.

a) 2×10^{-2} b) 2.2×10^{-2} c) 2.19×10^{-2} d) 2.194×10^{-2} e) 2.1939×10^{-2}

Answer: b

26. What is the correct answer to the following expression: $(53 + 63) \times 0.17051$? Carry out the addition operation first.

a) 20   b) 2.2×10^{2} c) 19.8   d) 19.78   e) 19.779

Answer: c

27. What is the correct answer to the following expression: $9.27 \times 10^{-4} + 6.3 \times 10^{-6}$?

a) 9×10^{-4} b) 9.3×10^{-4} c) 9.33×10^{-4} d) 9.333×10^{-4} e) 9.3330×10^{-4}

Answer: c

28. What is the correct answer to the following expression?
    $5.374 \times 10^{-3} + 4.1 \times 10^{-4} + 6 \times 10^{-5} =$

    a) 6×10^{-3} b) 5.8×10^{-3} c) 5.84×10^{-3} d) 5.844×10^{-3} e) 5.8440×10^{-3}

    Answer: c

29. The radius of a carbon atom is 77 pm. What is the volume (in m^3) of a carbon atom? The volume of a sphere is $(4/3)\pi r^3$.

    a) $1.9 \times 10^{-30}\ m^3$ b) $1.9 \times 10^{-21}\ m^3$ c) $1.9 \times 10^{-20}\ m^3$ d) $1.9 \times 10^{4}\ m^3$ e) $1.9 \times 10^{6}\ m^3$

    Answer: a

30. The wavelength of light emitted from a particular red diode laser is 6.72×10^{-7} m. What is the wavelength in nanometers?

    a) 0.672 nm   b) 6.72 nm   c) 67.2 nm   d) 672 nm   e) 6.72×10^{3} nm

    Answer: d

31. Light with a wavelength of 25 nm is in the x-ray region of the electromagnetic spectrum. What is the wavelength of this light in meters?

    a) 2.5×10^{-11} m   b) 2.5×10^{-10} m   c) 2.5×10^{-8} m   d) 2.5×10^{-7} m   e) 2.5×10^{10} m

    Answer: c

32. The volume of a cube is $3.0\ m^3$. What is the volume of the cube in liters?

    a) 3.0 L   b) 3.0×10^{1} L   c) 3.0×10^{2} L   d) 3.0×10^{3} L   e) 3.0×10^{4} L

    Answer: d

33. The dimensions of a box are 8.3 cm by 4.8 cm by 2.8 cm. Calculate the volume of the box in cubic feet. (2.54 cm = 1.00 inch, 12.0 inches = 1.00 foot)

    a) $3.9 \times 10^{-3}\ ft^3$ b) $1.1\ ft^3$ c) $3.6\ ft^3$ d) $24\ ft^3$ e) $1.1 \times 10^{3}\ ft^3$

    Answer: a

34. The volume of an object is 6.31×10^{-2} m^3. Calculate the volume in cubic feet. (2.54 cm = 1.00 inch, 12.0 inches = 1.00 foot)

a) 0.00179 ft^3 b) 0.0133 ft^3 c) 0.207 ft^3 d) 2.23 ft^3 e) 598 ft^3

Answer: d

35. A cylinder has a radius of 3.38 mm and a height of 275 mm. Calculate the volume of the cylinder in liters. (Volume = $\pi r^2 h$)

a) 9.87×10^{-3} L   b) 9.87×10^{-2} L   c) 0.987 L   d) 9.87 L   e) 9.87×10^3 L

Answer: a

36. The radius of a bromine atom is 114 pm. What is the volume of a bromine atom in cubic centimeters? The volume of a sphere is $(4/3)\pi r^3$.

a) 6.21×10^{-30} cm^3 b) 6.21×10^{-24} cm^3 c) 6.21×10^{-6} cm^3
d) 6.21×10^{-4} cm^3 e) 6.21×10^6 cm^3

Answer: b

37. If the fuel efficiency of an automobile is 21 miles per gallon, what is its fuel efficiency in kilometers per liter? (1 km = 0.621 mile, 1.000 L = 1.057 quarts, 4 quarts = 1 gallon)

a) 3.4 km/L   b) 8.0 km/L   c) 8.9 km/L   d) 49 km/L   e) 1.2×10^2 km/L

Answer: c

38. The speed of sound at one atmosphere of pressure is 769 miles per hour. What is the speed of sound in units of meters per second? (1 km = 0.621 mile)

a) 133 m/s   b) 344 m/s   c) 478 m/s   d) 4.46×10^4 m/s e) 4.78×10^5 m/s

Answer: b

39. The speed of light in a vacuum is 3.00×10^8 m/s. What is the speed of light in units of kilometers per hour?

a) 1.20×10^{-2} km/hr   b) 8.33×10^1 km/hr   c) 8.33×10^7 km/hr
d) 1.08×10^9 km/hr   e) 1.08×10^{15} km/hr

Answer: d

40. If a gallon (3.78 L) of latex paint can cover 365 ft^2 of the surface of a wall, what is the average thickness of one coat of paint (in millimeters)? (12 in = 1 ft, 2.54 cm = 1 in)

a) 1.11×10^{-1} mm   b) 2.83×10^{-1} mm   c) 4.64×10^{-1} mm
d) 4.89×10^{-1} mm   e) 3.40 mm

Answer: a

41. Silver iodide (AgI) is relatively insoluble in water. At 25 °C, 2.3×10^6 L of water are needed to dissolve 5.0 g of AgI. What mass (in micrograms) of AgI will dissolve in 1.0 L of water?

a) 0.43 μg   b) 2.2 μg   c) 4.6 μg   d) 12 μg   e) 4.6×10^6 μg

Answer: b

42. Calcium carbonate, or limestone, is relatively insoluble in water. At 25 °C, only 5.8 mg will dissolve in 1.0 liter of water. What volume of water is needed to dissolve 2.5 g of calcium carbonate?

a) 2.3×10^{-3} L   b) 1.5×10^{-2} L   c) 6.9×10^1 L   d) 1.7×10^2 L   e) 4.3×10^2 L

Answer: e

43. If the density of carbon dioxide in air is 0.65 mg/L, what volume is occupied by 25 kg of carbon dioxide?

a) 2.6×10^{-8} L   b) 1.6×10^1 L   c) 3.8×10^1 L   d) 1.6×10^7 L   e) 3.8×10^7 L

Answer: e

44. If 8.0 ounces of a liquid has a mass of 156 g, what is the density of the liquid in grams per cubic centimeter? (29.6 cm^3 = 1.00 oz.)

a) 0.051 g/cm^3 b) 0.66 g/cm^3 c) 1.5 g/cm^3 d) 5.3 g/cm^3 e) 42 g/cm^3

Answer: b

45. Which method is correct for determining the gallons of gasoline required to fill an automobile's 64 liter tank? (1.000 liter = 1.057 quarts, 4 quarts = 1 gallon)

a) $64 \text{ liters} \left(\dfrac{1 \text{ liter}}{1.057 \text{ quarts}} \right) \left(\dfrac{4 \text{ quarts}}{1 \text{ gallon}} \right) =$

b) $64 \text{ liters} \left(\dfrac{1 \text{ liter}}{1.057 \text{ quarts}} \right) \left(\dfrac{1 \text{ gallon}}{4 \text{ quarts}} \right) =$

c) $64 \text{ liters} \left(\dfrac{1.057 \text{ quarts}}{1 \text{ liter}} \right) \left(\dfrac{4 \text{ quarts}}{1 \text{ gallon}} \right) =$

d) $64 \text{ liters} \left(\dfrac{1.057 \text{ quarts}}{1 \text{ liter}} \right) \left(\dfrac{1 \text{ gallon}}{4 \text{ quarts}} \right) =$

e) none of the above

Answer: d

46. A barometer is filled with a cylindrical column of mercury that is 76.0 cm high and 1.000 cm in diameter. If the density of mercury is 13.53 g/cm^3, what is the mass of mercury in the column?

a) 0.227 g   b) 4.41 g   c) 808 g   d) 1.03×10^3 g   e) 3.23×10^3 g

Answer: c

47. What mass of zinc (in kg) is in a 2.50 kg roll of tin foil that is 93.50% tin and 6.50% zinc?

a) 0.163 kg   b) 0.385 kg   c) 1.63 kg   d) 2.34 kg   e) 2.50 kg

Answer: a

Short Answer Questions

48. There are ___________ cubic meters (m^3) in 1.0×10^3 cubic centimeters (cm^3).

Answer: 1.0×10^{-3}

49. What is the difference between the accuracy of measurements and the precision of measurements?

Answer: Accuracy is an indicator of how close a measurement, or the average of a set of measurements, is to the correct value. Precision is an indicator of how close two measurements are to each other, or how much numerical spread is present in a set of measurements.

50. Significant figures allow us to estimate uncertainty in calculated values. In some circumstances, following significant figure rules can lead to estimates of uncertainty that are too high or low. This is the case for the mathematical expression below.

$$99 \times 1.02 = 100.98$$

Following the rules governing significant figures in multiplication, the answer can be rounded to 1.0×10^2. What is wrong with rounding this answer to two significant figures?

Answer: Assume that absolute uncertainty in a value is equal to ±1 in the last place. An absolute uncertainty of ±1 in the value 99 is approximately a 1% relative uncertainty. Likewise, the absolute and relative uncertainty of 1.02 is ±0.01 and 1%, respectively. If both values have percent relative uncertainties of 1%, then their product ought to have an uncertainty close to 1%. But, by rounding to 2 significant figures, we have estimated the percent relative uncertainty to be 10%.

Chapter 2
ATOMS, MOLECULES AND IONS

CHAPTER OUTLINE

KEY TERMS: ball and stick model, space filling model

2.7 Ionic Compounds: Formulas, Names, and Properties(p. 70)

KEY TERMS: molecular compounds, ionic compounds, ions, cations, anions, monatomic ion, polyatomic ion

Formulas of Ionic Compounds (p. 74)

Table 2.4 Formulas and Names of Some Common Polyatomic Ions (p. 74)

Example 2.4: Ionic Compound Formulas (p. 75)

Example 2.5: Ionic Compound Formulas (p. 75)

Names of Ions (p. 76)

KEY TERMS: halide ions, oxoanions

Properties of Ionic Compounds (p. 78)

KEY TERMS: electrostatic forces, Coulomb's Law, crystal lattice, formula unit

PROBLEM SOLVING TIP 2.2: Is a Compound Ionic? (p. 80)

2.8 Molecular Compounds: Formulas and Names (p. 80)

KEY TERMS: binary compounds

2.9 Atoms, Molecules and The Mole (p. 82)

KEY TERMS: mole, Avogadro's number,

Historical Perspective: Amedeo Avogadro and His Number (p. 83)

Example 2.6: Mass, Moles and Atoms (p. 85)

Molecules, compounds and Molar Mass (p. 85)

KEY TERMS: molar mass, formula weight

Example 2.7: Molar Mass and Moles (p. 87)

2.10 Describing Compound Formulas (p. 88)

Percent Composition (p. 88)

KEY TERMS: percentage composition

Example 2.8: Using Percent Composition (p. 89)

Empirical and Molecular Formulas from Percent Composition (p. 90)

KEY TERMS: empirical formula

Example 2.9: Calculating a Formula from Percent Composition (p. 91)

PROBLEM SOLVING TIP 2.3 *Finding* Empirical and Molecular Formulas (p. 91)

Determining a Formula from Mass Data (p. 91)

Example 2.10: Formula of a Compound from Combining Masses (p. 93)

Example 2.11: Determining a Formula from Mass data (p. 93)

A Closer Look: Mass Spectrometry, Molar Mass, and Isotopes (p. 95)

2.11 Hydrated Compounds

KEY TERMS: hydrated compounds, anhydrous compound

Example: 2.12: Determining the Formula of a Hydrated Compound (p. 97)

Case Study: What's in Those French Fries? (p. 96)

CHAPTER OBJECTIVES

- Describe atomic structure and define atomic number and mass number (2.1-2.2)
- Understand the nature of isotopes, and calculate atomic weight from isotopic abundances and isotopic masses (2.2-2.4)
- Know the terminology of the periodic table (2.5)
- Interpret, predict, and write formulas for ionic and molecular compounds (2.6-2.7)

- Name ionic and molecular compounds (2.4, 2.7-2.8)
- Understand some properties of ionic compounds (2.3 and 2.7)
- Explain the concept of the mole, and use molar mass in calculations (2.9)
- Derive compound formulas from experimental data (2.10-2.1)

LESSON PLAN

AP Standards:

I.A1	Evidence for the atomic theory
I.A.2	Atomic masses: determination by chemical and physical means
I.A.3	Atomic number and mass number: isotopes
II.A.2.b	Avogadro's hypothesis and the mole concept
IV.2	Relationships in the periodic table: horizontal, vertical, and diagonal with examples from alkali metals, alkaline earth metals, halogens, and the first series of transition elements
I.B.2	Molecular models
I.B.3	Geometry of molecules and ions
III.B.3	Mass volume relations with emphasis on the mole concept, including empirical

formulas and limiting reagents.

Suggested Time:
Seven traditional class or 4 blocks. This chapter should be partly review however much of the information is critical for further study.

ASSESSMENT

Exercises:

2.1	Atomic Composition (p. 53)
2.2	Isotopes (p. 54)
2.3	Calculating Atomic Weight (p. 57)
2.4	The Periodic Table (p. 67)
2.5	Molecular Formulas (p. 69)
2.6	Predicting Ion Charges (p. 73)
2.7	Formulas of Ionic Compounds (p. 76)
2.8	Names of Ionic Compounds (p. 78)
2.9	Coulomb's Law (p. 80)
2.10	Naming Compounds of the Nonmetals (p. 82)
2.11	Mass/Mole Conversions (p. 85)
2.12	Atoms (p. 85)
2.13	Molar Mass and Moles-to-Mass Conversions (p. 88)
2.14	Percent Composition (p. 90)
2.15	Empirical and Molecular Formulas (p. 92)
2.16	Calculating a Formula from Percent Composition (p. 92)
2.17	Calculating a Formula from Percent Composition (p. 92)
2.18	Determining a Formula from Combining Masses (p. 94)
2.19	Determining the Formula of a Hydrated Compound (p. 98)

CHAPTER 2: ATOMS, Molecules and Ions

Answers to Exercises p. A41-A42
Practicing Skills pp. 100- 104, Answers IRM pp. 21-30
General Questions pp. 104-110, Answers IRM pp. 30-41
In the Laboratory Questions p. 110, Answers IRM pp. 42-43
Summary and Conceptual Questions: p. 111, Answers IRM p. 43

GLOSSARY

anhydrous compound The substance remaining after the water has been removed (usually by heating) from a hydrated compound.

allotrope Different forms of the same element that exist in the same physical state under the same conditions of temperature and pressure.

anion(s) An ion with a negative charge

actinides The series of elements between actinium and rutherfordium in the periodic table

alkali metals The metals in Group 1A of the periodic table

alkaline earth metals The metals in Group 2A of the periodic table

atomic mass unit (u) The unit of a scale of relative atomic masses of the elements: $1u = 1/12^{th}$ of the mass of a carbon atom with six protons and six neutrons

atomic number (Z) The number of protons in the nucleus of an atom of an element

atomic weight The average mass of an atom in a natural sample of the elements

Avogadro's Number The number of particles in one mole of any substance

Ball-and-stick model A diagram in which spheres represent atoms and sticks represent the bonds holding them together

binary compound(s) A compound formed from two elements

cation(s) An ion with a positive electrical charge

chalcogens The elements oxygen, sulfur, selenium, and tellurium of the periodic table.

condensed formula A variation of a molecular formula that shows groups of atoms

Coulomb's Law The force of attraction between the oppositely charged ions of an ionic compound is directly proportional to their charges and inversely proportional to the square of the distance between them

crystal lattice A solid, regular array of positive and negative ions

deuterium Hydrogen atoms consisting of one proton and one neutron, also known as heavy hydrogen.

electron (e⁻) A negatively charged subatomic particle found in the space about the nucleus

electrostatic forces Forces of attraction or repulsion caused by electric charges.

empirical formula A molecular formula showing the simplest possible ratio of atoms in a molecule

families in periodic table The vertical columns in the periodic table also known as groups

formula unit The simplest ratio of ions in an ionic compound, similar to the molecular formula of a molecular compound

formula weight The molar mass of an ionic compound.

Group 1A The alkali metals: Li, Na, K, Rb, Cs, Fr

Group 2A The alkaline earth metals: Be, Mg, Ca, Sr, Ba, Ra

Group 3A B, Al, Ga, In, Tl

Group 4A C, Si, Ge, Sn, Pb

Group 5A N, P, As, Sb, Bi

Group 6A O, S, Se, Te, Po

Group 7A The halogens: F, Cl, Br, I, At

Group 8A The noble gases: He, Ne, Ar, Kr, Xe, Rn

groups The vertical columns in the periodic table also known as chemical families.

halide ions Anions of the Group 7A elements

halogens The elements of Group 7A in the periodic table

hydrated compounds A compound in which molecules or water are associated with ions

ionic compounds A compound formed by the combination of positive and negative ions

ions atoms or groups of atoms that is not electrically neutral

CHAPTER 2: ATOMS, Molecules and Ions

isotopes Atoms with the same atomic number but different mass numbers, because of a difference in the numbers of neutrons.

lanthanides The series of elements between lanthanum and hafnium in the periodic table

law of chemical periodicity The properties of the elements are periodic functions of atomic number

main group elements The A groups in the periodic table.

mass number (A) The sum of the number of protons and neutrons in the nucleus of an atom of an element

metal(s) An element characterized by a tendency to give up electrons and be good thermal and electrical conductivity

metalloid(s) An element with properties of both metals and nonmetals also known as semi-metals

molar mass(M) The mass in grams of one mole of particles of any substance

mole (mol) The SI base unit for amount of substance

molecular compound(s) A compound formed by the combination of atoms without significant ionic character

molecular formula A formula that expresses the number of atoms of each type within one molecule of a compound

monatomic ion(s) An ion consisting of one atom bearing an electric charge

neutron An electrically neutral subatomic particle found in the nucleus

noble gases The elements in Group 8A of the periodic table

nonmetal(s) An element characterized by a lack of metallic properties

oxoanions Polyatomic anions containing oxygen

percent abundance The percentage of the atoms of a natural sample of the pure element represented by a particular isotope

percent composition The percentage of the mass of a compound represented by each of its constituent elements

periodicity The recurrence of properties as a function of atomic number

period(s) The horizontal rows in the periodic table of the elements

polyatmoic ion(s) An ion consisting of more than one atom

protium A hydrogen atom consisting of one proton and zero neutrons.

proton(s) A positively charged subatomic particle found in the nucleus

semi-metal(s) An element with properties of both metals and nonmetals also known as metalloids

space filling model Models that show the relative sizes of atoms and their proximity to each other

structural formula A variation of a molecular formula that expresses how that atoms in the compound are connected

subatomic particles A collective term for protons, neutrons, and electrons

transition elements Some elements that lie in rows 4-7 of the periodic table, comprising scandium through zinc, yttrium through cadmium, and lanthanum through mercury

tritium A radioactive hydrogen atom consisting of one proton and two neutrons

MEDIA MENU

The Media Menu section of this guide lists the interactive exercises, tutorials, active figures (animated, interactive figures) and additional resources available online for Chemistry Now, a unique web-based, assessment-centered personalized learning system for chemistry students. Go to www.cengagelearning.com/login.

Exercises
- Screen 2.13   Mass spectrometers and calculating atomic mass
- Screen 2.14   Periodic table organization
- Screen 2.15   Chemical periodicity
- Screen 2.17   Representations of molecules
- Screen 2.18   Ion Charge
- Screen 2.20   Ionic Compounds
- Screen 2.21   Ionic Compounds
- Screen 2.23   Properties of Ionic Compounds
- Screen 2.26   Naming alkanes
-

Tutorials
- Screen 2.11   The notation for symbolizing atoms
- Screen 2.12   Isotopes

CHAPTER 2: ATOMS, Molecules and Ions

- Screen 2.19   Polyatomic ions
- Screen 2.25   Naming compounds of the nonmetals
- Screen 2.25   Moles and atom conversions
- Screen 2.26   Molar mass conversions
- Screen 2.27   Determining molar mass
- Screen 2.28   Using molar mass
- Screen 2.29   Using percent composition
- Screen 2.30   Determining empirical formulas
- Screen 2.31   Determining molecular formulas

Active Figures
- 2.3   Mass Spectrometer
- 2.4   Some of the known 117 elements
- 2.12   Reaction of elements aluminum and bromine
- 2.15   Ways of depicting a molecule
- 2.17   Ions
- 2.19   Common ionic compounds based on polyatomic ions
- 2.29   Depicting hydrated cobalt (II) chloride hexa-hydrate

Additional Resources
- Screen 2.20   Relationship between cations and anions in ionic compounds
- Screen 2.23   Coulomb's Law
- Screen 2.27   Compounds and moles
- Podcast   Module 1

DIAGNOSTIC QUIZ QUESTIONS

Goal #1 Describe atomic structure and define atomic number and mass number

1.   Which of the following are true?

1. Electrons and protons have identical masses but opposite charges.
2. Most of the atom's mass is concentrated in a small, positively charged, nucleus.
3. Atoms have equal numbers of protons and neutrons.

a) 1 only   b) 2 only   c) 3 only
d) 1 and 2   e) 2 and 3

Answer: b

Goal #2 Understand the nature of isotopes, and calculate atomic weight from isotopic abundances and isotopic masses.

2.　　　Which of the following statements are correct?

1. Atomic number equals number of protons plus neutrons.
2. Mass number equals the number of neutrons.
3. An atomic mass unit equals $1/12^{th}$ the mass of a Carbon-12 atom

a) 1 only　　　　　　b) 2 only　　　　　　c) 3 only
d) 1 and 2　　　　　e) 2 and 3

Answer: c

3.　　　How many protons, neutrons, and electrons are in a fluorine-19 atom?

a) 9 protons, 9 neutrons, 10 electrons　　b) 9 protons, 10 neutrons, 9 electrons
c) 10 protons, 9 neutrons, 9 electrons　　d) 10 protons, 9 neutrons, 10 electrons
e) 10 protons; 10 neutrons; 9 electrons

Answer: b

4.　　　Which of the following atoms has the greatest number of protons?

a) ^{12}C　　　　　　b) ^{15}O　　　　　　c) ^{14}C
d) ^{15}N　　　　　　e) ^{13}C

Answer: b

5.　　　Which of the following atoms contains the greatest number of neutrons?

a) ^{28}Si　　　　　　b) ^{31}P　　　　　　c) ^{36}S
d) ^{35}Cl　　　　　　e) ^{34}S

Answer: c

6.　　　What is the identity of an atom that has 28 neutrons and a mass number of 52?

a) Cr　　　　　　b) Te　　　　　　c) Ni
d) Hg　　　　　　e) Ti

Answer: a

40

CHAPTER 2: ATOMS, Molecules and Ions

7. Silver has an average atomic mass of 107.9 u and is known to have only two naturally occurring isotopes. If 51.85% pf Ag exosts as Ag-107 (106.9051 u), what is the identify and the atomic mass of the other isotope?

   a) Ag-110; 110.1 u   b) Ag-110; 109.9 u   c) Ag-108; 107.9 u
   d) Ag-109; 109.0 u   e) Ag-107; 106.9 u

   Answer: d

8. An element consists of two isotopes. The abundance of one isotope is 60.40 % and its atomic mass is 68.9257 u. The atomic mass of the second isotope is 70.9249 u. What is the average atomic mass of the element?

   a) 69.72 u   b) 69.93 u   c) 70.13 um
   d) 139.9 u   e) 69.83 u

   Answer: a

Goal #3 Know the terminology of the periodic table

9. What alkaline earth metal is located in the fourth period?

   a) K   b) Ca   c) Ga
   d) Ge   e) Zr

   Answer: b

10. What halogen is located in the fourth period?

    a) Se   b) Kr   c) Cl
    d) Br   e) I

    Answer: d

11. Which chalcogen is located in the second period?

    a) P   b) N   c) O
    d) S   e) Si

    Answer: c

12.   Which grouping of elements is composed entirely of nonmetals?

a) iodine, indium, and xenon b) aluminum, silicon, and phosphorus
c) sulfur, neon, and bromine d) gallium, argon, and oxygen
e) silicon, bismuth, and xenon

Answer: c

13.   Which grouping of elements is composed entirely of metalloids?

a) B, As and Sb b) Si, P, and Ge c) As, Ge, and Pb
d) In, Sn, and Ge e) As, Te, and At

Answer: a

14.   The following lists of elements are all found in the human body. Which three elements are found in the highest concentrations?

a) sodium, oxygen, and magnesium b) oxygen, carbon, and hydrogen
c) selenium, oxygen, and potassium d) carbon, oxygen, and iron
e) hydrogen, sodium, and magnesium

Answer: b

Goal #4 Interpret, predict, and write formulas for ionic and molecular compounds

15.   The formula for acetic acid, CH_3CO_2H, is an example of a(n)

a) condensed formula b) mathematical formula c) structural formula
d) molecular formula e) ionic formula

Answer: a

16.   Which of the following statements are correct?

1.   Metals generally lose electrons to become cations.
2.   Nonmetals generally gain electrons to become anions.
3.   Group 2A elements form ions with a 2+ charge

a) 1 only b) 2 only c) 3 only
d) 1 and 3 e) 1, 2 and 3

Answer: e

17.	What charge is most commonly observed for an oxide ion?

a) 3-

b) 3+

c) 2-

d) 1-

e) 1+

Answer: c

18.	What is the formula for aluminum chloride?

a) $AlCl$

b) $AlCl_2$

c) $AlCl_3$

d) Al_2Cl_3

e) Al_3Cl_2

Answer: c

19.	What is the formula for silicon dioxide?

a) SiO

b) Si_2O

c) SiO_2

d) $(SiO)_2$

e) Si_3O_2

Answer: c

20.	What is the formula for lead (IV) sulfate?

a) PbS

b) $Pb(SO_3)_2$

c) $PbSO_4$

d) $Pb(SO_4)_2$

e) $Pb(SO_4)_4$

Answer: d

Goal #5 Name Ionic and Molecular Compounds

21. What is the name of $KHCO_3$?

a) hydrogen carbonate potassium ion

b) potassium hydrogen carbon trioxide

c) potassium bicarbonate

d) potassium hydrocarbonate

e) potassium carbonate

Answer: c

22.	What is the name of PCl_5?

a) phosphorus choride

b) phosphorus pentachlorine

c) phosphorus pentachloride

d) phosphorus (V) choride

e) phosphorus tetrachloride

Answer: c

23.   What is the name of Cu_2S?

a) copper sulfide
d) dicopper sulfide
b) copper (I) sulfide
e) copper disulfide
c) copper (II) sulfide

Answer: b

Goal #6 Understand some properties of ionic compounds

24.   Which of the following statements are correct?

1. The attractive forces between ions of opposite charges increase with increasing separation.
2. The greater the charges on ions, the greater the attractive forces between them.
3. Ions combine to form small molecules, such as $NaCl$.

a) 1 only
d) 1 and 2
b) 2 only
e) 1, 2 and 3
c) 3 only

Answer: b

Goal #7 Explain the concept of the mole, and use molar mass in calculations.

25.   Calculate the moles of Na in a 4.15 mg sampe.

a) 1.81×10^{-4} mol
d) 5.54×10^{3} mol
b) 9.54×10^{-2} mol
e) 9.54×10^{1} mol
c) 1.05×10^{-2} mol

Answer: a

26.   The molar mass of cesium is 132.9 g/mol. What is the mass of a single Cs atom?

a) 2.207×10^{-22} g
d) 4.531×10^{21} g
b) 1.249×10^{-26} g
e) 8.001×10^{-21} g
c) 2.763×10^{-23} g

Answer: a

27.   Calculate the moles present in 10.5 g $CaCO_3$.

a) 0.105 mol
d) 1.05×10^{3} mol
b) 0.419 mol
e) 10.5 mol
c) 9.53 mol

Answer: a

28.   How many oxygen atoms are in 0.20 g CO_2?

  a) 2.4×10^{23} oxygen atoms   b) 2.7×10^{21} oxygen atoms
  c) 5.5×10^{21} oxygen atoms   d) 1.2×10^{23} oxygen atoms
  e) 1.8×10^{21} oxygen atoms

  Answer: c

Goal #8 Derive compound formulas from experimental data

29.   Toluene is composed of 91.25% C and 8.75% H. Determine the empirical formula for
        toluene.

  a) CH   b) CH_3 c) C_4H_5
  d) C_7H_8 e) C_8H_7

  Answer: d

30.   You want to determine the value of x in a hydrated iron (II) sulfate, $FeSO_4$ x H_2O. In the
        laboratory you weigh out 1.983 g of the hydrate stale. After thorough heating, 1.084 g of
        the anhydrous salt remains. What is the value of x?

  a) 2   b) 3   c) 6
  d) 7   e) 9

  Answer: d

TESTBANK

2.1 Atomic Structure -Protons, Electrons, and Neutrons

1. Which of the following statements concerning atomic structure is/are correct?

    1. The nucleus contains most of an atom's mass.
    2. The nucleus contains all the positive charge of an atom.
    3. Electrons surround the nucleus and account for the majority of an atom's volume.

  a) 1 only   b) 2 only   c) 3 only   d) 2 and 3   e) 1, 2, and 3

  Answer: e

2. Rank the subatomic particles from least to greatest mass.

   a) electron mass = proton mass = neutron mass
   b) electron mass = neutron mass < proton mass
   c) electron mass = proton mass < neutron mass
   d) electron mass < proton mass < neutron mass
   e) electron mass < proton mass = neutron mass

Answer: d

2.2 Atomic Number and Atomic Mass

3. Atomic number is the number of __________ in the nucleus of an atom.

   a) electrons   b) protons   c) neutrons
   d) electrons and neutrons e) neutrons and protons

Answer: b

4. The atomic number of chromium is _____.

   a) 6   b) 24   c) 28   d) 48   e) 52

Answer: b

5. An atomic mass unit (u) is defined as

   a) the mass of one hydrogen-1 atom.   b) the mass of one carbon-12 atom.
   c) 1/12 the mass of one carbon-12 atom.   d) 1/8 the mass of one oxygen-16 atom.
   e) the sum of the masses of one proton, one neutron, and one electron.

Answer: c

6. What is the mass number of a nickel atom with 31 neutrons?

   a) 3   b) 28   c) 31   d) 58   e) 59

Answer: e

7. How many protons, neutrons, and electrons are in a yttrium-89 atom?

   a) 39 protons, 50 neutrons, 39 electrons   b) 39 protons, 89 neutrons, 39 electrons
   c) 39 protons, 50 neutrons, 50 electrons   d) 50 protons, 39 neutrons, 50 electrons
   e) 39 protons, 11 neutrons, 39 electrons

Answer: a

8. How many protons, neutrons, and electrons are in a hydrogen-3 atom?

   a) 1 proton, 1 neutron, 1 electron
   b) 1 proton, 2 neutrons, 1 electron
   c) 1 proton, 3 neutrons, 1 electron
   d) 2 protons, 1 neutron, 2 electrons
   e) 2 protons, 1 neutron, 1 electron

   Answer: b

9. What is the mass of sulfur-32 relative to carbon-12?

   a) 0.375
   b) 0.600
   c) 1.67
   d) 2.67
   e) 20

   Answer: d

10. Which of the following atoms contains the fewest protons?

   a) ^{128}Te
   b) ^{127}I
   c) ^{131}Xe
   d) ^{133}Cs
   e) ^{209}Bi

   Answer: a

11. Which of the following atoms contains the largest number of protons?

   a) ^{231}Pa
   b) ^{238}U
   c) ^{232}Th
   d) ^{244}Pu
   e) ^{251}Cf

   Answer: e

12. What is the atomic symbol for an element with 38 protons and 50 neutrons?

   a) $^{88}_{12}$Mg
   b) $^{50}_{38}$V
   c) $^{88}_{38}$Sr
   d) $^{88}_{50}$Sn
   e) $^{226}_{88}$Ra

   Answer: c

13. What is the identity of $^{54}_{24}$X ?

   a) Cr
   b) Xe
   c) Pt
   d) Zn
   e) Mn

   Answer: a

14. What is the atomic symbol for an element that has 24 neutrons and a mass number of 45?

   a) Tm
   b) Cr
   c) Rh
   d) Sc
   e) Dy

   Answer: d

15. How many neutrons are in an atom of arsenic-75?

   a) 33   b) 42   c) 75   d) 108   e) 117

Answer: b

16. Which of the following atoms contains the fewest protons?

   a) ^{103}Rh   b) ^{102}Pd   c) ^{106}Cd   d) ^{107}Ag   e) ^{102}Ru

Answer: e

17. Which of the following atoms contains the largest number of neutrons?

   a) $^{24}_{12}$Mg   b) $^{26}_{12}$Mg   c) $^{23}_{11}$Na   d) $^{27}_{13}$Al   e) $^{29}_{14}$Si

Answer: e

18. Which two of the atoms below have the same number of neutrons?
$$^{34}_{16}\text{S},\ ^{32}_{16}\text{S},\ ^{35}_{17}\text{Cl},\ ^{38}_{18}\text{Ar}$$

   a) $^{34}_{16}$S and $^{32}_{16}$S   b) $^{34}_{16}$S and $^{35}_{17}$Cl   c) $^{32}_{16}$S and $^{35}_{17}$Cl
   d) $^{32}_{16}$S and $^{38}_{18}$Ar   e) $^{35}_{17}$Cl and $^{38}_{18}$Ar

Answer: b

2.3 Isotopes

19. Two isotopes of a given element will have the same number of _________, but a different number of _________ in their nucleus.

   a) protons, electrons   b) electrons, protons   c) protons, neutrons
   d) neutrons, protons   e) neutrons, electrons

Answer: c

20. Which two of the following atoms are isotopes?
$$^{140}_{58}\text{Ce},\ ^{138}_{57}\text{La},\ ^{135}_{56}\text{Ba},\ ^{138}_{56}\text{Ba}$$

   a) $^{140}_{58}$Ce and $^{138}_{57}$La   b) $^{140}_{58}$Ce and $^{135}_{56}$Ba   c) $^{138}_{57}$La and $^{135}_{56}$Ba
   d) $^{138}_{57}$La and $^{138}_{56}$Ba   e) $^{135}_{56}$Ba and $^{138}_{56}$Ba

Answer: e

48

2.4 Atom Mass

21. Europium has two naturally occurring isotopes. The average mass of Eu is 151.965 u. If 47.8% of Eu is found as Eu-151 (150.919847 u), what is the mass of the other isotope?

    a) 1.04 u   b) 144.8 u   c) 152.9 u   d) 153.1 u   e) 158.1u

    Answer: c

22. An element consists of two isotopes. The abundance of one isotope is 60.1% and its atomic mass is 68.9256 u. The atomic mass of the second isotope is 70.9247 u. What is the average atomic mass of the element?

    a) 69.7 u   b) 69.9 u   c) 70.1 u   d) 84.1 u   e) 139.9 u

    Answer: a

23. Copper has an average atomic mass of 63.55 u. If 69.17% of Cu exists as Cu-63 (62.93960 u), what is the identity and the atomic mass of the other isotope?

    a) Cu-64; 63.82 u   b) Cu-64; 64.16 u   c) Cu-65; 64.16 u
    d) Cu-65; 64.92 u   e) Cu-66; 65.91 u

    Answer: d

24. Boron has two stable isotopes with masses of 10.01294 u and 11.00931 u. The average molar mass of boron is 10.81 u. What is the percent abundance of each isotope?

    a) 18.9% B-10 and 81.1% B-11   b) 20.0% B-10 and 80.0% B-11
    c) 48.6% B-10 and 51.4% B-11   d) 80.0% B-10 and 20.0% B-11
    e) 81.1% B-10 and 18.9% B-11

    Answer: b

2.5 The Periodic Table

25. How many elements are in the second period of the periodic table?

    a) 6   b) 8   c) 18   d) 32   e) 40

    Answer: b

26. How many elements are in the sixth period of the periodic table?

   a) 5   b) 6   c) 8   d) 18   e) 32

   Answer: e

27. What element is in the fifth period in Group 3A?

   a) Si   b) Ga   c) In   d) Sb   e) Tl

   Answer: c

28. What alkaline earth metal is in the fifth period?

   a) Rb   b) Sr   c) Zr   d) In   e) Bi

   Answer: b

29. What halogen is in the second period?

   a) N   b) O   c) F   d) Ne   e) Ar

   Answer: c

30. Which three elements are likely to have similar chemical and physical properties?

   a) boron, silicon, and germanium   b) sodium, magnesium, and aluminum
   c) magnesium, calcium, and strontium   d) uranium, plutonium, and americium
   e) carbon, nitrogen, and oxygen

   Answer: c

31. In which group of the periodic table do the elements all consist of diatomic molecules?

   a) 2A   b) 5A   c) 6A   d) 7A   e) 8A

   Answer: d

32. Which of the following elements is not a metalloid?

   a) boron   b) gallium   c) germanium   d) arsenic   e) tellurium

   Answer: b

2.6 Molecules, Compounds, and Formulas

33. The formula for acetic acid, CH_3CO_2H, is an example of a(n)

    a) condensed formula.   b) empirical formula.   c) structural formula.
    d) ionic compound formula. e) mass spectrum.

    Answer: a

34. $C_2H_2F_4$ is the formula for two possible molecules. $C_2H_2F_4$ is an example of a(n)

    a) structural formula.   b) empirical formula.   c) condensed formula.
    d) space-filling model.   e) molecular formula.

    Answer: e

2.7 Ionic Compounds: Formulas, Names, and Properties

35. A sulfide ion has _________ electrons.

    a) 14   b) 15   c) 16   d) 17   e) 18

    Answer: e

36. A magnesium ion has _________ electrons.

    a) 10   b) 11   c) 12   d) 13   e) 14

    Answer: a

37. Which atom is most likely to form a 1- ion?

    a) K   b) Mg   c) P   d) Br   e) Ar

    Answer: d

38. Which atom is most likely to form a 2+ ion?

    a) Sc   b) Sr   c) Ga   d) S   e) F

    Answer: b

39. Identify the ions present in $KClO_4$.

a) K^+, Cl^-, and O^{2-} b) KCl^+, and O_4^- c) K^+ and ClO_4^-
d) KCl^{2+} and O_4^{2-} e) K^{2+} and ClO_4^{2-}

Answer: c

40. Identify the ions in $CaHPO_4$.

a) Ca^{2+} and PO_4^{3-} b) Ca^{2+} and HPO_4^{2-} c) Ca^+ and HPO_4^-
d) Ca^{3+} and HPO_4^{3-} e) Ca^{2+}, H^+, P^{3-}, and O^{2-}

Answer: b

41. What charge is likely on a monatomic silver cation?

a) 2- b) 1- c) 1+ d) 2+ e) 3+

Answer: c

42. For a nonmetal in Group 6A of the periodic table, the most common monatomic ion will
have a charge of _________.

a) 3- b) 2- c) 1- d) 1+ e) 2+

Answer: b

43. What is the correct formula for an ionic compound that contains aluminum ions and sulfide
ions?

a) AlS b) AlS_2 c) Al_2S d) Al_2S_3 AlS_3

Answer: d

44. Which of the following formulas is not correct?

a) $Al_3(CO_3)_2$ b) $KClO_4$ c) BaO d) $Ca(NO_3)_2$ e) Na_2HPO_4

Answer: a

45. What is the correct formula for an ionic compound that contains barium ions and phosphate
ions?

a) $BaPO_4$ b) Ba_3P_2 c) $Ba_2(PO_4)_3$ d) $Ba(PO_4)_2$ e) $Ba_3(PO_4)_2$

Answer: e

46. Sodium sulfate has the chemical formula Na_2SO_4. Based on this information, the formula for titanium(IV) sulfate is ___________.

a) $TiSO_4$ b) $Ti(SO_4)_4$ c) $Ti(SO_4)_2$ d) Ti_2SO_4 e) $Ti_3(SO_4)_2$

Answer: c

47. What is the charge on the copper ion in Cu_3P?

a) 3-   b) 1-   c) 0   d) 1+   e) 3+

Answer: d

48. What is the correct formula for calcium nitrate?

a) CaN b) Ca_3N_2 c) $CaNO_2$ d) $Ca_3(NO_3)_2$ e) $Ca(NO_3)_2$

Answer: e

49. What is the correct formula for potassium hydrogen carbonate?

a) KCO_3 b) K_2CO_3 c) $KHCO_3$ d) K_2HCO_3 e) KH_2CO_3

Answer: c

50. What is the correct formula for ammonium fluoride?

a) NH_4F b) $(NH_4)_2F$ c) NH_3F d) NH_4F_2 e) NH_2F

Answer: a

51. What is the correct formula for iron(III) bromide?

a) $FeBr$ b) $FeBr_3$ c) Fe_2Br_3 d) Fe_3Br_2 e) Fe_3Br

Answer: b

52. What is the correct formula for cobalt(III) carbonate?

a) $CoCO_3$ b) $Co_3(CO_3)_2$ c) $Co_2(CO_3)_3$ d) $Co(CO_3)_3$ e) Co_3CO_3

Answer: c

53. What is the correct name for NH_4NO_3?

a) ammonia hydrogen nitrate
b) ammonia hydrogen nitride
c) ammonium nitric acid
d) ammonium nitrate
e) ammonium nitride

Answer: d

54. What is the correct name for CaI_2?

a) calcium diiodide
b) calcium diiodine
c) calcium(II) iodide
d) calcium iodide
e) iodine calcide

Answer: d

55. What is the correct name for Co_2S_3?

a) cobalt sulfide
b) dicobalt trisulfate
c) dicobalt trisulfide
d) cobalt(II) sulfate
e) cobalt(III) sulfide

Answer: e

56. Which of the following statements concerning ionic compounds is/are correct?

1. As ion charges increase, the attraction between oppositely charged ions increases.
2. Although not electrically conductive like metals, ionic compounds are malleable.
3. Positive and negative ions are attracted to each other by electrostatic forces.

a) 1 only
b) 2 only
c) 3 only
d) 1 and 3
e) 1, 2, and 3

Answer: d

57. Predict which ionic compound has the highest melting point.

a) KBr
b) MgO
c) RbI
d) $CaBr_2$
e) CsCl

Answer: b

<u>2.8 Molecular Compounds: Formulas and Names</u>

58. What is the correct name for SF_2?

a) sulfur fluoride
b) sulfur difluoride
c) disulfur fluoride
d) disulfur difluoride
e) sulfide difluoride

Answer: b

59 What is the correct name for N_2O_3?

    a) nitrogen oxide   b) oxygen nitride   c) dinitrogen trioxide
    d) nitrogen trioxide   e) trioxygen dinitride

Answer: c

60. What is the common name for PH_3?

    a) laughing gas   b) hydrazine   c) nitroglycerin
    d) ammonia   e) phosphine

Answer: e

2.9 Atoms, Molecules, and the Mole

61. You have 0.575 mole of each of the following elements: C, Cl, Ca, Cr, and Cd. Which sample has the greatest mass?

    a) C   b) Cl   c) Ca   d) Cr   e) Cd

Answer: e

62. You have 7.57 g of each of the following elements: Si, S, Se, Sr, and Sn. Which sample contains the largest number of atoms?

    a) Si   b) S   c) Se   d) Sr   e) Sn

Answer: a

63. Calculate the number of moles in 0.093 g Zn.

    a) 0.0014 mol   b) 0.16 mol   c) 6.1 mol   d) 7.0×10^2 mol   e) 8.6×10^{20} mol

Answer: a

64. What is the mass of 0.018 mol Mg?

    a) 4.0×10^{-23} g   b) 7.4×10^{-4} g   c) 0.44 g   d) 2.3 g   e) 1.4×10^3 g

Answer: c

65. A 0.10 g sample of silicon contains __________ Si atoms.

a) 5.9×10^{-27} b) 0.0036   c) 2.1×10^{21} d) 1.7×10^{24} e) 1.7×10^{26}

Answer: c

66. The molar mass of platinum is 195.08 g/mol. What is the mass of 1.00×10^2 Pt atoms?

a) 8.51×10^{-25} g   b) 3.24×10^{-24} g   c) 1.67×10^{-22} g
d) 3.24×10^{-22} g   e) 3.24×10^{-20} g

Answer: e

67. What mass of Ba contains the same number of atoms as 3.0 g Li?

a) 0.0031 g   b) 0.15 g   c) 3.0 g   d) 59 g   e) 2.9×10^3 g

Answer: d

68. A nail is coated with a 0.053 cm thick layer of zinc. The surface area of the nail is 8.59 cm^2. The density of zinc is 7.13 g/cm^3. How many zinc atoms are used in the coating?

a) 5.9×10^{20} atoms   b) 3.0×10^{22} atoms   c) 3.8×10^{22} atoms
d) 2.0×10^{24} atoms   e) 1.3×10^{26} atoms

Answer: b

69. What is the molar mass of calcium chloride hexahydrate?

a) 75.53 g/mol   b) 111.0 g/mol   c) 117.0 g/mol   d) 183.6 g/mol   e) 219.1 g/mol

Answer: e

70. What is the molar mass of potassium nitrate?

a) 39.10 g/mol   b) 53.11 g/mol   c) 85.11 g/mol   d) 101.1 g/mol   e) 163.1 g/mol

Answer: d

71. Calculate the number of moles in 0.197 g As_2O_3.

a) 9.96×10^{-4} mol   b) 5.05×10^{-3} mol   c) 39.0 mol
d) 198 mol   e) 1.00×10^3 mol

Answer: a

72. What is the mass of 8.04×10^{-3} mol I_2?

    a) 3.17×10^{-5} g   b) 0.490 g   c) 0.980 g   d) 1.02 g   e) 2.04 g

Answer: e

73. What is the mass of 0.50 mol chromium(III) sulfide?

    a) 2.5×10^{-3} g   b) 5.9×10^{-3} g   c) 42 g   d) 1.0×10^{2} g   e) 110 g

Answer: d

74. How many hydrogen atoms are in 1.0 g of CH_4?

    a) 6.2×10^{-2} atoms   b) 2.5×10^{-1} atoms   c) 3.8×10^{22} atoms
    d) 1.5×10^{23} atoms   e) 3.9×10^{25} atoms

Answer: d

75. How many bromide ions are in 0.55 g of iron(III) bromide?

    a) 1.1×10^{21} ions   b) 3.4×10^{21} ions   c) 3.3×10^{23} ions
    d) 9.9×10^{23} ions   e) 2.9×10^{26} ions

Answer: b

76. If 1.00 g of an unknown molecular compound contains 8.35×10^{21} molecules, what is its molar mass?

    a) 44.0 g/mol   b) 66.4 g/mol   c) 72.1 g/mol   d) 98.1 g/mol   e) 132 g/mol

Answer: c

2.10 Describing Compound Formulas

77. What is the mass percent of iodine in calcium iodide?

    a) 13.64%   b) 24.00%   c) 66.67%   d) 76.00%   e) 86.36%

Answer: e

78. What is the mass percent of each element in sulfuric acid, H_2SO_4?

a) 2.055% H, 32.69% S, 65.25% O
b) 1.028% H, 32.69% S, 66.28% O
c) 28.57% H, 14.29% S, 57. 17% O
d) 1.028% H, 33.72% S, 65.25% O
e) 2.016% H, 32.07% S, 65.91% O

Answer: a

79. Nitrogen and oxygen form an extensive series of oxides with the general formula N_xO_y. What is the empirical formula for an oxide that contains 36.85% by mass nitrogen?

a) N_2O　　　b) NO　　　c) NO_2　　　d) N_2O_3　　　e) N_2O_5

Answer: d

80. A molecule is found to contain 64.27% by mass C, 7.191% by mass H, and 28.54% by mass O. What is the empirical formula for this molecule?

a) C_2H_6O　　　b) C_3H_4O　　　c) $C_3H_8O_2$　　　d) $C_4H_6O_2$　　　e) $C_4H_8O_3$

Answer: b

81. An ionic compound has the formula MCl_2. The mass of 0.3011 mol of the compound is 62.69 grams. What is the identity of the metal?

a) Ni　　　b) Cu　　　c) Sn　　　d) Hg　　　e) Ba

Answer: e

2.11 Hydrated Salts

82. A 3.592 g sample of hydrated magnesium bromide, $MgBr_2 \cdot xH_2O$, is dried in an oven. When the anhydrous salt is removed from the oven, its mass is 2.263 g. What is the value of x?

a) 1　　　b) 3　　　c) 6　　　d) 8　　　e) 12

Answer: c

83. A 2.000 g sample of $FeF_2 \cdot xH_2O$ is dried in an oven. When the anhydrous salt is removed from the oven, its mass is 0.789 g. What is the value of x?

a) 1　　　b) 3　　　c) 6　　　d) 8　　　e) 12

Answer: d

Short Answer Questions

84. Buckminsterfullerene is a molecule composed of sixty _________ atoms.

   Answer: carbon

85. Pure oxygen can exist as O_2 or O_3. Elements that exist in more than one distinct form are called _________.

   Answer: allotropes

86. The group ___ elements are nonmetals and all exist as diatomic molecules.

   Answer: 7A or 17

87. What are the names of four metalloids?

   Answer: boron, silicon, germanium, arsenic, (antimony, and tellurium)

88. In reactions, metals generally lose electrons to become cations, and nonmetals gain electrons to become _________.

   Answer: anions

89. In which ionic compound, NaBr or KBr, is the force of attraction between anions and cations stronger?

   Answer: The force of attraction is stronger for NaBr. The electrostatic attraction between anions and cations decreases as the separation of the ions increases. The potassium ion will be farther from the bromide ion than the sodium ion due to its larger ionic radius.

90. The numerical quantity of a mole, 6.022×10^{23}, is defined as the number of atoms in a specific mass of an element. What is the mass and the identity of the element used to define one mole?

   Answer: A mole is equal to the number of atoms in 12.00 grams of carbon-12.

Chapter 3
CHEMICAL REACTIONS

CHAPTER OUTLINE

CHAPTER OBJECTIVES

- Balance equations for simple chemical reactions.
- Understand the nature and characteristics of chemical equilibria
- Understand the nature of ionic substances dissolved in water
- Recognize common acids and bases, and understand their behavior in aqueous solution
- Recognize the common types of reactions in aqueous solutions
- Write chemical equations for the common types of reactions in aqueous solutions
- Recognize common oxidizing and reducing agents, and identify oxidation-reduction reactions

LESSON PLAN

AP Standards

III.A.2	Precipitation reactions
III.A.3	Oxidation-Reduction reactions
III.A.3.a	Oxidation number
III.A.3.b	The role of the electron in oxidation-reduction
III.B.1	Ionic and molecular species present in chemical systems: net ionic equations
III.B.2	Balancing of equations including those for redox reactions
III.C.1	Concept of dynamic equilibrium

Suggested Time:
Five traditional classes or three blocks. You may want to spend extra time on section 3.6 since ionic equations have proven to be difficult for students. Sections 3.6, 3.7, and 3.10 cover the AP standards material. The remaining sections may be done briefly, skipped or moved to a later chapter in order to save time.

ASSESSMENT

Exercises:

Answers to Exercises pp A42-A43
Practicing Skills pp. 152-155, Answers IRM pp. 49-56
General Questions pp. 155-156, Answers IRM pp. 56-59
In the Laboratory Questions pp. 156-157, Answers IRM p. 59
Summary and Conceptual Questions pp. 157, Answers IRM pp. 59-60

GLOSSARY

acid A substance that, when dissolved in pure water, increases the concentration of hydrogen ions

acid-base reaction An exchange reaction between an acid and a base producing a salt and water

acidic oxide An oxide of a nonmetal that acts as an acid

amphiprotic A substance that can behave as either a Bronsted acid or Bronsted base

aqueous solution(s) A solution in which the solvent is water

balanced chemical equation A chemical equation showing the relative amounts of reactants and products

base(s) A substance that, when dissolved in pure water, increases the concentration of hydroxide ions

basic oxide(s) An oxide of a metal that acts as a base, basic oxygen furnace

chemical equilibrium A condition in which the forward and reverse reaction rates in a chemical system are equal

combustion reaction The reaction of a compound with molecular oxygen to form products in which all elements are combined with oxygen

complete ionic equation An equation representing the reactants and products in their ionized state in solution

dynamic equilibrium A reaction in which the forward and reverse processes are occurring

electrodes A device such as a metal plate or wire for conducting electrons into and out of solutions in electrochemical cells

electrolyte A substance that ionizes in water or on melting to form an electrically conducting solution

exchange reactions A chemical reaction that proceeds by the interchange of reactant cation-anion partners

half-reactions The two chemical equations into which the equation for an oxidation-reduction reaction can be divided, one representing the oxidation process and the other the reduction process

law of conservation of mass Matter can be neither created nor destroyed

net ionic reaction(s) A chemical equation involving only those substances undergoing chemical changes in the course of the reaction

neutralization reaction(s) An acid-base reaction that produces a neutral solution of a salt and water

nonelectrolyte A substance that dissolves in water to form an electrically nonconducting solution

oxidation The loss of electrons by an atom, ion, or molecule

oxidation number A number assigned to each element in a compound in order to keep track of the elecgtrons during a reaction

oxidation-reduction reaction A reaction involving the transfer of one or more electrons from one species to another

oxidizing agent The substance that accepts electrons and is reduced in an oxidation-reduction reaction

precipitate A water-insoluble solid product of a reaction, usually of water-soluble reactants

precipitation reaction An exchange reaction that produces an insoluble salt, or precipitate, from soluble reactants

product-favored reaction A system in which, when a reaction appears to stop, products predominate over reactants

product A substance formed in a chemical reaction

reactant-favored reaction A system in which, when a reaction appears to stop, reactants predominate over products

reactants A starting substance in a chemical reaction

redox reaction Abbreviation for oxidation-reduction reaction where electrons are transferred between substances

reducing agent The substance that donates electrons and is oxidized in an oxidation-reduction reaction

reduction The gain of electrons by an atom, ion, or molecule

solute The substance dissolved in a solvent to form a solution

solution A homogeneous mixture

solvent The medium in which a solute is dissolved to form a solution

spectator ion An ion that is present in a solution in which a reaction takes place, but that is not involved in the net process

stoichiometric coefficients the multiplying numbers assigned to the species in a chemical equation in order to balance the equation

stoichiometry The study of the quantitative relations between amounts of reactants and products

strong acid An acid that ionizes completely in aqueous solution

strong base A base that ionizes completely in aqueous solution

strong electrolyte A substance that dissolves in water to form a good conductor of electricity

weak acid An acid that is only partially ionized in aqueous solution

weak base A base that is only partially ionized in aqueous solution

weak electrolyte A substance that dissolves in water to form a poor conductor of electricity

MEDIA MENU

The Media Menu section of this guide lists the interactive exercises, tutorials, active figures (animated, interactive figures) and additional resources available online for Chemistry Now, a unique web-based, assessment-centered personalized learning system for chemistry students. Go to www.cengagelearning.com/login.

Exercises
- Screen 3.2   Conservation of mass in reactions
- Screen 3.3   Conservation of mass in reactions
- Screen 3.4   Balancing combustion chemical equations
- Screen 3.5   Dissolving of an ionic compound
- Screen 3.13   Acid-base reactions
- Screen 3.16   Oxidation numbers
- Screen 3.17   Redox reactions

Self-study modules
- Screen 3.6   Types of electrolytes
- Screen 3.7   Solubility of ionic compounds in water
- Screen 3.8   Types of reactions in solution
- Screen 3.12   Weak and strong bases
- Screen 3.8   Four reaction types

Tutorials
- Screen 3.9   Precipitation reactions
- Screen 3.10   Writing net ionic equations
- Screen 3.14   Identifying type of reaction

Active Figures
- 3.9   Classifying solutions based on conductivity
- 3.10   Guidelines to predict solubility
- 3.14   Acid-base reaction
- 3.19   Copper oxidation

Additional Resources
- Screen 3.11   Acid ionization simulation
- Screen 3.14   Four gases produced in reactions videos
- Podcast Module 5   Ions and molecules in aqueous solution

- Podcast Module 6   Precipitation reactions

DIAGNOSTIC QUIZ QUESTIONS

Goal #1 Balance equations for simple chemical reactions

1. Iron rusts according to the following equation:
$$\_\_ Fe(s) \ + \ \_\_ O_2(g) \ \rightarrow \ \_\_ Fe_2O_3(s)$$
What are the respective coefficients when the equation is balanced with the smallest integer values?

   a) 1, 1, 1   b) 1, 3, 1   c) 2, 3, 1   d) 3, 3, 2   e) 4, 3, 2

Answer: e

2. Balance the following chemical equation:
$Na_2S \ (s) + HCl \ (aq) \rightarrow H_2S \ (g) + NaCl \ (aq)$

   a) $Na_2S \ (s) + HCl \ (aq) \rightarrow H_2S \ (g) + NaCl \ (aq)$
   b) $Na_2S \ (s) + 2\ HCl \ (aq) \rightarrow H_2S \ (g) + 2\ NaCl \ (aq)$
   c) $NaS \ (s) + HCl \ (aq) \rightarrow HS \ (g) + NaCl \ (aq)$
   d) $Na_2S \ (s) + H_2Cl \ (aq) \rightarrow H_2S \ (g) + Na_2Cl \ (aq)$
   e) $Na_2S \ (s) + HCl \ (aq) \rightarrow H_2S \ (g) + 2\ NaCl \ (aq)$

Answer: b

Goal #2 Understand the nature of and characteristics of chemical equilibria.

3. Consider the reaction $A(aq) \rightleftharpoons 2\ B(aq)$ at 25 °C. If 0.50 moles of A are initially present in a 1.0 L flask at 25 °C, what change in concentrations (if any) will occur in time?

   a) [A] will decrease and [B] will decrease.
   b) [A] will decrease and [B] will increase.
   c) [A] will increase and [B] will decrease.
   d) [A] will increase and [B] will increase.
   e) [A] and [B] remain unchanged.

Answer: c

CHAPTER 3: Chemical Reactions

Goal #3 Understand the nature of ionic substances dissolved in water.

4. Which of the following statements is/are correct?

> 1. Water soluble ionic compounds, such as NaCl, are strong electrolytes.
> 2. Some molecular compounds, such as HCl, are strong electrolytes.
> 3. Some molecular compounds, such as acetic acid, are weak electrolytes.

   a) 1 only   b) 2 only   c) 3 only   d) 1 and 2   e) 1, 2, and 3

   Answer: e

5. What ions are formed when $NaHCO_3$ dissolves in water?

   a) Na^+, H^+, and CO_3^- b) NaH^+, and CO_3^{-2} c) Na^+ and HCO_3^-
   d) Na^-, H^+, and CO_3^- e) Na^{+2}, HCO_3^{-2}

   Answer: c

Goal # 4 Recognize common acids and bases, and understand their behavior in aqueous solution.

6. Which of the following compounds is a weak acid?

   a) HCl   b) CH_3CO_2H c) HNO_3 d) $HClO_4$ e) H_2SO_4

   Answer: b

7. Which of the following compounds is a weak base?

   a) KOH   b) H_2CO_3 c) LiCl   d) NH_3 e) HNO_3

   Answer: d

8. What is the net ionic equation for the reaction of aqueous perchloric acid and aqueous potassium hydroxide?

   a) $HClO_4(aq) + OH^-(aq) \rightarrow H_2O(\ell) + ClO_4^-(aq)$
   b) $ClO_4^-(aq) + K^+(aq) \rightarrow KClO_4(s)$
   c) $HClO_4(aq) + KOH(aq) \rightarrow KClO_4(aq) + H_2O(\ell)$
   d) $ClO_4^-(aq) + K^+(aq) \rightarrow KClO_4(aq)$
   e) $H^+(aq) + OH^-(aq) \rightarrow H_2O(\ell)$

   Answer: e

9. Which of the following compounds will produce an acidic solution when dissolved in water?

   a) CaO   b) Na_2SO_4 c) Fe_2O_3 d) SO_2 e) Li_2O

   Answer: d

Goal #5 Recognize and write equations for the common types of reactions in aqueous solution

Goal #6 Write chemical equations for the common types of reactions in aqueous solution.

10. The chemical equation below,

   $$AlCl_3 \text{ (aq)} + 3\ NaOH \text{ (aq)} \rightarrow Al(OH)_3 \text{ (s)} + 3\ NaCl \text{ (aq)}$$

   is an example of a(n) ___________________ reaction.

   a) precipitation   b) oxidation-reduction   c) strong acid-strong base
   d) gas-forming   e) combustion

   Answer: a

11. What is the net ionic equation for the reaction of aqueous sodium hydroxide and aqueous iron(II) chloride?

   a) $Na^+(aq) + OH^-(aq) \rightarrow NaOH(s)$ b) $Na^+(aq) + Cl^-(aq) \rightarrow NaCl(s)$
   c) $Fe^{2+}(aq) + 2\ OH^-(aq) \rightarrow Fe(OH)_2(s)$ d) $Fe^{2+}(aq) + OH^-(aq) \rightarrow FeOH^+(s)$
   e) $Fe^{2+}(aq) + 2\ Cl^-(aq) \rightarrow FeCl_2(s)$

   Answer: c

12. Write a balanced equation for the reaction between zinc carbonate with nitric acid.

   a) $ZnCO_3(s) + 2\ HNO_3(aq) \rightarrow Zn(NO_3)_2 \text{ (aq)} + H_2O\ (\ell) + CO_2(g)$
   b) $ZnCO_3(s) + HNO_3(aq) \rightarrow ZnO \text{ (aq)} + HNO_3(aq) + CO_2(g)$
   c) $ZnCO_3(s) + 2\ HNO_3(aq) \rightarrow Zn(OH)_2 \text{ (aq)} + N_2O_5 \text{ (s)} + CO_2(g)$
   d) $ZnCO_3(s) + 2\ HNO_3(aq) \rightarrow Zn(NO_3)_2 \text{ (aq)} + H_2CO_3(aq)$
   e) $ZnCO_3(s) + 2\ HNO_2(aq) \rightarrow Zn(NO_2)_2 \text{ (aq)} + H_2CO_3(aq)$

   Answer: a

Goal #7 Recognize common oxidizing and reducing agents and identify oxidation-reduction reactions.

13. Which species is oxidized in the reaction below?

$$I^-(aq) + ClO^-(aq) \rightarrow IO^-(aq) + Cl^-(aq)$$

a) I^- b) H_2O c) Cl^- d) IO^- e) ClO^-

Answer: a

14. What is the oxidation number of iodine in potassium periodate, KIO_4?

a) -1   b) 0   c) +3   d) +5   e) +7

Answer: e

15. Which molecule in the reaction below is the reducing agent?

$$2\ C_2H_6(g) + 7\ O_2(g) \rightarrow 4\ CO_2(g) + 6\ H_2O(g)$$

a) C_2H_6 b) O_2 c) H_2O d) CO_2 e) None

Answer: a

16. All of the following are oxidation-reduction reactions EXCEPT

a) $CaCO_3(s) \rightarrow CaO(s) + CO_2(g)$ b) $2\ Na(s) + Br_2(g) \rightarrow 2\ NaBr(g)$
c) $Fe(s) + 2\ HCl(aq) \rightarrow FeCl_2(aq) + H_2(g)$ d) $C(s) + O_2(g) \rightarrow 2\ CO(g)$
e) $2\ H_2O(\ell) \rightarrow 2\ H_2(g) + O_2(g)$

Answer: a

TESTBANK

<u>3.2 Balancing Chemical Equations</u>

1. Nitroglycerin decomposes violently according to the balanced chemical equation below.

$$2\, C_3H_5(NO_3)_3(\ell) \rightarrow 3\, N_2(g) + \tfrac{1}{2}\, O_2(g) + 6\, CO_2(g) + 5\, H_2O(g)$$

Which one of the following statements concerning this reaction is correct?

a) Two grams of nitroglycerine will produce one gram of nitrogen and one gram of water.
b) Two grams of nitroglycerine will produce three grams of nitrogen and six grams of water.
c) One mole of nitroglycerine will produce one mole of each gaseous product.
d) Two moles of nitroglycerine will produce a total of two moles of gaseous products.
e) Two moles of nitroglycerine will produce three moles of nitrogen and five moles of water.

Answer: e

2. Iron rusts according to the following equation:

$$\_\_\ Fe(s) + \_\_\ O_2(g) \rightarrow \_\_\ Fe_2O_3(s)$$

What are the respective coefficients when the equation is balanced with the smallest integer values?

a) 1, 1, 1   b) 1, 3, 1   c) 2, 3, 1   d) 3, 3, 2   e) 4, 3, 2

Answer: e

3. When ethanol undergoes complete combustion, the products are carbon dioxide and water.

$$\_\_\ C_2H_5OH(\ell) + \_\_\ O_2(g) \rightarrow \_\_\ CO_2(g) + \_\_\ H_2O(g)$$

What are the respective coefficients when the equation is balanced with the smallest whole numbers?

a) 1, 1, 1, 1   b) 1, 2, 1, 3   c) 2, 3, 4, 6   d) 1, 3, 2, 3   e) 2, 7, 4, 6

Answer: d

4. The reaction of elemental chlorine with potassium iodide yields elemental iodine and potassium chloride. Write a balanced chemical equation for this reaction.

a) $Cl_2(g) + KI(s) \rightarrow I(s) + KCl_2(s)$ b) $Cl_2(g) + 2\,KI(s) \rightarrow I_2(s) + 2\,KCl(s)$
c) $Cl_2(g) + KI_2(s) \rightarrow I_2(s) + KCl_2(s)$ d) $Cl(g) + KI(s) \rightarrow I(s) + KCl(s)$
e) $Cl_2(g) + 2\,K_2I(s) \rightarrow I_2(s) + 2\,K_2Cl(s)$

Answer: b

5. Metals react with oxygen gas to produce oxides with the general formula M_xO_y. Write a balanced chemical equation for the reaction of copper to yield copper(I) oxide.

a) $4\,Cu(s) + O_2(g) \rightarrow 2\,Cu_2O(s)$ b) $Cu + O_2(g) \rightarrow CuO_2(s)$
c) $2\,Cu(s) + O_2(g) \rightarrow 2\,CuO(s)$ d) $Cu(s) + O(g) \rightarrow CuO(s)$
e) $2\,Cu(s) + O(g) \rightarrow 2\,Cu_2O(s)$

Answer: a

6. The products of the complete combustion of octane, C_8H_{18}, are carbon dioxide and water. Write a balanced chemical equation for this reaction.

a) $C_8H_{18}(\ell) \rightarrow 8\,C(s) + 9\,H_2(g)$

b) $C_8H_{18}(\ell) + 25\,O_2(g) \rightarrow 8\,CO_2(g) + 9\,H_2O(g)$

c) $2\,C_8H_{18}(\ell) + 25\,O_2(g) \rightarrow 16\,CO_2(g) + 18\,H_2O(g)$

d) $C_8H_{18}(\ell) + 16\,O_2(g) \rightarrow 8\,CO_2(g) + 9\,H_2(g)$

e) $2\,C_8H_{18}(\ell) + 17\,O_2(g) \rightarrow 16\,CO(g) + 18\,H_2O(g)$

Answer: c

7. What is the balanced chemical equation for the complete combustion of methanol, CH_3OH?

a) $CH_3OH(\ell) \rightarrow CO(g) + 2\,H_2(g)$

b) $CH_3OH(\ell) \rightarrow CH_2(g) + H_2O(g)$

c) $CH_3OH(\ell) + O_2(g) \rightarrow CO_2(g) + H_2O(g)$

d) $2\,CH_3OH(\ell) + 3\,O_2(g) \rightarrow 2\,CO_2(g) + 4\,H_2O(g)$

e) $2\,CH_3OH(\ell) + 4\,O_2(g) \rightarrow 2\,CO_2(g) + 4\,H_2O(g)$

Answer: d

8. Which of the following statements is/are correct?
    1. A solute is a mixture of a solvent and a soluble compound.
    2. A solution is a homogeneous mixture of a solvent and a solute.
    3. Water is a solvent that is commonly used by chemists.

    a) 1 only   b) 2 only   c) 3 only   d) 1 and 2   e) 2 and 3

    Answer: e

3.5 Ions and Molecules in Aqueous Solutions

9. Which of the following statements is/are correct?
    1. All ionic compounds that are soluble in water are electrolytes.
    2. All ionic compounds dissolve in water.
    3. Molecular compounds are never soluble in water.

    a) 1 only   b) 2 only   c) 3 only   d) 1 and 2   e) 2 and 3

    Answer: a

10. Which of the following statements is/are correct?
    1. Water soluble ionic compounds, such as NaCl, are strong electrolytes.
    2. Some molecular compounds, such as HCl, are strong electrolytes.
    3. Some molecular compounds, such as acetic acid, are weak electrolytes.

    a) 1 only   b) 2 only   c) 3 only   d) 1 and 2   e) 1, 2, and 3

    Answer: e

11. Which one of the following compounds is a nonelectrolyte when dissolved in water?

    a) HCl   b) $MgBr_2$ c) Cl_2 d) $Zn(NO_3)_2$ e) KI

    Answer: c

12. Which of the following statements is/are correct?
    1. Most ionic compounds containing ammonium ion are insoluble in water.
    2. Most ionic compounds containing nitrate ion are insoluble in water.
    3. Most ionic compounds containing carbonate ion are insoluble in water.

    a) 1 only   b) 2 only   c) 3 only   d) 1 and 2   e) 1, 2, and 3

    Answer: c

13. Which of the following statements is/are correct?
    1. Most ionic compounds containing phosphate ion are soluble in water.
    2. Most ionic compounds containing chloride ion are soluble in water.
    3. Most ionic compounds containing hydroxide ion are soluble in water.

    a) 1 only   b) 2 only   c) 3 only   d) 1 and 2   e) 1, 2, and 3

    Answer: b

14. Which of the following compounds are soluble in water: K_2CO_3, $CaCO_3$, $NiCO_3$, and $Fe_2(CO_3)_3$?

    a) K_2CO_3 only   b) K_2CO_3 and $CaCO_3$
    c) $CaCO_3$ and $NiCO_3$ d) $NiCO_3$ and $Fe_2(CO_3)_3$
    e) $CaCO_3$, $NiCO_3$, and $Fe_2(CO_3)_3$

    Answer: a

15. Which of the following compounds are soluble in water: $Cu(NO_3)_2$, Al_2S_3, $Ba_3(PO_4)_2$, and $CaBr_2$?

    a) $Cu(NO_3)_2$ only   b) $Cu(NO_3)_2$ and Al_2S_3 c) Al_2S_3 and $Ba_3(PO_4)_2$
    d) $Ba_3(PO_4)_2$ and $CaBr_2$ e) $Cu(NO_3)_2$ and $CaBr_2$

    Answer: e

16. All of the following compounds are insoluble in water except __________.

    a) $BaSO_4$ b) AgI c) CuS d) $Ca(ClO_4)_2$ e) $PbCrO_4$

    Answer: d

<u>3.6 Precipitation Reactions</u>

17. A precipitate will form when aqueous $Pb(NO_3)_2$ is added to an aqueous solution of __________.

    a) $Cu(NO_3)_2$ b) $CaBr_2$ c) $NaCH_3CO_2$ d) $Ca(ClO_4)_2$ e) $NaNO_3$

    Answer: b

18. A precipitate will form when aqueous nickel(II) chloride is added to an aqueous solution of
_________.

a) SrI_2 b) $Cu(NO_3)_2$ c) KOH   d) Na_2SO_4 e) NaF

Answer: c

19. If an aqueous solution of _________ is added to a mixture of Ba^{2+} and Fe^{3+}, the barium ion
will precipitate, but the iron ion will remain in solution.

a) NaOH   b) Na_2SO_4 c) K_3PO_4 d) KCl   e) $Pb(NO_3)_2$

Answer: b

20. If an aqueous solution of _________ is added to a mixture of F^- and SO_4^{2-}, the fluoride ion
will precipitate, but the sulfate ion will remain in solution.

a) LiBr   b) HNO_3 c) $Pb(ClO_4)_2$ d) $AgNO_3$ e) $AlCl_3$

Answer: d

21. Write a balanced chemical equation for the reaction of aqueous solutions of sodium sulfide
and zinc(II) chloride.

a) $Na_2S(aq) + ZnCl_2(aq) \rightarrow ZnS(s) + 2\,NaCl(aq)$
b) $Na_2S(aq) + ZnCl_2(aq) \rightarrow ZnS(s) + 2\,NaCl(s)$
c) $Na_2S(aq) + ZnCl_2(aq) \rightarrow Na_2Zn(s) + SCl_2(aq)$
d) $Na_2S(aq) + ZnCl_2(aq) \rightarrow Na_2Zn(aq) + SCl_2(s)$
e) No reaction occurs.

Answer: a

22. Write a balanced chemical equation for the reaction of aqueous solutions of aluminum nitrate
and potassium phosphate.

a) $Al(NO_3)_3(aq) + K_3PO_4(aq) \rightarrow K_3Al(s) + PO_4(NO_3)_3(aq)$
b) $3\,Al(NO_3)_2(aq) + 2\,K_3PO_4(aq) \rightarrow Al_3(PO_4)_2(s) + 6\,KNO_3(aq)$
c) $AlNO_3(aq) + KPO_4(aq) \rightarrow AlPO_4(s) + KNO_3(aq)$
d) $AlNO_3(aq) + KPO_4(aq) \rightarrow AlPO_4(aq) + KNO_3(s)$
e) $Al(NO_3)_3(aq) + K_3PO_4(aq) \rightarrow AlPO_4(s) + 3\,KNO_3(aq)$

Answer: e

23. What is the net ionic equation for the reaction of aqueous calcium acetate and aqueous sodium carbonate?

    a) $Ca^{2+}(aq) + 2\ CH_3CO_2^-(aq) \rightarrow Ca(CH_3CO_2)_2(s)$
    b) $Na^+(aq) + CH_3CO_2^-(aq) \rightarrow NaCH_3CO_2(aq)$
    c) $Na^+(aq) + CH_3CO_2^-(aq) \rightarrow NaCH_3CO_2(s)$
    d) $Ca^{2+}(aq) + CO_3^{2-}(aq) \rightarrow CaCO_3(s)$
    e) $Ca^{2+}(aq) + 2\ Na^+(aq) \rightarrow CaNa_2(s)$

    Answer: d

24. What is the net ionic equation for the reaction of aqueous sodium hydroxide and aqueous iron(II) chloride?

    a) $Na^+(aq) + OH^-(aq) \rightarrow NaOH(s)$
    b) $Na^+(aq) + Cl^-(aq) \rightarrow NaCl(s)$
    c) $Fe^{2+}(aq) + 2\ OH^-(aq) \rightarrow Fe(OH)_2(s)$
    d) $Fe^{2+}(aq) + OH^-(aq) \rightarrow FeOH^+(s)$
    e) $Fe^{2+}(aq) + 2\ Cl^-(aq) \rightarrow FeCl_2(s)$

    Answer: c

<u>3.7 Acids and Bases</u>

25. Which of the following compounds is a weak acid?

    a) HCl b) CH_3CO_2H c) HNO_3 d) $HClO_4$ e) H_2SO_4

    Answer: b

26. Which of the following compounds is a weak base?

    a) KOH b) H_2CO_3 c) $LiCl$ d) NH_3 e) HNO_3

    Answer: d

27. Sulfuric acid is the product of the reaction of ________ and H_2O.

    a) SO_3 b) SO_2 c) S_8 d) H_2S e) SO_4^{2-}

    Answer: a

28. What is the balanced equation for the reaction that occurs when bicarbonate ion donates a proton to water?

a) $HCO_3^-(aq) + H_2O(\ell) \rightleftharpoons H_2CO_3(aq) + H_3O^+(aq)$

b) $CO_3^{2-}(aq) + H_2O(\ell) \rightleftharpoons HCO_3^-(aq) + OH^-(aq)$

c) $HCO_3^-(aq) + H_2O(\ell) \rightleftharpoons CO_3^{2-}(aq) + H_3O^+(aq)$

d) $HCO_3^-(aq) + H_2O(\ell) \rightleftharpoons H_2CO_3(aq) + OH^-(aq)$

e) $H_2CO_3(aq) + H_2O(\ell) \rightleftharpoons HCO_3^-(aq) + H_3O^+(aq)$

Answer: c

29. What is the balanced equation for hydrogen phosphate ion (HPO_4^{2-}) acting as a Brønsted base in a reaction with water?

a) $HPO_4^{2-}(aq) + H_2O(\ell) \rightleftharpoons PO_4^{3-}(aq) + H_3O^+(aq)$

b) $HPO_4^{2-}(aq) + H_2O(\ell) \rightleftharpoons H_2PO_4^-(aq) + OH^-(aq)$

c) $HPO_4^{2-}(aq) + H_2O(\ell) \rightleftharpoons H_2PO_4^-(aq) + H_3O^+(aq)$

d) $HPO_4^{2-}(aq) + H_2O(\ell) \rightleftharpoons PO_4^{3-}(aq) + OH^-(aq)$

e) $HPO_4^{2-}(aq) + H_2O(\ell) \rightleftharpoons H_3PO_5^{2-}(aq)$

Answer: b

30. What is the net ionic equation for the reaction of aqueous perchloric acid and aqueous potassium hydroxide?

a) $HClO_4(aq) + OH^-(aq) \rightarrow H_2O(\ell) + ClO_4^-(aq)$

b) $ClO_4^-(aq) + K^+(aq) \rightarrow KClO_4(s)$

c) $HClO_4(aq) + KOH(aq) \rightarrow KClO_4(aq) + H_2O(\ell)$

d) $ClO_4^-(aq) + K^+(aq) \rightarrow KClO_4(aq)$

e) $H^+(aq) + OH^-(aq) \rightarrow H_2O(\ell)$

Answer: e

31. What are the spectator ions in the reaction between aqueous hydrobromic acid and aqueous sodium hydroxide?

a) Na^+ only

b) H^+ and OH^-

c) Na^+ and Br^-

d) Br^- only

e) H^+, Br^-, Na^+, and OH^-

Answer: c

32. What are the spectator ions in the reaction between aqueous hydrochloric acid and ammonia?

    a) H^+ only

    b) Cl^- only

    c) H^+ and NH_4^+

    d) Cl^- and NH_4^+

    e) H^+, Cl^-, and NH_4^+

    Answer: b

33. Formic acid, HCO_2H, is a weak acid. Write a net ionic equation for the reaction of aqueous formic acid and aqueous potassium hydroxide.

    a) $HCO_2H(aq) + KOH(aq) \rightarrow K^+(aq) + HCO_2^-(aq) + H_2O(\ell)$

    b) $HCO_2H(aq) + H_2O(aq) \rightarrow HCO_2^-(aq) + H_3O^+(\ell)$

    c) $H^+(aq) + OH^-(aq) \rightarrow H_2O(\ell)$

    d) $HCO_2H(aq) + OH^-(aq) \rightarrow HCO_2^-(aq) + H_2O(\ell)$

    e) $H^+(aq) + KOH(aq) \rightarrow K^+(aq) + H_2O(\ell)$

    Answer: d

34. Which of the following compounds will produce an acidic solution when dissolved in water?

    a) CaO b) Na_2SO_4 c) Fe_2O_3 d) SO_2 e) Li_2O

    Answer: d

35. Metal oxides react with water to produce __________.

    a) bascs

    b) hydrogen gas

    c) oxygen gas

    d) acids

    e) hydronium ions

    Answer: a

3.8 <u>Gas-Forming Reactions</u>

36. Write a balanced net ionic equation for the reaction of barium carbonate and aqueous hydrochloric acid.

    a) $BaCO_3(s) + 2 H^+(aq) \rightarrow Ba^{2+}(aq) + CO_3^{2-}(aq) + H_2(g)$

    b) $BaCO_3(s) + 2 H^+(aq) \rightarrow Ba^{2+}(aq) + CO_2(g) + H_2O(\ell)$

    c) $BaCO_3(s) + 2 HCl(aq) \rightarrow BaCl_2(aq) + H_2CO_3(aq)$

    d) $BaCO_3(s) + 2 H^+(aq) \rightarrow Ba^{2+}(aq) + H_2CO_3(s)$

    e) $BaCO_3(s) + 2 H^+(aq) \rightarrow BaO(s) + CO_2(g) + H_2(g)$

    Answer: b

37. Write a balanced net ionic equation for the reaction of aqueous solutions of baking soda
($NaHCO_3$) and acetic acid.

a) $HCO_3^-(aq) + CH_3CO_2H(aq) \rightarrow CH_3CO_2^-(aq) + H_2O(\ell) + CO_2(g)$

b) $2\,NaHCO_3(aq) + CH_3CO_2H(aq) \rightarrow 2\,Na_2CO_3(aq) + CH_4(aq) + 2H_2O(\ell) + CO_2(g)$

c) $NaHCO_3(aq) + H^+(aq) \rightarrow H_2CO_3(s) + Na^+(aq)$

d) $HCO_3^-(aq) + H^+(aq) \rightarrow H_2O(\ell) + CO_2(g)$

e) $HCO_3^-(aq) + H^+(aq) \rightarrow H_2CO_3(aq)$

Answer: a

38. Write a balanced chemical equation for the reaction of aqueous solutions of potassium
sulfide and nitric acid.

a) $K_2S(aq) + 2\,HNO_3(aq) \rightarrow 2\,KH(aq) + S(NO_3)_2(g)$
b) $K_2S(aq) + HNO_3(aq) \rightarrow HS(g) + K_2NO_3(aq)$
c) $K_2S(aq) + 2\,HNO_3(aq) \rightarrow S(s) + H_2(g) + 2\,KNO_3(aq)$
d) $K_2S(aq) + 2\,HNO_3(aq) \rightarrow 2\,K(s) + H_2(g) + S(NO_3)_2(g)$
e) $K_2S(aq) + 2\,HNO_3(aq) \rightarrow H_2S(g) + 2\,KNO_3(aq)$

Answer: e

3.9 Oxidation-Reduction Reactions

39. Which molecule in the reaction below is the reducing agent?

$2\,C_2H_6(g) + 7\,O_2(g) \rightarrow 4\,CO_2(g) + 6\,H_2O(g)$

a) C_2H_6 b) O_2 c) H_2O d) CO_2 e) None

Answer: a

40. Which species in the reaction below undergoes reduction?

$2\,Na(s) + 2\,H_2O(aq) \rightarrow 2\,Na^+(aq) + 2\,OH^-(aq) + H_2(g)$

a) Na   b) H_2O c) Na^+ d) OH^- e) H_2

Answer: b

41. Which species is oxidized in the reaction below?

$I^-(aq) + ClO^-(aq) \rightarrow IO^-(aq) + Cl^-(aq)$

   a) I^- b) H_2O c) Cl^- d) IO^- e) ClO^-

Answer: a

42. What is the oxidation number of iodine in potassium periodate, KIO_4?

   a) -1   b) 0   c) +3   d) +5   e) +7

Answer: e

43. What is the oxidation number of sulfur in sulfur in SCl_2?

   a) -2   b) 0   c) +2   d) +4   e) +6

Answer: c

44. What is the oxidation number of each atom in sulfurous acid, H_2SO_3?

   a) H = +1, S = -2, O = -2
   b) H = 0, S = +6, O = -2
   c) H = 0, S = 0, O = 0
   d) H = +1, S = +4, O = -2
   e) H = -1, S = +8, O = -2

Answer: d

45. What is the oxidation number of each atom in sodium phosphate, Na_3PO_4?

   a) Na = +1, P = -3, O = -2
   b) Na = +1, P = +5, O = -2
   c) Na = +1, P = -3, O = +2
   d) Na = -1, P = +5, O = -2
   e) Na = 0, P = 0, O = 0

Answer: b

46. Which of the following elements generally acts as an oxidizing agent?

   a) Cl_2 b) H_2 c) F^- d) C   e) Na

Answer: a

47. Which of the following chemical equations show oxidation-reduction reactions?
   1. $Mg(s) + I_2(aq) \rightarrow MgI_2(s)$
   2. $Pb(ClO_4)_2(aq) + 2\,KI(aq) \rightarrow PbI_2(s) + 2\,NaClO_4(aq)$
   3. $Fe_2O_3(s) + 3\,CO(g) \rightarrow 2\,Fe(s) + 3\,CO_2(g)$

   a) 1 only   b) 2 only   c) 1 and 2   d) 1 and 3   e) 2 and 3

Answer: d

3.10 Classifying Reactions in Aqueous Solution

48. Which of the following chemical equations is an acid-base reaction?

   a) $Ba(OH)_2(aq) + K_2SO_4(aq) \rightarrow BaSO_4(s) + 2\,KOH(aq)$
   b) $3\,NaOH(aq) + AlCl_3(aq) \rightarrow Al(OH)_3(s) + 3\,NaCl(aq)$
   c) $2\,HCl(aq) + Zn(s) \rightarrow H_2(g) + ZnCl_2(aq)$
   d) $2\,HCl(aq) + Pb(NO_3)_2(aq) \rightarrow PbCl_2(s) + 2\,HNO_3(aq)$
   e) $HF(aq) + NH_3(aq) \rightarrow NH_4F(aq)$

Answer: e

49. All of the following are oxidation-reduction reactions EXCEPT

   a) $CaCO_3(s) \rightarrow CaO(s) + CO_2(g)$
   b) $2\,Na(s) + Br_2(g) \rightarrow 2\,NaBr(g)$
   c) $Fe(s) + 2\,HCl(aq) \rightarrow FeCl_2(aq) + H_2(g)$
   d) $C(s) + O_2(g) \rightarrow 2\,CO(g)$
   e) $2\,H_2O(\ell) \rightarrow 2\,H_2(g) + O_2(g)$

Answer: a

Short Answer Questions

50. A(n) _________ agent loses electrons in an oxidation-reduction reaction.

Answer: reducing

51. _________ acid is produced in a larger quantity than any other chemical in the United States. This chemical is used in the production of fertilizers, pigments, alcohol, paper and detergents.

Answer: Sulfuric

52. The net ionic equation for the reaction of barium(II) chloride and sodium sulfate is shown below.
$$Ba^{2+}(aq) + SO_4^{2-}(aq) \rightarrow BaSO_4(s)$$
Chloride and sodium ions are referred to as _________ ions because they are not involved in the reaction.

    Answer: spectator

53. Species that can function as either acids or bases are called _____________.

    Answer: amphiprotic

54. Give the name of a basic oxide and write a balanced chemical equation for the reaction of the oxide with water.

    Answer: Calcium oxide. $CaO(aq) + H_2O(\ell) \rightarrow Ca(OH)_2(aq)$. Other basic oxides include the group 1A and 2A metal oxides.

55. If an aqueous sodium hydroxide solution is left in contact with air, the concentration of hydroxide ion gradually decreases. The process can be hastened if a person exhales over a sodium hydroxide solution. Write a balanced chemical equation that describes the process by which the hydroxide concentration decreases.

    Answer: $OH^-(aq) + CO_2(aq) \rightarrow HCO_3^-(aq)$

Chapter 4
STOICHIOMETRY: QUANTITATIVE INFORMATION ABOUT CHEMICAL RELATIONSHIPS

CHAPTER OUTLINE

CHAPTER 4: Stoichiometry: Quantitative Information about Chemical Relationships

CHAPTER OBJECTIVES

- Perform stoichiometry calculations using balanced chemical equations. (4.1)
- Understand the meaning of a limiting reactant in a chemical reaction (4.2)
- Calculate theoretical and percent yields of a chemical reaction (4.3)
- Use stoichiometry to analyze a mixture of compounds or to determine the formula of a compound (4.4)
- Define and use concentrations in solution stoichiometry (4.5-4.8)

LESSON PLAN

AP Standards

III.B.3   Mass and volume relationships with emphasis on the mole concept, including empirical formulas and limiting reagents

II.C.2   Methods of expressing solutions concentration (use of normalities is not tested)

Suggested Time:

Four traditional classes or two blocks. You may want to spend extra time on section 4.2 as stoichiometric factor in critical for understanding chemical equations. Section 4.8 may be skipped if necessary however sectrophotometry is part of the lab component of AP. This basic introduction to pH will be expanded upon in later chapters.

ASSESSMENT

Exercises:
    4.1   Mass Relations in Chemical Reactions (p. 162)
    4.2   A Reaction with a Limiting Reagent (p. 166)

GLOSSARY

absorbance The negative logarithm of the transmittance.

absorbance spectrum A plot of the intensity of light absorbed by a sample as a function of the wavelength of the light.

actual yield The measured amount of product obtained from a reaction

amounts table A table indicating the mole and mass relationships of reactants and products in a reaction.

Beer-Lambert law The absorbance of a sample is proportional to the path length and the concentration

calibration curve Plot of absorbance versus concentration for a series of standard solutions whose concentrations are accurately known.

chemical analysis The determination of the amounts or identities of the components of a mixture

equivalence point The point in a titration at which one reactant has been exactly consumed by addition of the other reactant

indicator A substance used to signal the equivalence point of a titration by a change in some physical property such as color

CHAPTER 4: Stoichiometry: Quantitative Information about Chemical Relationships

limiting reagent The reactant present in limited supply that determines the amount of product formed.

molarity The number of moles of solute per liter of solution

mole ratio A conversion factor relating moles of one species in a reaction to moles of another species in the same reaction also known as stoichiometric factor

percent yield The actual yield of a chemical reaction as a percentage of its theoretical yield.

pH The negative of the base-10 logarithm of the hydrogen ion concentration

primary standard A pure, solid acid or base that can be accurately weighed for preparation of a titrating reagent

spectrophotometry The quantitative measurement of light absorption and its relation to the concentration of the dissolved solute

standardization The accurate determination of the concentration of an acid, base, or other reagent for use in a titration

stoichiometric factor A conversion factor relating moles of one species in a reaction to moles of another species in the same reaction also known as mole ratio

theoretical yield The maximum amount of product that can be obtained from the given amounts of reactants in a chemical reaction

titration A procedure for quantitative analysis of a substance by an essentially complete reaction in solution with a measured quantity of a reagent of known concentration

transmittance (T) The ratio of the amount of light passing through the sample to the amount of light that initially fell on the sample

MEDIA MENU

The Media Menu section of this guide lists the interactive exercises, tutorials, active figures (animated, interactive figures) and additional resources available online for Chemistry Now, a unique web-based, assessment-centered personalized learning system for chemistry students. Go to www.cengagelearning.com/login.

Exercises
- Screen 4.3   Reaction between chlorine and phosphorus
- Screen 4.5   Zinc and hydrochloric acid in solution
- Screen 4.11   Direct addition method of preparing a solution
- Screen 4.11   Dilution method of preparing a solution

- ▪ Screen 4.12   Solution stoichiometry

Tutorials
- ▪ Screen 4.3   Yield
- ▪ Screen 4.6   Determining the theoretical yield of a reaction
- ▪ Screen 4.6   Determining the percent yield of a reaction
- ▪ Screen 4.7   Chemical analysis
- ▪ Screen 4.9   Solution concentration
- ▪ Screen 4.9   Ion concentration
- ▪ Screen 4.11   Direct addition method of preparing a solution
- ▪ Screen 4.11   Dilution method of preparing a solution
- ▪ Screen 4.11   Determining the pH of a solution
- ▪ Screen 4.12   Determining the mass of product
- ▪ Screen 4.12   Determining the volume of a reactant
- ▪ Screen 4.13   Volume of titrant used
- ▪ Screen 4.13   Determining the concentration of acid solution
- ▪ Screen 4.13   Determine the concentration of an unknown acid

Active Figures
- ▪ 4.2   Oxidation of ammonia
- ▪ 4.6   Analysis for the sulfate content of a sample
- ▪ 4.7   Combustion analysis of a hydrocarbon
- ▪ 4.9   Making a solution
- ▪ 4.11   pH of common substances
- ▪ 4.14   Titration of an acid in aqueous solution with a base

Additional Resources
- ▪ Screen 4.2   Phosphorus and chlorine video and animation
- ▪ Screen 4.4   Limiting reagent in the methanol and oxygen reaction video and animation
- ▪ Screen 4.5   Limiting reagent simulation
- ▪ Podcast Module 7   Mass relationships in chemical reactions
- ▪ Podcast Module 8   Limiting reagents
- ▪ Podcast Module 9   pH of acids and bases

DIAGNOSTIC QUIZ QUESTIONS

Goal #1 Perform stoichiometry calculations using balanced chemical equations.

1. Magnesium reacts with chlorine gas to produce magnesium chloride. How many moles of Mg will react with 2.6 moles of Cl_2?
    a) 1.3 mol   b) 2.6 mol   c) 3.9 mol   d) 5.2 mol   e) 7.8 mol

    Answer: b

2. When strongly heated, boric acid breaks down to boric oxide and water. What mass of boric oxide is formed from the decomposition of 10.0 g $B(OH)_3$?

$$2\ B(OH)_3(s)\ \rightarrow\ B_2O_3(s)\ +\ 3\ H_2O(g)$$

a) 4.37 g   b) 4.44 g   c) 5.63 g   d) 11.3 g   e) 22.5 g

Answer: c

Goal #2 Understand the meaning of a limiting reactant in a chemical reaction

3. How many moles of $Cu_2O(s)$ can be produced from the reaction of 0.66 mol Cu(s) with 0.40 mol $O_2(g)$?

a) 0.33 mol   b) 0.40 mol   c) 0.66 mol   d) 0.80 mol   e) 1.06 mol

Answer: a

4. If 5.00 g Br_2 and 3.00 g NH_3 react according to the equation below, what is the maximum mass of ammonium bromide produced?

$$3\ Br_2(\ell)\ +\ 8\ NH_3(g)\ \rightarrow\ 6\ NH_4Br(s)\ +\ N_2(g)$$

a) 3.06 g   b) 6.13 g   c) 12.9 g   d) 18.4 g   e) 34.5 g

Answer: b

Goal #3 Calculate theoretical and percent yields of a chemical reaction

5. Under certain conditions the formation of ammonia from nitrogen and hydrogen has a 9.82% yield. Under these conditions, what mass of NH_3 will be produced from the reaction 15.0 g N_2 with 2.00 g H_2?

$$N_2(g)\ +\ 3\ H_2(g)\ \rightarrow\ 2\ NH_3(g)$$

a) 1.11 g   b) 1.67 g   c) 1.79 g   d) 3.32 g   e) 11.3 g

Answer: a

6. Sulfur hexafluoride is produced by reacting elemental sulfur with fluorine gas.

$$S_8(s)\ +\ 24\ F_2(g)\ \rightarrow\ 8\ SF_6(g)$$

What is the percent yield if 18.3 g SF_6 is isolated from the reaction of 10.0 g S_8 and 30.0 g F_2?

a) 40.2%   b) 45.8%   c) 47.6%   d) 54.6%   e) 61.0%

Answer: c

Goal #4 Use stoichiometry to analyze a mixture of compounds or to determine the formula of a compound

7. A 5.00 g sample of an oxide of platinum, when heated in a stream of hydrogen, produces $Pt(s)$ and 0.794 g H_2O. What is formula for the compound?

   a) Pt_2O b) PtO c) Pt_2O_3 d) Pt_2O_5 e) PtO_2

   Answer: e

8. An unknown hydrocarbon contains carbon, hydrogen, and oxygen. Combustion of 1.6000 g of the hydrocarbon produces 1.0286 g H_2O and 3.7681 g CO_2. What is the empirical formula of the hydrocarbon?

   a) CHO b) C_2H_3O c) C_3H_4O d) $C_4H_6O_5$ e) C_5H_5O

   Answer: c

Goal #5 Define and use concentrations in solution stoichiometry

9. If 8.19 g KIO_3 is dissolved in enough water to make 500.0 mL of solution, what is the molarity of the potassium iodate solution? The molar mass of KIO_3 is 214 g/mol.

   a) 1.64×10^{-2} M
   b) 1.91×10^{-2} M
   c) 7.65×10^{-2} M
   d) 3.51 M
   e) 16.4 M

   Answer: c

10. Iron reacts with hydrochloric acid.
    $$Fe(s) + 2\ HCl(aq) \rightarrow FeCl_2(aq) + H_2(g)$$
    What volume of 2.55 M HCl(aq) will react with 35.0 g Fe(s)?

   a) 0.246 L   b) 0.492 L   c) 1.60 L   d) 3.20 L   e) 27.5 L

TESTBANK

4.1 Mass Relationships in Chemical Reactions: Stoichiometry

1. Aluminum reacts with oxygen to produce aluminum oxide.
$$4\ Al(s)\ +\ 3\ O_2(g)\ \rightarrow\ 2\ Al_2O_3(s)$$
If 4.0 moles of Al react with excess O_2, how many moles of Al_2O_3 can be formed?

a) 2.0 mol   b) 4.0 mol   c) 6.0 mol   d) 7.0 mol   e) 8.0 mol

Answer: a

2. Magnesium reacts with chlorine gas to produce magnesium chloride. How many moles of Mg will react with 2.6 moles of Cl_2?

a) 1.3 mol   b) 2.6 mol   c) 3.9 mol   d) 5.2 mol   e) 7.8 mol

Answer: b

3. The complete combustion of 0.20 moles of pentane, C_5H_{12}, will

a) consume 0.20 mol O_2 and produce 0.20 mol CO_2.
b) consume 1.0 mol O_2 and produce 1.0 mol CO_2.
c) consume 1.6 mol O_2 and produce 1.0 mol CO_2.
d) consume 2.0 mol O_2 and produce 1.0 mol CO_2.
e) consume 2.0 mol O_2 and produce 1.6 mol CO_2.

Answer: c

4. Iron reacts with hydrochloric acid to produce iron(II) chloride and hydrogen gas.
$$Fe(s)\ +\ 2\ HCl(aq)\ \rightarrow\ FeCl_2(aq)\ +\ H_2(g)$$
What mass of $H_2(g)$ is produced from the reaction of 5.2 g Fe(s) with excess hydrochloric acid?

a) 0.094 g   b) 0.19 g   c) 5.2 g   d) 6.8 g   e) 1.4×10^2 g

Answer: b

5. The reaction of coal and water at a high temperature produces a mixture of hydrogen and carbon monoxide gases. This mixture is known as synthesis gas (or syngas). What mass of carbon monoxide can be formed from the reaction of 71.3 g of carbon with excess water?
$$C(s)\ +\ H_2O(g)\ \rightarrow\ H_2(g)\ +\ CO(g)$$

a) 5.94 g   b) 12.0 g   c) 71.3 g   d) 107 g   e) 166 g

Answer: e

6. When strongly heated, boric acid breaks down to boric oxide and water. What mass of boric oxide is formed from the decomposition of 10.0 g $B(OH)_3$?

$$2\, B(OH)_3(s) \rightarrow B_2O_3(s) + 3\, H_2O(g)$$

a) 4.37 g b) 4.44 g c) 5.63 g d) 11.3 g e) 22.5 g

Answer: c

7. If the complete combustion of an unknown mass of ethylene produces 16.0 g CO_2, what mass of ethylene is combusted?

$$C_2H_4(g) + 3\, O_2(g) \rightarrow 2\, CO_2(g) + 2\, H_2O(g)$$

a) 0.182 g b) 0.364 g c) 5.10 g d) 8.00 g e) 12.6 g

Answer: c

8. Nitroglycerine decomposes violently according to the <u>unbalanced</u> chemical equation below. What mass of nitrogen gas is produced from the decomposition of 0.300 mol $C_3H_5(NO_3)_3$?

$$C_3H_5(NO_3)_3(\ell) \rightarrow CO_2(g) + N_2(g) + H_2O(g) + O_2(g)$$

a) 0.0370 g b) 0.0555 g c) 8.41 g d) 12.6 g e) 25.2 g

Answer: d

9. The compound P_4S_3 is used in matches. It reacts with oxygen to produce P_4O_{10} and SO_2. The <u>unbalanced</u> chemical equation is shown below.

$$P_4S_3(s) + O_2(g) \rightarrow P_4O_{10}(s) + SO_2(g)$$

What mass of SO_2 is produced from the combustion of 0.331 g P_4S_3?

a) 0.00150 g b) 0.00451 g c) 0.0321g d) 0.0964 g e) 0.289 g

Answer: e

10. What mass of water is produced by the complete combustion of 6.21 grams of hexane, C_6H_{14}?

a) 0.186 g b) 1.30 g c) 9.09 g d) 43.5 g e) 208 g

Answer: c

11. Hydrogen peroxide decomposes into oxygen and water. What mass of oxygen is formed from the decomposition of 125 g of H_2O_2?

a) 58.8 g b) 66.4 g c) 107 g d) 118 g e) 125 g

Answer: a

<u>4.2 Reactions in which One Reactant is Present in Limited Supply</u>

12. How many moles of lithium fluoride can be produced from the reaction of 1.73 moles of lithium with 1.58 moles of fluorine gas?

$$2\ Li(s) + F_2(g) \rightarrow 2\ LiF(s)$$

a) 1.58 mol b) 1.73 mol c) 3.16 mol d) 3.31 mol e) 3.46 mol

Answer: b

13. How many moles of $Cu_2O(s)$ can be produced from the reaction of 0.66 mol Cu(s) with 0.40 mol $O_2(g)$?

a) 0.33 mol b) 0.40 mol c) 0.66 mol d) 0.80 mol e) 1.06 mol

Answer: a

14. Nitric oxide is made from the oxidation of ammonia. What mass of nitric oxide can be made from the reaction of 8.00 g NH_3 with 17.0 g O_2?

$$4\ NH_3(g) + 5\ O_2(g) \rightarrow 4\ NO(g) + 6\ H_2O(g)$$

a) 4.54 g b) 12.8 g c) 14.1 g d) 15.9 g e) 25.0 g

Answer: b

15. If 5.00 g Br_2 and 3.00 g NH_3 react according to the equation below, what is the maximum mass of ammonium bromide produced?

$$3\ Br_2(\ell) + 8\ NH_3(g) \rightarrow 6\ NH_4Br(s) + N_2(g)$$

a) 3.06 g b) 6.13 g c) 12.9 g d) 18.4 g e) 34.5 g

Answer: b

16. What mass of iron can be produced from the reaction of 175 kg Fe_2O_3 with 385 kg CO?

$$Fe_2O_3(s) + 3\ CO(g) \rightarrow 2\ Fe(s) + 3\ CO_2(g)$$

a) 2.19 kg b) 30.6 kg c) 61.2 kg d) 122 kg e) 512 kg

Answer: d

17. Magnesium reacts with iodine gas at high temperatures to form magnesium iodide. What mass of MgI_2 can be produced from the reaction of 5.15 g Mg and 50.0 g I_2?

a) 29.5   b) 44.9 g   c) 54.8 g   d) 55.2 g   e) 58.9 g

Answer: c

18. Aspirin ($C_9H_8O_4$) is produced by the reaction of salicylic acid ($M = 138.1$ g/mol) and acetic anhydride ($M = 102.1$ g/mol).
$$C_7H_6O_3(s) \; + \; C_4H_6O_3(\ell) \; \rightarrow \; C_9H_8O_4(s) \; + \; C_2H_4O_2(\ell)$$
If you mix 2.00 grams of each reactant, what mass of aspirin ($M = 180.2$ g/mol) can theoretically by obtained?
a) 2.00 g   b) 2.61 g   c) 2.71 g   d) 3.53 g   e) 4.00 g

Answer: b

<u>4.3 Percent Yield</u>

19. Aspirin is produced by the reaction of salicylic acid ($M = 138.1$ g/mol) and acetic anhydride ($M = 102.1$ g/mol).
$$C_7H_6O_3(s) \; + \; C_4H_6O_3(\ell) \; \rightarrow \; C_9H_8O_4(s) \; + \; C_2H_4O_2(\ell)$$
If 2.04 g of $C_9H_8O_4$ ($M = 180.2$ g/mol) is produced from the reaction of 3.03 g $C_7H_6O_3$ and 4.01 g $C_4H_6O_3$, what is the percent yield?

a) 28.8%   b) 29.0%   c) 50.9%   d) 51.6%   e) 67.3%

Answer: d

20. Under certain conditions the reaction of ammonia with excess oxygen will produce a 24.8% yield of NO. What mass of NH_3 must react with excess oxygen to yield 12.5 g NO?
$$4\,NH_3(g) \; + \; 5\,O_2(g) \; \rightarrow \; 4\,NO(g) \; + \; 6\,H_2O(g)$$

a) 1.76 g   b) 7.10 g   c) 28.6 g   d) 50.4 g   e) 88.8 g

Answer: c

21. Under certain conditions the formation of ammonia from nitrogen and hydrogen has a 9.82% yield. Under these conditions, what mass of NH_3 will be produced from the reaction 15.0 g N_2 with 2.00 g H_2?
$$N_2(g) \; + \; 3\,H_2(g) \; \rightarrow \; 2\,NH_3(g)$$

a) 1.11 g   b) 1.67 g   c) 1.79 g   d) 3.32 g   e) 11.3 g

Answer: a

22. Sulfur hexafluoride is produced by reacting elemental sulfur with fluorine gas.
$$S_8(s) + 24 F_2(g) \rightarrow 8 SF_6(g)$$
What is the percent yield if 18.3 g SF_6 is isolated from the reaction of 10.0 g S_8 and 30.0 g F_2?

a) 40.2%   b) 45.8%   c) 47.6%   d) 54.6%   e) 61.0%

Answer: c

4.6 Chemical Equations and Chemical Analysis

23. A mass of 3.051 g of a metal carbonate, MCO_3, is heated to drive off carbon dioxide. The remaining metal oxide has a mass of 1.458 g.
$$MCO_3(s) \rightarrow MO(s) + CO_2(g)$$
What is the identity of the metal?

a) Ni   b) Ba   c) Ca   d) Co   e) Mg

Answer: e

24. A mixture of $MnSO_4$ and $MnSO_4{\cdot}4H_2O$ has a mass of 2.005 g. After heating to drive off all the water the mass is 1.780 g. What is the mass percent of $MnSO_4{\cdot}4H_2O$ in the mixture?

a) 11.2%   b) 34.7%   c) 65.3%   d) 69.6%   e) 88.8%

Answer: b

25. A 1.500 g mixture of $CaCO_3$ and CaO is heated. The $CaCO_3$ decomposes to CaO and CO_2. What was the mass percentage of $CaCO_3$ in the mixture if the mass after heating is 1.103 g?

a) 26.5%   b) 33.7%   c) 47.2%   d) 60.2%   e) 73.5%

Answer: d

26. A 5.00 g sample of an oxide of platinum, when heated in a stream of hydrogen, produces $Pt(s)$ and 0.794 g H_2O. What is formula for the compound?

a) Pt_2O b) PtO c) Pt_2O_3 d) Pt_2O_5 e) PtO_2

Answer: e

27. The reaction of 0.779 g K with O_2 forms 1.417 g potassium superoxide, a substance used in self-contained breathing devices. Determine the formula for potassium superoxide.

a) KO_2 b) KO c) K_2O_3 d) K_2O e) KO_4

Answer: a

28. A 2.007 g sample of a hydrocarbon is combusted to give 1.389 g of H_2O and 6.785 g of CO_2. What is the empirical formula of the compound?

a) CH b) CH_2 c) C_2H_3 d) C_3H_4 e) C_5H_{12}

Answer: a

29. The combustion of 0.3037 mole of a hydrocarbon produces 16.414 g H_2O and 26.731 g CO_2. What is the molar mass of the hydrocarbon?

a) 16.04 g/mol b) 30.07 g/mol c) 44.08 g/mol d) 72.15 g/mol e) 92.13 g/mol

Answer: b

30. Polyethylene is a polymer consisting of only carbon and hydrogen. If 2.300 g of the polymer is burned in oxygen it produces 2.955 g H_2O and 7.217 g CO_2. What is the empirical formula of polyethylene?

a) CH b) CH_2 c) C_2H_3 d) C_5H_8 e) C_7H_8

Answer: b

31. Soft drink bottles are made of polyethylene terephthalate (PET), a polymer composed of carbon, hydrogen, and oxygen. If 2.8880 g PET is burned in oxygen it produces 1.0000 g H_2O and 6.1058 g CO_2. What is the empirical formula of PET?

a) CHO b) CH_7O_5 c) C_5H_7O d) $C_8H_{10}O$ e) $C_{10}H_8O_5$

Answer: e

32. An unknown hydrocarbon contains carbon, hydrogen, and oxygen. Combustion of 1.6000 g of the hydrocarbon produces 1.0286 g H_2O and 3.7681 g CO_2. What is the empirical formula of the hydrocarbon?

a) CHO b) C_2H_3O c) C_3H_4O d) $C_4H_6O_5$ e) C_5H_5O

Answer: c

33. Cyclohexane, a hydrocarbon, has an approximate molar mass of 84 g/mole. If the combustion of 0.6000 g cyclohexane produces 0.7709 g H_2O and 1.883 g CO_2, what is the molecular formula of this compound?

a) CH_2 b) C_5H_{12} c) C_5H_{24} d) C_6H_{12} e) C_7H_{10}

Answer: d

4.7 Measuring Concentrations of Compounds in Solution

34. If 8.19 g KIO_3 is dissolved in enough water to make 500.0 mL of solution, what is the molarity of the potassium iodate solution? The molar mass of KIO_3 is 214 g/mol.

   a) 1.64×10^{-2} M
   b) 1.91×10^{-2} M
   c) 7.65×10^{-2} M
   d) 3.51 M
   e) 16.4 M

   Answer: c

35. How many liters of 0.2805 M $C_6H_{12}O_6$(aq) contain 1.000 g of $C_6H_{12}O_6$?

   a) 0.001557 L   b) 0.01979 L   c) 0.2805 L   d) 3.565 L   e) 50.5 L

   Answer: b

36. If 5.15 g $Fe(NO_3)_3$ is dissolved in enough water to make exactly 150.0 mL of solution, what is the molar concentration of nitrate ion?

   a) 0.00319 M   b) 0.0343 M   c) 0.142 M   d) 0.313 M   e) 0.426 M

   Answer: e

37. What is the mass of sodium iodide in 50.0 mL of 2.63×10^{-2} M NaI(aq)?

   a) 0.00132 g   b) 0.00877 g   c) 0.0788 g   d) 0.197 g   e) 78.8 g

   Answer: d

38. If 25.00 mL of 4.50 M NaOH(aq) is diluted with water to a volume of 750.0 mL, what is the molarity of the diluted NaOH(aq)?

   a) 0.0333 M   b) 0.150 M   c) 0.155 M   d) 6.67 M   e) 1.35×10^{3} M

   Answer: b

39. What volume of 0.15 M HCl(aq) must be diluted to make 2.0 L of 0.050 M HCl(aq)?
   a) 0.015 L   b) 0.10 L   c) 0.30 L   d) 0.67 L   e) 6.0 L

   Answer: d

40. What is the pH of 0.51 M HCl(aq)?

a) -0.29 b) 0.29 c) 0.31 d) 0.51 e) 0.67

Answer: b

41. The pH of an aqueous NaOH solution is 13.17. What is the hydrogen ion concentration of this solution?

a) 6.8×10^{-14} M b) 1.9×10^{-6} M c) 0.89 M d) 1.1 M e) 1.5×10^{13} M

Answer: a

42. The pH of a vinegar solution is 4.15. What is the H_3O^+ concentration of the solution?

a) 7.1×10^{-5} M b) 1.6×10^{-2} M c) 0.62 M d) 1.4 M e) 1.4×10^4 M

Answer: a

43. An aqueous nitric acid solution has a pH of 1.15. What mass of HNO_3 is present in 2.0 L of this solution?

a) 0.071 g b) 0.14 g c) 2.2 g d) 4.5 g e) 8.9 g

Answer: e

4.7 Stoichiometry of Reactions in Aqueous Solution

44. Iron reacts with hydrochloric acid.
$$Fe(s) + 2\,HCl(aq) \rightarrow FeCl_2(aq) + H_2(g)$$
What volume of 2.55 M HCl(aq) will react with 35.0 g Fe(s)?

a) 0.246 L b) 0.492 L c) 1.60 L d) 3.20 L e) 27.5 L

Answer: b

45. What volume of 0.200 M Na_2SO_4(aq) will completely react with 50.0 mL of 0.135 M $Ba(NO_3)_2$(aq)?
$$Na_2SO_4(aq) + Ba(NO_3)_2(aq) \rightarrow BaSO_4(s) + 2\,NaNO_3(aq)$$

a) 33.8 mL b) 67.5 mL c) 74.1 mL d) 148 mL e) 540. mL

Answer: a

46. A mass of 0.4113 g of an unknown acid, HA, is titrated with NaOH. If the acid reacts with 28.10 mL of 0.1055 M NaOH(aq), what is the molar mass of the acid?

   a) 2.965×10^{-3} g/mol
   b) 9.128 g/mol
   c) 138.7 g/mol
   d) 337.3 g/mol
   e) 820.7 g/mol

   Answer: c

47. A 25.00 mL sample of KOH is titrated with 21.83 mL of 0.2120 M HCl(aq). What is the concentration of the KOH solution?

   a) 0.0006720 M   b) 0.002574 M   c) 0.1851 M   d) 0.2428 M   e) 4.119 M

   Answer: c

48. If 0.1800 g of impure soda ash (Na_2CO_3) is titrated with 15.66 mL of 0.1082 M HCl, what is the percent purity of the soda ash?
$$Na_2CO_3(aq) + 2\,HCl(aq) \rightarrow 2\,NaCl(aq) + H_2O(\ell) + CO_2(g)$$

   a) 17.96%   b) 49.89%   c) 50.11%   d) 94.13%   e) 99.77%

   Answer: b

4.8 Spectrophotometry, Another method of Analysis

49. Transmittance, T, is defined as the intensity of light transmitted through a sample divided by the intensity of light incident on the sample. The absorbance, A, is defined as

   a) $A = -\log T$
   b) $A = \log (-T)$
   c) $A = 10^{T}$
   d) $A = 10^{-T}$
   e) $A = 10^{-1/T}$

   Answer: a

50. Which of the following statements is/are CORRECT?
   1. Absorbance is directly proportional to the intensity of the incident light.
   2. Absorbance is inversely proportional to the analyte concentration.
   3. Absorbance is directly proportional to the path length of the light.

   a) 1 only   b) 2 only   c) 3 only   d) 2 and 3   e) 1,2, and 3

   Answer: c

51. The numbers preceding the formulas in chemical equations are referred to as the __________ coefficients.

Answer: stoichiometric

52. The __________ yield of a chemical reaction is calculated by dividing the actual yield by the theoretical yield.

Answer: percent

53. The pH of purified water (or of a neutral solution) is ______.

Answer: 7.00

54. The following equation: $A = \varepsilon \times \ell \times C$ (where ℓ is path length and C is concentration) is known as the ______-Lambert law.

Answer: Beer

55. PCl_3 can be produced from the reaction of P_4 with Cl_2.
$$P_4(s) + 6\ Cl_2(g) \rightarrow 4\ PCl_3(g)$$
If 1.00 g P_4 reacts with 1.00 g Cl_2, which reactant is the limiting reactant? What mass of product may be produced?

Answer: Chlorine is the limiting reactant and 1.29 g PCl_3 may be produced.

56. The reaction of 2.75 g N_2 with excess H_2 produces 1.77 g NH_3. The percent yield of this reaction is __________.

Answer: 52.9%

Chapter 5
PRINCIPLES OF CHEMICAL REACTIVITY: ENERGY AND CHEMICAL REACTIONS

CHAPTER OUTLINE

CHAPTER OBJECTIVES

- Assess the transfer of energy as heat associated with changes in temperature and changes of state (5.1-5.3)
- Understand and apply the first law of thermodynamics (5.4)
- Define and understand state functions (enthalpy and internal energy) (5.4)
- Learn how energy changes are measured (5.5-5.6)
- calculate the energy evolved or required for physical changes and chemical reactions using tables of thermodynamic data (5.7)

LESSON PLAN

AP Standards

III.E.1	State Functions
III.E.2	First law of Thermodynamics: change in enthalpy; heat of formation; heat of reaction; Hess's law; heats of vaporization and fusion; Calorimetry

Suggested Time:
Three traditional classes or two blocks. Sections 5.4-5.7 should take up most of the time allotted. Sections 5.1 and 5.2 have parts that accompany the first law of thermodynamics which need to be included. Also, section 5.7 has essential material about Hess's law. Section 5.8 may be delayed for further discussion in thermodynamics.

CHAPTER 5: Principles of Chemical Reactivity: Energy and Chemical Reactions

ASSESSMENT

Exercises:

Answers to Exercises p. 44
Practicing Skills pp. 242-247, Answers IRM pp. 91-98
General Questions pp. 247-249, Answers IRM pp. 99-104
In the Laboratory Questions pp. 249-250, Answers IRM pp. 105-107
Summary and Conceptual Questions pp. 251-253, Answers IRM pp. 107-112

GLOSSARY

calorie (cal) The quantity of energy required to raise the temperature of 1.00 g of pure liquid water from $14.5^{\circ}C$ to $15.5^{\circ}C$

calorimetry The experimental determination of the enthalpy changes of reactions

endothermic process A thermodynamic process in which heat flows into a system from its surroundings

energy The capacity to do work and transfer heat

enthalpy change (ΔH) Heat energy transferred at constant pressure

exothermic process A thermodynamic process in which heat flows from a system to its surroundings

first law of thermodynamics The total energy of the universe is constant

fusion The state change from solid to liquid

heat of fusion The quantity of heat required to convert a solid to a liquid at constant temperature

heat of vaporization The quantity of heat required to convert a liquid to a gas at its normal boiling point

Hess's Law If a reaction is the sum of two or more other reactions, the enthalpy change for the overall process is the sum of the enthalpy changes for the constituent reactions

internal energy The sum of the potential and kinetic energies of the particles in the system

joule (J) The SI unit for energy

kilocalorie(kcal) A unit of energy equivalent to 1000 calories

kilojoule (kJ) A unit of energy equivalent to 1000 joules

kinetic energy The energy of a moving object, dependent on its mass and velocity

law of conservation of energy Energy can be neither created nor destroyed

potential energy The energy that results from an object's position

specific heat capacity(C) The quantity of heat required to raise the temperature of 1.00 g of a substance by $1.00^{\circ}C$

standard molar enthalpy of formation($\Delta_f H^o$) The enthalpy change of a reaction for the formation of 1 mol of a compound directly from its elements, all in their standard states

standard reaction enthalpy ($\Delta_r H^o$) The enthalpy change or a reaction that occurs with all reactants and products in their standard state. It is read as per mol-rxn, indicating that the reaction has progressed one mole of times.

standard state The most stable form of anelement or compound in the physical state in which it exists at 1 bar and the specified temperature.

state function A quantity whose value is determined only by the state of the system

sublimation The direct conversion of a solid to a gas

surroundings Everything outside of the system in a thermodynamic process

system The substance being evaluated for energy content in a thermodynamic process

thermal equilibrium A condition in which the system and its surroundings are at the same temperature and heat transfer stops

thermodynamics The science of heat or energy flow in chemical processes

vaporization The state change from liquid to gas

CHAPTER 5: Principles of Chemical Reactivity: Energy and Chemical Reactions

MEDIA MENU

The Media Menu section of this guide lists the interactive exercises, tutorials, active figures (animated, interactive figures) and additional resources available online for Chemistry Now, a unique web-based, assessment-centered personalized learning system for chemistry students. Go to www.cengagelearning.com/login.

Exercises
- Screen 5.3   Energy transfer
- Screen 5.7   Energy transfer as heat
- Screen 5.8   Energy transfers as heat between substances
- Screen 5.10   Calculating energy transfer
- Screen 5.14   Reactions in a constant volume calorimeter
- Screen 5.15   Hess's law

Tutorials
- Screen 5.5   Different Energy Units
- Screen 5.8   Energy transfers as heat between substances
- Screen 5.10   Calculating energy transfer
- Screen 5.12   Enthalpy changes
- Screen 5.14   Calculating the enthalpy change for a reaction from a calorimetry experiment
- Screen 5.16   Calculating the standard enthalpy change for a reaction

Self-study module
- Screen 5.11   Energy changes in a physical process
- Screen 5.12   Enthalpy changes
- Screen 5.13   Enthalpy changes
-

Active Figures
- 5.1   Energy and its conversion
- 5.6   Exothermic and endothermic processes
- 5.10   Changes of state
- 5.11   Energy changes in a physical process
- 5.13   The exothermic combustion of hydrogen in air
- 5.15   Constant volume calorimeter
- 5.16   Energy level diagrams

Additional Resources
- Screen 5.4   Endothermic and exothermic animation
- Scree. 5.7   Energy transfer as heat animation
- Screen 5.8   Energy transfers as heat between substances simulations
- Screen 5.10   Calculating energy transfer simulations
- Screen 5.15   Hess's law simulation
- Podcast Module 10   Enthalpy calculations

DIAGNOSTIC QUIZ QUESTIONS

Goal #1   Assess the transfer of energy as heat associated with changes in temperature and changes of state.

1. A battery-operated power tool, such as a cordless drill, converts

    a) electrostatic energy to chemical potential energy.
    b) mechanical energy to electrostatic energy.
    c) thermal energy to mechanical energy.
    d) thermal energy to gravitational energy.
    e) chemical potential energy to mechanical energy.

    Answer: e

2. Which of the following statements is/are CORRECT?
    1. A system is defined as an object or collection of objects being studied.
    2. Surroundings are defined as everything outside of the system being studied.
    3. In an exothermic reaction, heat is transferred from the system to the surroundings.

    a) 1 only   b) 2 only   c) 3 only   d) 1 and 3   e) 1, 2, and 3

    Answer: e

3. How many joules are equivalent to 75 calories?

    a) 0.056 J   b) 0.075 J   c) 18 J   d) 3.1×10^2 J   e) 7.5×10^4 J

    Answer: d

4. If 245 J is required to change the temperature of 14.4 g of chromium by 38.0 K, what is the specific heat capacity of chromium?

    a) 0.448 J/g·K   b) 2.23 J/g·K   c) 4.18 J/g·K   d) 4.68 J/g·K   e) 92.8 J/g·K

    Answer: a

5. The heat of vaporization of benzene, C_6H_6, is 30.8 kJ/mol at its boiling point of 80.1 °C. How much energy in the form of heat is required to vaporize 102 g benzene at its boiling point?

    a) 0.302 kJ   b) 23.6 kJ   c) 24.2 kJ   d) 40.2 kJ   e) 3.14×10^3 kJ

    Answer: d

CHAPTER 5: Principles of Chemical Reactivity: Energy and Chemical Reactions

Goal #2   Understand and apply the first law of thermodynamics.

Goal #3   Define and understand state functions (enthalpy and internal energy).

6. One statement of the first law of thermodynamics is that

a) the amount of work done on a system is dependent of pathway.
b) the total work done on a system must equal the heat absorbed by the system.
c) the total work done on a system is equal in magnitude, but opposite in sign of the heat absorbed by the system.
d) the total energy change for a system is equal to the sum of the heat transferred to or from the system and the work done by or on the system.
e) in any chemical process the heat flow must equal the change in enthalpy.

Answer: d

Goal #4   Learn how energy changes are measured.

7. When 10.0 g KOH is dissolved in 100.0 g of water in a coffee-cup calorimeter, the temperature rises from 25.18 °C to 47.53 °C. What is the enthalpy change per gram of KOH dissolved in the water? Assume that the solution has a specific heat capacity of 4.18 J/g·K.

a) -116 J/g
b) -934 J/g
c) -1.03 × 10^3 J/g
d) -2.19 × 10^3 J/g
e) -1.03 × 10^4 J/g

Answer: c

8. A chemical reaction in a bomb calorimeter evolves 4.66 kJ of heat. If the heat capacity of the calorimeter is 1.38 kJ/°C, what is the temperature change of the calorimeter?

a) 0.296 °C   b) 3.28 °C   c) 3.38 °C   d) 6.04 °C   e) 6.43 °C

Answer: c

Goal #5   Calculate the energy evolved or required for physical changes and chemical reactions using tables of thermodynamic data.

9. Calculate $\Delta_f H°$ for sulfur dioxide given the thermochemical equations below.

$$2\ S(s) + 3\ O_2(g) \rightarrow 2\ SO_3(g) \qquad \Delta_r H° = -791.5\ kJ$$
$$2\ SO_2(g) + O_2(g) \rightarrow 2\ SO_3(g) \qquad \Delta_r H° = -197.9\ kJ$$

a) -296.8 kJ/mol   b) -395.7 kJ/mol   c) -494.7 kJ/mol   d) -593.6 kJ/mol   e) -989.4 kJ/mol

Answer: a

10. Hydrazine, N_2H_4, is a liquid used as a rocket fuel. It reacts with oxygen to yield nitrogen gas and water.

$$N_2H_4(\ell) + O_2(g) \rightarrow N_2(g) + 2\,H_2O(\ell)$$

The reaction of 6.50 g N_2H_4 evolves 126.2 kJ of heat. Calculate the enthalpy change per mole of hydrazine combusted.

a) -19.4 kJ/mol   b) -25.6 kJ/mol   c) -126 kJ/mol   d) -622 kJ/mol   e) -820. kJ/mol

Answer: d

11. Calculate $\Delta_r H°$ for the combustion of gaseous dimethyl ether,

$$CH_3OCH_3(g) + 3\,O_2(g) \rightarrow 2\,CO_2(g) + 3\,H_2O(\ell)$$

using standard molar enthalpies of formation.

molecule	$\Delta_f H°$ (kJ/mol)
$CH_3OCH_3(g)$	-184.1
$CO_2(g)$	-393.5
$H_2O(\ell)$	-285.8

a) -76.4 kJ   b) -495.2 kJ   c) -863.4 kJ   d) -1460.3 kJ   e) -1828.5 kJ

Answer: d

TESTBANK

<u>5.1. Energy: Some Basic Principles</u>

1. Many homes are heated using natural gas. The combustion of natural gas converts

   a) chemical potential energy to thermal energy.
   b) thermal energy to mechanical energy.
   c) mechanical energy to thermal energy.
   d) electrostatic energy to mechanical energy.
   e) thermal energy to acoustic energy.

   Answer: a

2. A battery-operated power tool, such as a cordless drill, converts

   a) electrostatic energy to chemical potential energy.
   b) mechanical energy to electrostatic energy.
   c) thermal energy to mechanical energy.
   d) thermal energy to gravitational energy.
   e) chemical potential energy to mechanical energy.

   Answer: e

3. Which of the following lists contains only forms of kinetic energy?

    a) electrostatic, gravitational, and mechanical energy
    b) gravitational, mechanical, and electrical energy
    c) thermal, acoustic, and mechanical energy
    d) chemical, thermal, and acoustic energy
    e) gravitational, chemical, and electrostatic energy

    Answer: c

4. Of the following types of energy, which is/are classified as potential energy?
    1. chemical energy
    2. electrostatic energy
    3. gravitational energy

    a) 1 only   b) 2 only   c) 3 only   d) 2 and 3   e) 1, 2, and 3

    Answer: e

5. Which of the following statements is/are CORRECT?
    1. A system is defined as an object or collection of objects being studied.
    2. Surroundings are defined as everything outside of the system being studied.
    3. In an exothermic reaction, heat is transferred from the system to the surroundings.

    a) 1 only   b) 2 only   c) 3 only   d) 1 and 3   e) 1, 2, and 3

    Answer: e

6. Which of the following statements is/are CORRECT?
    1. A calorie of heat is defined as the energy required to change 1.00 g of water from
        14.5 °C to 15.5 °C.
    2. One calorie (c) equals 4.184 J, where the joule (J) is the SI unit of energy.
    3. A calorie (c) is equal to 1000 dietary calories (C).

    a) 1 only   b) 2 only   c) 3 only   d) 1 and 2   e) 1, 2, and 3

    Answer: d

7. Which one of the following statements is INCORRECT?
    a) Energy is neither created nor destroyed in chemical reactions.
    b) Potential energy is the energy associated with motion.
    c) Endothermic processes transfer heat from the surroundings into the system.
    d) Increasing the thermal energy of a gas increases the motion of its atoms.
    e) Energy is the capacity to do work.

    Answer: b

8. How many joules are equivalent to 75 calories?

a) 0.056 J   b) 0.075 J   c) 18 J   d) 3.1×10^2 J   e) 7.5×10^4 J

Answer: d

9. How many nutritional calories are equivalent to 25 kJ?

a) 0.10 Cal   b) 0.17 Cal   c) 6.0 Cal   d) 1.0×10^2 Cal   e) 6.0×10^3 Cal

Answer: c

5.2 Specific Heat Capacity: Heating and Cooling

10. Specific heat capacity is

a) the quantity of heat needed to change the temperature of 1.00 g of a substance by 1.00 K.
b) the quantity of heat needed to change the temperature of 1.00 g of a substance by 4.184 K.
c) the mass of a substance that 1.00 J of energy will heat by 1.00 K.
d) the temperature change undergone when 1.00 g of a substance absorbs 4.184 J.
e) the maximum amount of heat that 1.00 g of a substance may absorb without decomposing.

Answer: a

11. Heat capacity is defined as

a) the amount of heat required to raise the temperature of 1 gram of substance by 1 K.
b) the amount of heat required to raise the temperature of a substance by 1 K.
c) the amount of heat required to vaporize a solid or liquid.
d) the maximum amount of heat that a substance may absorb without decomposing.
e) 4.18 cal/g·K.

Answer: b

12. Which one of the following statements is CORRECT?

a) When heat is transferred from the surroundings to the system, q is negative.
b) Heat is transferred from the system to the surroundings in an endothermic process.
c) The SI unit of specific heat capacity is calories per gram (cal/g).
d) Specific heat capacity is a positive value for liquids and a negative value for solids.
e) The larger the heat capacity of an object, the more thermal energy it can store.

Answer: e

13. If 245 J is required to change the temperature of 14.4 g of chromium by 38.0 K, what is the specific heat capacity of chromium?

a) 0.448 J/g·K   b) 2.23 J/g·K   c) 4.18 J/g·K   d) 4.68 J/g·K   e) 92.8 J/g·K

Answer: a

14. If the same amount of energy in the form of heat is added to 5.00 g samples of each of the metals below, which metal will undergo the largest temperature change?

Metal	Specific Heat Capacity (J/g·K)
Ag	0.235
Al	0.897
Cu	0.385
Fe	0.449
Mg	1.017

a) Ag   b) Al   c) Cu   d) Fe   e) Mg

Answer: a

15. If 50.0 g of ethanol, CH_3CH_2OH, at 20.0 °C absorbs 1.45 kJ of heat, what is the final temperature of the ethanol? The specific heat capacity of ethanol is 2.44 J/g·K.

a) 8.1 °C   b) 21.2°C   c) 31.9 °C   d) 47.7 °C   e) 90.8 °C

Answer: c

16. How much energy (in kJ) is required to change the temperature of 1.00 kg Fe from 25.0 °C to 1515 °C? The specific heat capacity of iron is 0.449 J/g·K.

a) 1.49 kJ   b) 3.32 kJ   c) 66.9 kJ   d) 669 kJ   e) 3.32×10^6 kJ

Answer: d

17. If 25.0 g H_2O at 16.0 °C is combined with 70.0 g H_2O at 85.5 °C, what is the final temperature of the mixture? The specific heat capacity of water is 4.184 J/g·K.

a) 14.4 °C   b) 34.3 °C   c) 50.8 °C   d) 67.2 °C   e) 70.0 °C

Answer: d

18. If 46.1 g Al at 21.0 °C is placed in 80.0 g H_2O at 88.0 °C, what is the final temperature of the mixture? The specific heat capacities of aluminum and water are 0.897 J/g·K and 4.184 J/g·K, respectively.

a) 28.4 °C   b) 39.2 °C   c) 54.5 °C   d) 69.8 °C   e) 80.6 °C

Answer: e

19. When 27.0 g of an unknown metal at 18.4 °C is placed in 70.0 g H_2O at 79.5 °C, the water temperature decreases to 76.8 °C. What is the specific heat capacity of the metal? The specific heat capacity of water is 4.184 J/g·K.

a) 0.34 J/g·K   b) 0.50 J/g·K   c) 0.74 J/g·K   d) 0.94 J/g·K   e) 1.4 J/g·K

Answer: b

5.3 Energy and Changes of State

20. Calculate the energy in the form of heat (in kJ) required to convert 225 grams of liquid water at 21.0 °C to steam at 115 °C. (Heat of fusion = 333 J/g; heat of vaporization = 2256 J/g; specific heat capacities: liquid water = 4.184 J/g·K, steam = 1.92 J/g·K)

a) 88.5 kJ   b) 158 kJ   c) 588 kJ   d) 596 kJ   e) 663 kJ

Answer: c

21. Calculate the energy in the form of heat (in kJ) required to change 50.0 g ice at -15.0 °C to liquid at 65.0 °C. (Heat of fusion = 333 J/g; heat of vaporization = 2256 J/g; specific heat capacities: ice = 2.06 J/g·K, liquid water = 4.184 J/g·K)

a) 15.5 kJ   b) 16.7 kJ   c) 31.8 kJ   d) 128 kJ   e) 145 kJ

Answer: c

22. 65.0 g of ice at -17.0 °C is mixed with 256 g of water at 29.2 °C. Calculate the final temperature of the mixture. Assume that no energy in the form of heat is transferred to the environment. (Heat of fusion = 333 J/g; specific heat capacities: ice = 2.06 J/g·K, liquid water = 4.184 J/g·K)

a) 0.0 °C   b) 5.5 °C   c) 6.1 °C   d) 19.8 °C   e) 21.6 °C

Answer: b

23. Ice at 0.0 °C is used to cool water. What is the minimum mass of ice required to cool 325 g of water from 30.5 °C to 4.0 °C? (Heat of fusion = 333 J/g; specific heat capacities: ice = 2.06 J/g·K, liquid water = 4.184 J/g·K)

   a) 108 g   b) 125 g   c) 325 g   d) 605 g   e) 1.75×10^4 g

   Answer: a

24. The heat of vaporization of benzene, C_6H_6, is 30.8 kJ/mol at its boiling point of 80.1 °C. How much energy in the form of heat is required to vaporize 102 g benzene at its boiling point?

   a) 0.302 kJ   b) 23.6 kJ   c) 24.2 kJ   d) 40.2 kJ   e) 3.14×10^3 kJ

   Answer: d

5.4 The First Law of Thermodynamics

25. One statement of the first law of thermodynamics is that

   a) the amount of work done on a system is dependent of pathway.
   b) the total work done on a system must equal the heat absorbed by the system.
   c) the total work done on a system is equal in magnitude, but opposite in sign of the heat absorbed by the system.
   d) the total energy change for a system is equal to the sum of the heat transferred to or from the system and the work done by or on the system.
   e) in any chemical process the heat flow must equal the change in enthalpy.

   Answer: d

26. Which of the following statements is CORRECT?

   a) If a reaction occurs at constant pressure, $w = \Delta E$.
   b) If a reaction occurs at constant pressure, $q = \Delta H$.
   c) If a reaction occurs at constant pressure, $q = \Delta E$.
   d) If a reaction occurs at constant volume, $\Delta E > \Delta H$.
   e) If a reaction occurs at constant volume, $w = \Delta E$.

   Answer: b

27. Which of the following thermodynamic quantities are state functions: heat (q), work (w), enthalpy change (ΔH), and/or internal energy change (ΔE)?

   a) q only   b) w only   c) ΔH only   d) ΔE only   e) ΔH and ΔE

   Answer: e

<u>5.5 Enthalpy Changes for Chemical Reactions</u>

28. The thermochemical equation for the combustion of benzene is shown below.

$2\ C_6H_6(\ell) + 15\ O_2(g) \rightarrow 12\ CO_2(g) + 6\ H_2O(g)$ $\Delta_rH° = -3909.9$ kJ/mol-rxn

What is the enthalpy change for the combustion of 12.5 g C_6H_6?

a) -313 kJ   b) -626 kJ   c) -1.22×10^4 kJ   d) -2.44×10^4 kJ   e) -4.89×10^4 kJ

Answer: a

29. The thermochemical equation for the combustion of propane is shown below.

$C_3H_8(g) + 5\ O_2(g) \rightarrow 3\ CO_2(g) + 4\ H_2O(\ell)$ $\Delta_rH° = -2220$ kJ/mol-rxn

What is the enthalpy change for the following reaction?

$$6\ CO_2(g) + 8\ H_2O(\ell) \rightarrow 2\ C_3H_8(g) + 10\ O_2(g)$$

a) -1.33×10^4 kJ   b) +2220 kJ   c) +370 kJ   d) +2220 kJ   e) +4440 kJ

Answer: e

30. Hydrazine, N_2H_4, is a liquid used as a rocket fuel. It reacts with oxygen to yield nitrogen gas and water.

$$N_2H_4(\ell) + O_2(g) \rightarrow N_2(g) + 2\ H_2O(\ell)$$

The reaction of 6.50 g N_2H_4 evolves 126.2 kJ of heat. Calculate the enthalpy change per mole of hydrazine combusted.

a) -19.4 kJ/mol   b) -25.6 kJ/mol   c) -126 kJ/mol   d) -622 kJ/mol   e) -820. kJ/mol

Answer: d

<u>5.6 Calorimetry</u>

31. MgO reacts with water to form $Mg(OH)_2$. If 5.00 g MgO is combined with 100.0 g H_2O in a coffee cup calorimeter, the temperature of the resulting solution increases from 22.3 °C to 32.9 °C. Calculate the enthalpy change for the reaction per mole of MgO. Assume that the specific heat capacity of the solution is 4.18 J/g·K.

a) -37.5 kJ   b) -93.0 kJ   c) -577 kJ   d) -1.11×10^3 kJ   e) -4.65×10^3 kJ

Answer: a

32. Commercial cold packs consist of solid ammonium nitrate and water. NH_4NO_3 absorbs 25.69 kJ of heat per mole dissolved in water. In a coffee-cup calorimeter, 3.40 g NH_4NO_3 is dissolved in 100.0 g of water at 21.0 °C. What is the final temperature of the solution? Assume that the solution has a specific heat capacity of 4.18 J/g·K.

    a) 0.0 °C   b) 18.5 °C   c) 20.2 °C   d) 21.0 °C   e) 23.5 °C

    Answer: b

33. When 10.0 g KOH is dissolved in 100.0 g of water in a coffee-cup calorimeter, the temperature rises from 25.18 °C to 47.53 °C. What is the enthalpy change per gram of KOH dissolved in the water? Assume that the solution has a specific heat capacity of 4.18 J/g·K.

    a) -116 J/g
    b) -934 J/g
    c) -1.03×10^3 J/g
    d) -2.19×10^3 J/g
    e) -1.03×10^4 J/g

    Answer: c

34. A chemical reaction in a bomb calorimeter evolves 4.66 kJ of heat. If the heat capacity of the calorimeter is 1.38 kJ/°C, what is the temperature change of the calorimeter?

    a) 0.296 °C   b) 3.28 °C   c) 3.38 °C   d) 6.04 °C   e) 6.43 °C

    Answer: c

35. A 1.994 g sample of ethanol, CH_3CH_2OH, is combusted in a bomb calorimeter. The temperature of the calorimeter increases by 10.91 K. If the heat capacity of the bomb is 615.5 J/K and it contains 1.150 kg of water, what is the enthalpy change per mole of ethanol combusted? The specific heat capacity of water is 4.184 J/g·K and the molar mass of ethanol is 46.07 g/mol.

    a) -5.921×10^1 kJ/mol
    b) -2.563×10^3 kJ/mol
    c) -1.181×10^5 kJ/mol
    d) -1.368×10^3 kJ/mol
    e) -122.7×10^3 kJ/mol

    Answer: d

36. Acetylene, C_2H_2, is a gas used in welding. The molar enthalpy of combustion for acetylene is -2599 kJ. A mass of 0.250 g $C_2H_2(g)$ is combusted in a bomb calorimeter. If the heat capacity of the calorimeter is 715 J/K and it contains 1.200 kg of water, what is the temperature increase of the bomb calorimeter? The specific heat capacity of water is 4.184 J/g·K and the molar mass of acetylene is 26.04 g/mol.

a) 4.35 K   b) 4.72 K   c) 13.0 K   d) 11.3 K   e) 17.4 K

Answer: a

5.7 Enthalpy Calculations

37. Determine the heat of condensation of titanium(IV) chloride,

$$TiCl_4(g) \rightarrow TiCl_4(\ell)$$

given the enthalpies of reaction below.

$$Ti(s) + 2\,Cl_2(g) \rightarrow TiCl_4(\ell) \qquad \Delta_rH° = -804.2 \text{ kJ}$$
$$Ti(s) + 2\,Cl_2(g) \rightarrow TiCl_4(g) \qquad \Delta_rH° = -763.2 \text{ kJ}$$

a) -1567.4 kJ   b) -41.0 kJ   c) +1.054 kJ   d) +41.0 kJ   e) +1567.4 kJ

Answer: b

38. Determine the enthalpy change for the oxidation of iron,

$$4\,Fe(s) + 3\,O_2(g) \rightarrow 2\,Fe_2O_3(s)$$

given the thermochemical equations below.

$$Fe(s) + 3\,H_2O(\ell) \rightarrow Fe(OH)_3(s) + 3/2\,H_2(g) \qquad \Delta_rH° = +160.9 \text{ kJ}$$
$$H_2(g) + \tfrac{1}{2}\,O_2(g) \rightarrow H_2O(\ell) \qquad \Delta_rH° = -285.8 \text{ kJ}$$
$$Fe_2O_3(s) + 3\,H_2O(\ell) \rightarrow 2\,Fe(OH)_3(s) \qquad \Delta_rH° = +288.6 \text{ kJ}$$

a) -1648.4 kJ   b) -1182.1 kJ   c) -219.4 kJ   d) +162.8 kJ   e) +1447.1 kJ

Answer: a

39. Determine the enthalpy change for the oxidation of ethanol to acetic acid,

$$CH_3CH_2OH(\ell) + O_2(g) \rightarrow CH_3COOH(\ell) + H_2O(\ell)$$

given the thermochemical equations below.

$$2\,CH_3CH_2OH(\ell) + O_2(g) \rightarrow 2\,CH_3CHO(\ell) + 2\,H_2O(\ell) \qquad \Delta_rH° = -400.8 \text{ kJ}$$
$$2\,CH_3CHO(\ell) + O_2(g) \rightarrow 2\,CH_3COOH(\ell) \qquad \Delta_rH° = -584.4 \text{ kJ}$$

a) -985.2 kJ   b) -492.6 kJ   c) -183.6 kJ   d) +183.6 kJ   e) +492.6 kJ

Answer: b

40. Determine $\Delta_r H$ for the following reaction,

$$4\ NH_3(g)\ +\ 5\ O_2(g)\ \rightarrow\ 4\ NO(g)\ +\ 6\ H_2O(g)$$

given the thermochemical equations below.

$N_2(g)\ +\ O_2(g)\ \rightarrow\ 2\ NO(g)$	$\Delta_r H^\circ = +180.8$ kJ
$N_2(g)\ +\ 3\ H_2(g)\ \rightarrow\ 2\ NH_3(g)$	$\Delta_r H^\circ = -91.8$ kJ
$2\ H_2(g)\ +\ O_2(g)\ \rightarrow\ 2\ H_2O(g)$	$\Delta_r H^\circ = -483.6$ kJ

a) -1272.8 kJ   b) -905.6 kJ   c) -394.6 kJ   d) -211.0 kJ   e) +1272.8 kJ

Answer: b

41. Calculate $\Delta_r H^\circ$ for the following reaction,

$$CaCO_3(s)\ \rightarrow\ CaO(s)\ +\ CO_2(g)$$

given the thermochemical equations below.

$2\ Ca(s)\ +\ O_2(g)\ \rightarrow\ 2\ CaO(s)$	$\Delta_r H^\circ = -1270.2$ kJ
$C(s)\ +\ O_2(g)\ \rightarrow\ CO_2(g)$	$\Delta_r H^\circ = -393.5$ kJ
$2\ Ca(s)\ +\ 2\ C(s)\ +\ 3\ O_2(g)\ \rightarrow 2\ CaCO_3(s)$	$\Delta_r H^\circ = -2413.8$ kJ

a) -4077.3 kJ   b) -2870.6 kJ   c) -750.1 kJ   d) -456.8 kJ   e) +178.3 kJ

Answer: e

42. Which of the following chemical equations corresponds to the standard molar enthalpy of formation of SO_3?

a) $SO_2(g)\ +\ \frac{1}{2}\ O_2(g)\ \rightarrow\ SO_3(g)$
b) $2\ SO_2(g)\ +\ O_2(g)\ \rightarrow\ 2\ SO_3(g)$
c) $S_8(s)\ +\ 12\ O_2(g)\ \rightarrow\ 8\ SO_3(g)$
d) $2\ S(s)\ +\ 3\ O_2(g)\ \rightarrow\ 2\ SO_3(g)$
e) $1/8\ S_8(s)\ +\ 3/2\ O_2(g)\ \rightarrow\ SO_3(g)$

Answer: e

43. Which of the following chemical equations does not correspond to a standard molar enthalpy of formation?

a) $Ca(s)\ +\ C(s)\ +\ 3/2\ O_2(g)\ \rightarrow\ CaCO_3(s)$
b) $C(s)\ +\ O_2(g)\ \rightarrow\ CO_2(g)$
c) $NO(g)\ +\ \frac{1}{2}\ O_2(g)\ \rightarrow\ NO_2(g)$
d) $N_2(g)\ +\ 2\ O_2(g)\ \rightarrow\ N_2O_4(g)$
e) $H_2(g)\ +\ \frac{1}{2}\ O_2(g)\ \rightarrow\ H_2O(\ell)$

Answer: c

44. Calculate $\Delta_f H°$ for sulfur dioxide given the thermochemical equations below.

$$2\,S(s) + 3\,O_2(g) \rightarrow 2\,SO_3(g) \qquad \Delta_r H° = -791.5 \text{ kJ}$$
$$2\,SO_2(g) + O_2(g) \rightarrow 2\,SO_3(g) \qquad \Delta_r H° = -197.9 \text{ kJ}$$

a) -296.8 kJ/mol   b) -395.7 kJ/mol   c) -494.7 kJ/mol   d) -593.6 kJ/mol   e) -989.4 kJ/mol

Answer: a

45. Calculate $\Delta_r H°$ for the combustion of gaseous dimethyl ether,

$$CH_3OCH_3(g) + 3\,O_2(g) \rightarrow 2\,CO_2(g) + 3\,H_2O(\ell)$$

using standard molar enthalpies of formation.

molecule	$\Delta_f H°$ (kJ/mol)
$CH_3OCH_3(g)$	-184.1
$CO_2(g)$	-393.5
$H_2O(\ell)$	-285.8

a) -76.4 kJ   b) -495.2 kJ   c) -863.4 kJ   d) -1460.3 kJ   e) -1828.5 kJ

Answer: d

46. The standard enthalpy change for the combustion of 1 mole of ethene is -1323.1 kJ.

$$C_2H_4(g) + 3\,O_2(g) \rightarrow 2\,CO_2(g) + 2\,H_2O(g)$$

Calculate $\Delta_f H°$ for ethene based on the following standard molar enthalpies of formation.

molecule	$\Delta_f H°$ (kJ/mol)
$CO_2(g)$	-393.5
$H_2O(g)$	-241.8

a) -1958.4 kJ   b) -687.8 kJ   c) -52.5 kJ   d) +52.5 kJ   e) +687.8 kJ

Answer: d

47. The standard molar enthalpy of formation of $NH_3(g)$ is -45.9 kJ/mol. What is the enthalpy change if 6.31 g $N_2(s)$ and 1.96 g $H_2(g)$ react to produce $NH_3(g)$?

a) -10.3 kJ   b) -20.7 kJ   c) -29.8 kJ   d) -43.7 kJ   e) -65.6 kJ

Answer: b

Short Answer Questions

48. ________ energy is often referred to as the energy of motion. Examples of this type of energy include thermal, mechanical, and electrical energies.

Answer: Kinetic

49. _______ energy is associated with the motion of atoms, molecules, or ions at the particulate level.

Answer: Thermal

50. In thermodynamics, a _________ is defined as the object, or collection of objects, being studied. The surroundings include everything outside.

Answer: system

51. The heat required to convert a liquid at its boiling point to a gas is called the heat of _________.

Answer: vaporization

52. The enthalpy change is the heat absorbed or evolved in a reaction that occurs at constant _________.

Answer: pressure

53. Why are you at greater risk from being burned by steam at 100 °C than from liquid water at the same temperature?

Answer: At 100 °C, steam has approximately 40 kJ/mole more potential energy than liquid water at the same temperature.

54. Internal energy and enthalpy are state functions. What is meant by this statement?

Answer: In any process, the change in internal energy or the change in enthalpy is independent of the path between the two states. These values depend only on the initial and final states.

Chapter 6
THE STRUCTURE OF ATOMS

CHAPTER OUTLINE

CHAPTER 6: The Structure of Atoms

d Orbitals (p. 291)
f Orbitals (p. 291)

6.7   One More Electron Property: Electron Spin
The Electron Spin Quantum Number, m_s (p. 291)
    KEY TERMS: electron spin quantum number (m_s),
Diamagnetism and Paramagnetism
    KEY TERMS: paramagnetic, ferromagnetic, diamagnetic
    A Closer Look: Paramagnetism and Ferromagnetism

CHAPTER OBJECTIVES

- Describe the properties of electromagnetic radiation (6.1-6.2)
- Understand the origin of light from excited atoms and its relationship to atomic structure (6.3)
- Describe the experimental evidence for particle-wave duality (6.4)
- Describe the basic ideas of quantum numbers (6.5)
- Define the four quantum numbers (n, ℓ, m_ℓ, m_s), and recognize their relationship to electronic structure (6.5-8.7)

LESSON PLAN

AP Standards
I.A.4   Electron energy levels: atomic spectra, quantum numbers, atomic orbitals

Suggested Time:
Four traditional classes or 2 blocks. Chapters 6 and 7 are key elements in understanding chemistry and can be taught together. The AP material for this chapter is covered in sections 6.4-6.7. If time allows, then equal time should be spent on both chapters. If you are behind, suggest cutting back on sections 6.1, 6.2 and 6.4.

ASSESSMENT

Exercises:

6.1	Radiation, Wavelength (p. 271)	
6.2	Photon Energies (p. 275)	
6.3	Electron Energies (p. 278)	
6.4	Energy of an Atomic Spectral Line (p. 280)	
6.5	Ionization of Hydrogen (p. 280)	
6.6	De Broglie's Equation (p. 283)	
6.7	Using Quantum Numbers (p. 287)	
6.8	Orbital Shapes (p. 291)	

Answers to Exercises p. A44-A45
Practicing Skills pp. 297-299, Answers IRM pp. 114-122
General Questions pp. 299-301, Answers IRM pp. 122-126
In the Laboratory Questions pp. 301-302, Answers IRM pp. 126
Summary and Conceptual Questions pp. 302-303, Answers IRM pp. 126-130

GLOSSARY

amplitude The maximum height of a wave, as measured from the axis of propagation

azimuthal quantum number (ℓ) The quantum number that defines the shape of an orbital

atomic orbital The matter wave for an allowed energy state of an electron in an atom

Balmer equation An equation which relates the wavelength of the visible spectral lines for hydrogen to the change in energy level of the electron

Balmer series A series of spectral line that have energies in the visible region

boundary surface The surface of the orbital within which the electron is found 90% of the time

continuous spectrum The spectrum of white light emitted by a heated object, consisting of light of all wavelengths

surface density plot (radial distribution plot) A plot of the surface density for an orbital ($4\pi r^2 \psi^2$) as a function of distance from the nucleus

electromagnetic radiation Radiation that consists of wave-like electric and magnetic fields, including light, microwaves, radio signals, and x-rays

electron density The probability of finding an atomic electron within a give region of space, related to the square of the electron's wave function

electron shell The region in space corresponding to a particular principle quantum number

excited state The state of an atom in which a least one electron is not in the lowest possible energy level

frequency (v) The number of complete waves passing a point in a given amount of time

ground state The state of an atom in which all electrons are in the lowest possible energy level

hertz The unit of frequency, or cycles per second: $1 \text{ Hz} = 1 \text{ s}^{-1}$

line emission spectrum The spectrum of light emitted by excited atoms in the gas phase, consisting of discrete wavelengths

Lyman series A series of spectral line that have energies in the ultraviolet region

magnetic quantum number (m_ℓ) The quantum number related to the orientation in space of the orbitals within a subshell

CHAPTER 6: The Structure of Atoms

nodal surface A surface on which there is zero probability of finding an electron

node A point of zero amplitude of a wave

orbitals The matter wave for an allowed energy state of an electron in an atom or molecule

photoelectric effect The ejection of electrons from a metal bombarded with light of at least a minimum frequency

photons A "particle" of electromagnetic radiation having zero mass and an energy given by Planck's law

Planck's constant (h) the proportionality constant that relates the frequency of radiation to its energy

Planck's equation An equation which relates the energy of an electron to its wave frequency

principle quantum number (n) The quantum number related to the energy of the orbital

probability Likelihood of finding

quantization A situation in which only certain energies are allowed

quantum (or wave) mechanics A general theoretical approach to atomic behavior that describes the electron in an atom as a matter wave

quantum numbers A set of numbers with integer values that define the properties of an atomic orbital

Rydberg constant The proportionality constant between the inverse wavelength of a hydrogen electron and its corresponding change in energy levels

standing waves A single frequency wave having fixed points of zero amplitude

subshell Regions within an energy shell which correspond to particular orbitals and equivalent energy

uncertainty principle It is impossible to determine both the position and the momentum of an electron in an atom simultaneously with great certainty

wave function (Ψ) A set of equations that characterize the electron as a matter wave

wavelength (λ) The distance between successive crests or troughs in a wave

wave-particle duality The idea that the electron has properties of both a wave and a particle

MEDIA MENU

The Media Menu section of this guide lists the interactive exercises, tutorials, active figures (animated, interactive figures) and additional resources available online for Chemistry Now, a unique web-based, assessment-centered personalized learning system for chemistry students. Go to www.cengagelearning.com/login.

Exercises

- Screen 6.5   Using Planck's equation to calculate wavelength
- Screen 6.6   The radiation emitted when electrons of excited hydrogen atoms return to ground state
- Screen 6.13   Orbital shapes, quantum numbers, and nodes
- Screen 6.16   Spinning electrons and magnetism

Tutorials

- Screen 6.3   Calculating the frequency of ultraviolet light
- Screen 6.3   Calculating the frequency of visible light
- Screen 6.6   Calculating the wavelength of radiation emitted when electrons change energy levels
- Screen 6.8   Calculating the wavelength of a moving electron
- Screen 6.12   Determining values for the quantum numbers of an orbital

Self-study module

- Screen 6.15   Electron spin

Active Figures

- 6.2   The electromagnetic spectrum
- 6.6   The line emission spectrum of hydrogen
- 6.8   Energy levels for the H atom in the Bohr model
- 6.9   Absorption of energy by the atom as the electron moves to an excited state
- 6.10   Some of the electronic transitions that can occur in an excited H atom
- 6.13   Different views of a $1s$ orbital
- 6.14   Atomic orbitals
- 6.18   Observing and measuring paramagnetism

Additional Resources

- Screen 6.4   Exploring the wavelength and frequency of visible light simulation
- Screen 6.5   Exploring the relationship between wavelength, frequency, and photon energy simulation
- Screen 6.6   The radiation emitted when electrons of excited hydrogen atoms return to ground state simulation
- Screen 6.9   Quantum mechanical view of the atom animation

DIAGNOSTIC QUIZ QUESTIONS

Goal # 1 Describe the properties of electromagnetic radiation

1. Which of the following types of radiation has the highest energy?

    a) microwave   b) radio-wave   c) infrared
    d) visible   e) ultraviolet

    Answer: e

2. Which of the following types of radiation has the longest wavelength?

    a) x-ray   b) ultraviolet   c) microwave
    d) visible   e) γ-ray

    Answer: c

3. A typical laser pointer emits light with a wavelength of 6.50×10^2 nm. What is the frequency of this light?

    a) 2.17×10^{-15} s^{-1} b) 5.13×10^{-3} s^{-1} c) 1.95×10^2 s^{-1}
    d) 4.61×10^{14} s^{-1} e) 3.05×10^{-19} s^{-1}

    Answer: d

4. What is the energy (in Joules) of a 455 nm photon?

    a) 4.84×10^{-36} J   b) 9.04×10^{-32} J   c) 4.37×10^{-19} J
    d) 1.46×10^{-18} J   e) 9.10×10^{-7} J

    Answer: c

Goal #2 Understand the origin of light from excited atoms and its relationship to atomic structure

5. Which of the following transitions in a hydrogen atom would **emit** the highest energy photon?

    a) $n = 1$ to $n = 2$ b) $n = 3$ to $n = 2$ c) $n = 5$ to $n = 1$
    d) $n = 2$ to $n = 8$ e) $n = 6$ to $n = 5$

    Answer: c

6. For a hydrogen atom, calculate the energy of a photon in the Balmer series that results from the transition $n = 3$ to $n = 2$. The Rydberg constant is 1.0974×10^7 m^{-1}. (h = 6.626×10^{-34} J·s and c = 2.998×10^8 m/s)

a) 3.03×10^{-19} J
b) 3.63×10^{-19} J
c) 4.36×10^{-19} J
d) 2.18×10^{-18} J
e) 1.09×10^{-17} J

Answer: a

Goal #3 Describe the experimental evidence for particle-wave duality

7. If the de Broglie wavelength of an electron is 15 nm, what is its velocity? The mass of an electron is 9.1×10^{-31} kg.

a) 2.1×10^{-5} m/s   b) 1.2×10^{-2} m/s   c) 9.2×10^1 m/s   d) 4.9×10^4 m/s   e) 9.2×10^{10} m/s

Answer: d

Goal #4 Describe the basic ideas of quantum numbers

8. Which of the following statements is INCORRECT?

a) It is not possible to know the exact location of an electron and its exact energy simultaneously.
b) The energies of an atom's electrons are quantized.
c) Quantum numbers define the energy states and the orbitals available to an electron.
d) The behavior of an atom's electrons can be described by circular orbits around a nucleus.
e) Electrons have both wave and particle properties.

Answer: d

Goal #5 Define the four quantum numbers (n, ℓ, m_ℓ, m_s), and recognize their relationship to electronic structure

9. What type of orbital is designated $n = 2$, $\ell = 3$, $m_\ell = -2$?

a) $4s$ b) $4p$ c) $4d$ d) $4f$ e) none

Answer: e

10. What shell contains a total of 16 orbitals?

a) $n = 1$ b) $n = 2$ c) $n = 3$ d) $n = 4$ e) $n = 5$

Answer: d

11. Which of the following statements is/are CORRECT?
    1. A diamagnetic substance is attracted to a magnetic field.
    2. An atom with no unpaired electrons is ferromagnetic.
    3. Atoms with one or more unpaired electrons are paramagnetic.

    a) 1 only   b) 2 only   c) 3 only   d) 1 and 2   e) 1, 2, and 3

    Answer: c

TESTBANK

6.1 Electromagnetic Radiation

1. Which of the following regions of the electromagnetic spectrum have shorter wavelengths than visible light?
    1. infrared radiation
    2. ultraviolet radiation
    3. microwave radiation

    a) 1 only   b) 2 only   c) 3 only   d) 1 and 2   e) 1 and 3

    Answer: b

2. Which of the following regions of the electromagnetic spectrum has the lowest frequency?

    a) microwave   b) gamma ray   c) ultraviolet   d) infrared   e) visible

    Answer: a

3. Which of the following colors of visible light has the highest frequency?

    a) green   b) orange   c) red   d) yellow   e) blue

    Answer: e

4. Which of the following regions of the electromagnetic spectrum has the longest wavelength?

    a) ultraviolet   b) x-ray   c) infrared   d) visible   e) gamma ray

    Answer: c

5. An argon ion laser emits light at 457.9 nm. What is the frequency of this radiation?

    a) 4.338×10^{-19} s^{-1} b) 1.527×10^{-15} s^{-1} c) 1.373×10^{11} s^{-1}
    d) 6.547×10^{14} s^{-1} e) 2.305×10^{18} s^{-1}

    Answer: d

6. If an FM radio station broadcasts at 92.1 MHz, what is the wavelength of this radiation?

    a) 0.0326 m   b) 0.307 m   c) 2.76 m   d) 3.26 m   e) 30.7 m

Answer: d

7. A plutonium-239 atom may emit a photon of light with a wavelength of 24.0 pm when it undergoes nuclear fission. What is the frequency of this radiation?

    a) 8.01×10^{-20} s^{-1} b) 8.28×10^{-15} s^{-1} c) 1.25×10^{7} s^{-1} d) 7.20×10^{9} s^{-1} e) 1.25×10^{19} s^{-1}

Answer: e

8. When heated in a flame, sodium atoms emit light with a frequency of 5.09×10^{14} s^{-1}. What is the wavelength of this radiation?

    a) 153 nm   b) 170 nm   c) 589 nm   d) 1.70×10^{6} nm e) 1.53×10^{14} nm

Answer: c

6.2 Planck, Einstein, Energy, and Photons

9. The _________ of a photon of light is _________ proportional to its frequency and _________ proportional to its wavelength.

    a) energy, directly, inversely   b) energy, inversely, directly c) velocity, directly, inversely
    d) intensity, inversely, directly   e) amplitude, directly, inversely

Answer: a

10. According to experiments concerned with the photoelectric effect, which of the following will increase the kinetic energy of an electron ejected from a metal surface?
    1. decreasing the wavelength of the light striking the surface
    2. decreasing the frequency of the light striking the surface
    3. increasing the number of photons of light striking the surface

    a) 1 only   b) 2 only   c) 3 only   d) 1 and 2   e) 1 and 3

Answer: a

11. Excited hydrogen atoms emit light in the infrared at 1.87×10^{-6} m. What is the energy of a single photon with this wavelength?

    a) 1.24×10^{-39} J b) 4.13×10^{-20} J c) 1.06×10^{-19} J d) 6.24×10^{-15} J e) 1.60×10^{14} J

Answer: c

12. A common infrared laser operates at a frequency of 2.82×10^{14} s^{-1}. What is the energy of a photon with this frequency?

    a) 7.04×10^{-40} J   b) 6.25×10^{-28} J   c) 1.87×10^{-19} J   d) 3.54×10^{-15} J   e) 1.06×10^{-6} J

    Answer: c

13. A red laser pointer emits light at a wavelength of 652 nm. If the laser emits 9.0×10^{-4} J of energy per second in the form of visible radiation, how many photons per second are emitted from the laser?

    a) 3.4×10^{-6} photons/sec   b) 4.1×10^{11} photons/sec   c) 8.9×10^{14} photons/sec
    d) 3.0×10^{15} photons/sec   e) 5.1×10^{17} photons/sec

    Answer: d

14. What is the energy (in kJ) of 1.00 mole of photons of green light with a wavelength of 507 nm?

    a) 4.24 kJ   b) 60.7 kJ   c) 91.6 kJ   d) 152 kJ   e) 236 kJ

    Answer: e

15. If the energy of 1.00 mole of photons is 346 kJ, what is the wavelength of the light?

    a) 289 nm   b) 346 nm   c) 574 nm   d) 787 nm   e) 867 nm

    Answer: b

16. What is the binding energy of an electron in a photosensitive metal (in kJ/mol) if the longest wavelength of light that can eject electrons from the metal is 238 nm?

    a) 139 kJ/mol   b) 503 kJ/mol   c) 835 kJ/mol
    d) 1.26×10^{12} kJ/mol   e) 8.35×10^{22} kJ/mol

    Answer: b

17. The energy required to break one mole of hydrogen-hydrogen bonds in H_2 is 436 kJ/mol. What is the longest wavelength of light capable of breaking a single hydrogen-hydrogen bond?

    a) 0.688 nm   b) 119 nm   c) 132 nm   d) 274 nm   e) 688 nm

    Answer: d

18. According to the Bohr model for the hydrogen atom, the energy necessary to excite an electron from $n = 4$ to $n = 5$ is _________ the energy necessary to excite an electron from $n = 3$ to $n = 4$.

    a) less than   b) greater than   c) equal to
    d) either equal to or greater than   e) either less than or equal to

    Answer: a

19. Which of the following transitions in a hydrogen atom would **emit** the highest energy photon?

    a) $n = 1$ to $n = 2$ b) $n = 3$ to $n = 2$ c) $n = 5$ to $n = 1$ d) $n = 2$ to $n = 8$ e) $n = 6$ to $n = 5$

    Answer: c

20. For which of the following transitions would a hydrogen atom absorb a photon with the longest wavelength?

    a) $n = 1$ to $n = 2$ b) $n = 2$ to $n = 4$ c) $n = 5$ to $n = 1$ d) $n = 7$ to $n = 6$ e) $n = 5$ to $n = 6$

    Answer: e

21. For a hydrogen atom, calculate the energy of a photon in the Balmer series that results from the transition $n = 3$ to $n = 2$. The Rydberg constant is 1.0974×10^7 m^{-1}. ($h = 6.626 \times 10^{-34}$ J·s and $c = 2.998 \times 10^8$ m/s)

    a) 3.03×10^{-19} J   b) 3.63×10^{-19} J   c) 4.36×10^{-19} J
    d) 2.18×10^{-18} J   e) 1.09×10^{-17} J

    Answer: a

22. If a hydrogen atom in the excited $n = 4$ state relaxes to the ground state, what is the maximum number of possible emission lines?

    a) 1   b) 4   c) 6   d) 8   e) infinite

    Answer: c

CHAPTER 6: The Structure of Atoms

<u>6.4 Particle-Wave Duality: Prelude to Quantum Mechanics</u>

23. For a neutron (mass = 1.675×10^{-27} kg) moving with a velocity of 3.10×10^4 m/s, what is the de Broglie wavelength (in pm)?

   a) 0.128 pm   b) 1.23 pm   c) 12.8 pm   d) 78.4 pm   e) 123 pm

   Answer: c

24. If the de Broglie wavelength of an electron is 15 nm, what is its velocity? The mass of an electron is 9.1×10^{-31} kg.

   a) 2.1×10^{-5} m/s   b) 1.2×10^{-2} m/s   c) 9.2×10^{1} m/s   d) 4.9×10^{4} m/s   e) 9.2×10^{10} m/s

   Answer: d

25. What is the de Broglie wavelength of a 120 g golf ball traveling at 84 km/hr?

   a) 2.4×10^{-34} m   b) 1.9×10^{-33} m   c) 3.3×10^{-33} m   d) 1.3×10^{-31} m   e) 4.2×10^{33} m

   Answer: a

<u>6.5 The Modern View of Electronic Structure: Wave or Quantum Mechanics</u>

26. Which type of experiment demonstrates that an electron has the properties of a wave?

   a) nuclear fission   b) electron diffraction   c) light emission from atomic gases
   d) mass spectroscopy   e) photoelectric effect

   Answer: b

27. Which type of experiment demonstrates that light has the properties of a particle?

   a) nuclear fission   b) electron diffraction   c) light emission from atomic gases
   d) mass spectroscopy   e) photoelectric effect

   Answer: e

28. The Schrödinger wave equation

   a) calculates the exact position and momentum of an electron at any given time.
   b) can be solved to determine the probability of finding an electron in a region of space.
   c) proves that energy is equal to mass times the speed of light squared.
   d) incorrectly predicts circular orbits of electrons around nuclei.
   e) is used to calculate the velocity of an electron.

   Answer: b

29. Which of the following statements is INCORRECT?

a) It is not possible to know the exact location of an electron and its exact energy
   simultaneously.
b) The energies of an atom's electrons are quantized.
c) Quantum numbers define the energy states and the orbitals available to an electron.
d) The behavior of an atom's electrons can be described by circular orbits around a nucleus.
e) Electrons have both wave and particle properties.

Answer: d

30. How many orbitals have the following set of quantum numbers: $n = 6$, $\ell = 3$, $m_\ell = +1$?

a) 0   b) 1   c) 3   d) 6   e) 7

Answer: b

31. All of the following sets of quantum numbers are allowed EXCEPT

a) $n = 1$, $\ell = 0$, $m_\ell = 0$ b) $n = 3$, $\ell = 2$, $m_\ell = +2$ c) $n = 4$, $\ell = 3$, $m_\ell = -1$
d) $n = 5$, $\ell = 1$, $m_\ell = 0$ e) $n = 6$, $\ell = 2$, $m_\ell = +3$

Answer: e

32. All of the following sets of quantum numbers are allowed EXCEPT

a) $n = 5$, $\ell = 3$, $m_\ell = +2$ b) $n = 4$, $\ell = 2$, $m_\ell = -1$ c) $n = 3$, $\ell = 3$, $m_\ell = 0$
d) $n = 2$, $\ell = 1$, $m_\ell = +1$ e) $n = 2$, $\ell = 0$, $m_\ell = 0$

Answer: c

33. What is the total number of orbitals having $n = 5$ and $\ell = 3$?

a) 3   b) 5   c) 7   d) 9   e) 10

Answer: c

34. Which of the following properties is associated with the value of the ℓ quantum number?

a) the shape of an orbital   b) the size of an orbital
c) the number of electrons in an orbital   d) the energy of an orbital
e) the orientation in space of an orbital

Answer: a

35. Which of the following properties is associated with the value of the n quantum number?

    a) the number of electrons in an orbital   b) the size of an orbital
    c) the orientation in space of an orbital   d) the energy of an orbital
    e) the shape of an orbital

    Answer: d

6.6 The Shapes of Atomic Orbitals

36. What type of orbital is designated $n = 3$, $\ell = 1$, $m_\ell = -1$?

    a) *3s*   b) *3p*   c) *3d*   d) *2f*   e) *2d*

    Answer: b

37. What type of orbital is designated $n = 4$, $\ell = 0$, $m_\ell = 0$?

    a) *4s*   b) *4p*   c) *4d*   d) *4f*   e) none

    Answer: a

38. What type of orbital is designated $n = 2$, $\ell = 3$, $m_\ell = -2$?

    a) *4s*   b) *4p*   c) *4d*   d) *4f*   e) none

    Answer: e

39. What shell contains a total of 16 orbitals?

    a) $n = 1$ b) $n = 2$ c) $n = 3$ d) $n = 4$ e) $n = 5$

    Answer: d

40. Which of the following sets of quantum numbers refers to a *4d* orbital?

    a) $n = 2, \ell = 1, m_\ell = -1$ b) $n = 2, \ell = 4, m_\ell = -1$ c) $n = 4, \ell = 2, m_\ell = -1$
    d) $n = 4, \ell = 3, m_\ell = 0$ e) $n = 4, \ell = 3, m_\ell = +2$

    Answer: c

41. Which of the following orbitals might have $m_\ell = -2$?

a) s b) s and p c) p and d d) d and f e) p, d, and f

Answer: d

42. How many spherical nodes exist for a $3p$ orbital?

a) 0   b) 1   c) 2   d) 3   e) 4

Answer: b

43. The $n = $ ______ shell is the lowest that may contain d-orbitals.

a) 2   b) 3   c) 4   d) 5   e) 6

Answer: b

44. Which of the following diagrams represent d-orbitals?

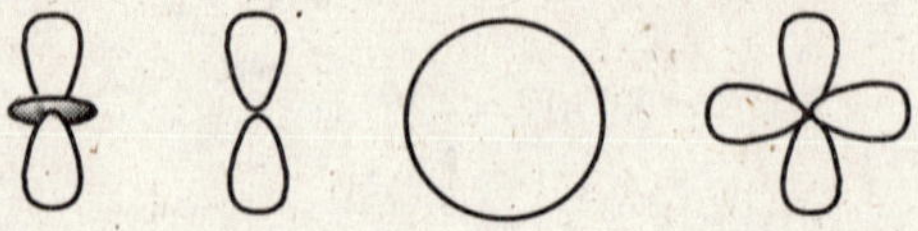

1   2   3   4

a) 1 only   b) 2 only   c) 3 only   d) 4 only   e) 1 and 4

Answer: e

45. Which of the following diagrams represent p-orbitals?

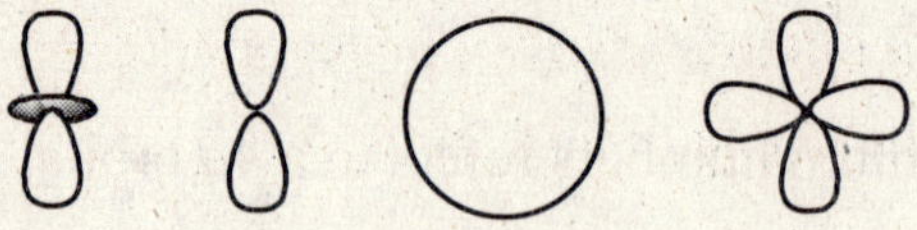

1   2   3   4

a) 1 only   b) 2 only   c 3 only   d) 4 only   e) 1 and 2

Answer: b

6.7 One More Electron Property: Electron Spin

46. Which of the following statements is/are CORRECT?
    1. A diamagnetic substance is attracted to a magnetic field.
    2. An atom with no unpaired electrons is ferromagnetic.
    3. Atoms with one or more unpaired electrons are paramagnetic.

    a) 1 only b) 2 only c) 3 only d) 1 and 2 e) 1, 2, and 3

    Answer: c

47. Which of the following statements is/are CORRECT?
    1. Substances that retain their magnetism after they are withdrawn from a magnetic field are called ferromagnetic.
    2. A paramagnetic substance is unaffected by a magnetic field.
    3. An atom with an even number of electrons must be diamagnetic.

    a) 1 only b) 2 only c) 3 only d) 1 and 3 e) 1, 2, and 3

    Answer: a

Short Answer Questions

48. According to Heisenberg's _________ principle, it is impossible to simultaneously measure the exact location and energy of an electron.

    Answer: uncertainty

49. A point in a standing wave that has zero amplitude is called a _________.

    Answer: node

50. The _________ momentum quantum number is given the symbol ℓ. This quantum number is the primary factor in determining the shape of an orbital.

    Answer: angular

51. The _________ of a photon is equal to the speed of light divided by the wavelength of the photon.

    Answer: frequency

52. The Bohr model predicts that the energy of an atom's electron is _________, meaning that the electron can only occupy orbitals of specific energies.

    Answer: quantized

53. What evidence does the photoelectric effect provide that photons are not only waves?

Answer: The photoelectric effect provides evidence that an electron is ejected when a single particle of light (as opposed to a wave of light) interacts with a metal surface. When photons of sufficient energy strike a metal surface, electrons are ejected. The number of ejected electrons is proportional to the number of photons striking the surface (which is consistent with both wave and particle theories of light). However, the energy of an ejected electron is independent of the number of photons striking the surface. If the interaction of matter and light was dependent on wave properties, then increasing the intensity of the light would increase the energy striking the surface and thus increase the energy of an ejected electron. Furthermore, there is a minimum wavelength of light required to eject an electron from a metal surface. At longer wavelengths, no photons are ejected from a metal regardless of the intensity of the light.

54. The size of an electron orbital is often chosen to be the distance within which 90% of the electron density is found. Why is the orbital radius not chosen to be the distance within which 100% of the electron density is found?

Answer: The probability of finding an electron decreases exponentially as distance from the nucleus increases. However, the probability does not reach zero until the electron is an infinite distance from the nucleus.

Chapter 7
THE STRUCTURE OF ATOMS AND PERIODIC TRENDS

CHAPTER OUTLINE

Case Study : Metals in Biochemistry and Medicine (p. 327)
7.6   Periodic Trends and Chemical Properties (p. 328)

CHAPTER OBJECTIVES

- Recognize the relationship of the four quantum numbers to atomic structure (7.1)
- Write the electron configuration for atoms and monatomic ions (7.2-7.4)
- Rationalize trends in atom and ion sizes, ionization energy, and electron affinity (7.5-7.6)

LESSON PLAN

AP Standards

I.A.5   Periodic relationships including, for example, atomic radii, ionization energies, electron affinities, oxidation numbers

IV.2   Relationships in the periodic table: horizontal, vertical, diagonal with examples from alkali metals, alkaline earth metals, halogens, and the first series of transition elements

Suggested Time:

Four traditional classes or 2 blocks. See Chapter 6 suggested time. The entire chapter is essential for the AP material.

ASSESSMENT

Exercises:

Answers to Exercises p. A45-A46
Practicing Skills pp. 332-333, Answers IRM pp. 130-135
General Questions pp. 333-335, Answers IRM pp. 135-138
In the Laboratory Questions pp. 335, Answers IRM pp. 138-139
Summary and Conceptual Questions pp. 335-337, Answers IRM pp. 140-143

GLOSSARY

actinide(s) The series of elements between actinium and rutherfordium in the periodic table

core electrons The electrons in an atom's completed set of shells

effective nuclear charge($Z*$) The nuclear charge experienced by an electron in a multi-electron atom, as modified by the other electrons

electron affinity The energy change occurring when an atom of that element in the gas phase gains an electron

electron configurations The arrangements of electrons in the atoms

Hund's rule The most stable arrangement of electrons is that the maximum number of unpaired electrons, all with the same spin direction

inner transition elements The lanthanide and actinide series of the periodic table constituting the *f*-block

ionization energy The energy required to remove an electron from an atom or ion in the gas phase

isoelectronic Molecules or ions that have the same number of valence electrons and comparable Lewis structures

lanthanides The series of elements between lanthanum and hafnium in the periodic table

noble gas notation An abbreviated form of *spdf* notation that replaces the completed electron shells with the symbol of the corresponding noble gas in brackets

orbital box diagram A notation for the electron configuration of an atom in which each orbital is shown as a box and the number and spin direction of the electrons are shown by arrows

Pauli exclusion principle No two electrons in an atom can have the same set of four quantum numbers

***p*-block elements** Main group elements in Groups 3A-8A

***s*-block elements** Main group elements in Groups 1A – 2A

***spdf* notation** A notation for the electron configuration of an atom in which the number of electrons assigned to a subshell is shown as a superscript after the subshell's symbol

transition elements Elements in groups 1B-8B where the *d*-orbitals are being filled

valence electron(s) The outermost and most reactive electrons of an atom

MEDIA MENU

The Media Menu section of this guide lists the interactive exercises, tutorials, active figures (animated, interactive figures) and additional resources available online for Chemistry Now, a unique web-based, assessment-centered personalized learning system for chemistry students. Go to www.cengagelearning.com/login.

Exercises

- Screen 7.2   Effective nuclear charge
- Screen 7.3   Shielding value
- Screen 7.4   Effective nuclear charge and shielding value

Tutorials

- Screen 7.5   Determining an element's box and spdf notations
- Screen 7.6   Determining an ion's box notation
- Screen 7.6   Determining whether an element is diamagnetic or paramagnetic

Active Figures

- 7.1   Order of subshell energies
- 7.4   Electron configurations and the periodic table
- 7.8   Atomic radii in picometers for main group elements
- 7.10   First ionization energies of the main group elements in these first four periods
- 7.11   Electron affinity
- 7.12   Relative sizes of some common ions
- 7.13   Examples of the periodicity of Group 1A and Group 7A elements

Additional Resources

- Screen 7.4   Effective nuclear charge and shielding value simulation
- Screen 7.5   The relationship between an element's electron configuration and its position in the periodic table simulation
- Screen 7.6   Changes to an element's configuration when it ionizes simulation
- Screen 7.8   The trends in atomic size moving across and down the periodic table simulation
- Screen 7.9   Energy levels of orbitals and the ability to retain electrons simulation
- Screen 7.9-7.10   Periodic trends in these properties simulations
- Screen 7.12   The relationship between ion formation and orbital energies in main group elements simulation
- Screen 7.12   The relationship between orbital energies and electron configurations on the size of main group elements simulation
- Screen 7.13   Relationship of atomic electron configurations and orbital energies on periodic trends videos
- Podcast Module 11   Periodic trends

CHAPTER 7: The Structure of Atoms and Periodic Trends

DIAGNOSTIC QUIZ QUESTIONS

Goal #1 Recognize the relationship of the four quantum numbers to atomic structure

1. The Pauli exclusion principle states that

    a) no two electrons in an atom can have the same four quantum numbers.
    b) electrons can have either $\pm\frac{1}{2}$ spins.
    c) electrons with opposing spins are attracted to each other.
    d) no two electrons in an atom can have the same spin.
    e) two electrons occupying the same orbital must have identical spins.

    Answer: a

2. How many electrons can be described by the following quantum numbers: $n = 3$, $\ell = 1$, $m_\ell = 0$, $m_s = -1/2$?

    a) 0   b) 1   c) 2   d) 3   e) 6

    Answer: b

3. Which one of the following sets of quantum numbers is NOT allowed?

    a) $n = 7$, $\ell = 0$, $m_\ell = 0$, $m_s = +1/2$ b) $n = 5$, $\ell = 3$, $m_\ell = -2$, $m_s = +1/2$
    c) $n = 4$, $\ell = 2$, $m_\ell = 0$, $m_s = -1/2$ d) $n = 3$, $\ell = 1$, $m_\ell = -1$, $m_s = +1/2$
    e) $n = 2$, $\ell = 2$, $m_\ell = 0$, $m_s = -1/2$

    Answer: e

Goal #2 Write the electron configuration for atoms and monatomic ions

4. Which of the following atoms is diamagnetic?

    a) Rb   b) N   c) Br   d) Cu   e) Hg

    Answer: e

5. Which element has the electron configuration $1s^2 2s^2 2p^6 3s^2 3p^2$?

    a) Mg   b) Si   c) Ge   d) Ti   e) Ga

    Answer: b

6. Which element has the electron configuration $[Ar]3d^{10}4s^2$?

    a) Cu   b) Zn   c) Ge   d) Ag   e) Cd

Answer: b

7. What is the electron configuration for a bismuth atom?

    a) $[Xe]6s^2 6p^3$ b) $[Kr]5d^{10}6s^2 6p^3$ c) $[Xe]4f^{14}5d^{10}6s^2 6p^3$
    d) $[Xe]5d^{10}6s^2 6p^3$ e) $[Kr]4d^{10}4f^{14}5d^3$

Answer: c

8. Which element has the following ground state electron configuration?

$$[Ar] \quad \boxed{\uparrow\downarrow} \, \boxed{\uparrow\downarrow} \, \boxed{\uparrow\downarrow} \, \boxed{\uparrow\downarrow} \, \boxed{\uparrow\downarrow} \qquad \boxed{\uparrow\downarrow} \qquad \boxed{\uparrow\downarrow} \, \boxed{\uparrow} \, \boxed{\uparrow}$$
$$\qquad\qquad 3d \qquad\qquad\qquad 4s \qquad\quad 4p$$

    a) Se   b) As   c) S   d) Cl   e) Ge

Answer: a

9. Which of the following cations has the same number of unpaired electrons as Cr^{3+}?

    a) Ni^{2+} b) Fe^{2+} c) Zn^{2+} d) Mn^{2+} e) Co^{2+}

Answer: e

Goal #3 Rationalize trends in atom and ion sizes, ionization energy, and electron affinity

10. Place the following atoms in order of increasing atomic radii: K, Mg, Ca, and Rb?

    a) K < Mg < Ca < Rb   b) K < Mg < Rb < Ca   c) Mg < Ca < K < Rb
    d) K < Rb < Mg < Ca   e) Mg < K < Ca < Rb

Answer: c

11. Which one of the following statements is INCORRECT?

    a) Ionization energy is a positive value for all elements.
    b) Ionization energy is the energy required to remove an electron from a gaseous atom.
    c) For any element, the second ionization energy is larger than the first ionization energy.
    d) Ionization energy increases across a periodic of the periodic table.
    e) Ionization energy increases down a group of the periodic table.

Answer: e

TESTBANK

7.1 The Pauli Exclusion Principle

1. The Pauli exclusion principle states that

    a) no two electrons in an atom can have the same four quantum numbers.
    b) electrons can have either $\pm\frac{1}{2}$ spins.
    c) electrons with opposing spins are attracted to each other.
    d) no two electrons in an atom can have the same spin.
    e) two electrons occupying the same orbital must have identical spins.

    Answer: a

2. How many electrons can be described by the following quantum numbers: $n = 3$, $\ell = 1$, $m_\ell = 0$, $m_s = -1/2$?

    a) 0 b) 1 c) 2 d) 3 e) 6

    Answer: b

3. How many electrons can be described by the quantum numbers $n = 5$, $\ell = 2$, $m_s = +1/2$?

    a) 1 b) 3 c) 5 d) 7 e) 9

    Answer: c

4. How many electrons can be described by the quantum numbers $n = 5$, $\ell = 2$?

    a) 0 b) 2 c) 6 d) 10 e) 14

    Answer: d

5. What is the maximum number of electrons that can occupy the $n = 4$ shell?

    a) 10 b) 16 c) 25 d) 32 e) 50

    Answer: d

6. What is the maximum number of electrons that can occupy the $n = 3$ shell?

    a) 4 b) 8 c) 18 d) 32 e) 50

    Answer: c

7. Which one of the following sets of quantum numbers is NOT allowed?

    a) $n = 7$, $\ell = 0$, $m_\ell = 0$, $m_s = +1/2$ b) $n = 5$, $\ell = 3$, $m_\ell = -2$, $m_s = +1/2$

    c) $n = 4$, $\ell = 2$, $m_\ell = 0$, $m_s = -1/2$ d) $n = 3$, $\ell = 1$, $m_\ell = -1$, $m_s = +1/2$

    e) $n = 2$, $\ell = 2$, $m_\ell = 0$, $m_s = -1/2$

Answer: e

8. Which of the following sets of quantum numbers is allowed?

    a) $n = 2$, $\ell = 1$, $m_\ell = +1/2$, $m_s = -1/2$ b) $n = 3$, $\ell = 2$, $m_\ell = +1$, $m_s = +1$

    c) $n = 4$, $\ell = 2$, $m_\ell = -2$, $m_s = -1/2$ d) $n = 4$, $\ell = 4$, $m_\ell = -1$, $m_s = +1/2$

    e) $n = 5$, $\ell = 2$, $m_\ell = +2$, $m_s = -1$

Answer: c

7.3 Atomic Subshell Energies and Electron Assignments

9. The procedure by which electrons are assigned to (or built up into) orbitals is known as the
_______ principle.

    a) aufbau   b) Bohr   c) Planck   d) Hund   e) Pauli

Answer: a

10. Which of the following statements regarding subshell filling order for a neutral atom is/are
CORRECT?
    1. Electrons are assigned to the $4s$ subshell before they are assigned to the $3d$ subshell.
    2. Electrons are assigned to the $4f$ subshell before they are assigned to the $6s$ subshell.
    3. Electrons are assigned to the $4d$ subshell before they are assigned to the $5p$ subshell.

    a) 1 only   b) 2 only   c) 3 only   d) 1 and 3   e) 1, 2, and 3

Answer: d

11. Which of the following statements is/are CORRECT for a carbon atom?
    1. The effective nuclear charge felt by a $2s$ electron is greater than that felt by a $1s$
electron.
    2. The effective nuclear charge felt by a $2p$ electron is less than that felt by a $2s$ electron.
    3. The effective nuclear charges felt by $2s$ and $2p$ electrons are identical.

    a) 1 only   b) 2 only   c) 3 only   d) 1 and 3   e) 1, 2, and 3

Answer: b

12. Which of the following statements concerning ground state electron configurations is/are CORRECT?
    1. For a hydrogen atom with one electron, the $2s$ and $2p$ orbitals have identical energies.
    2. For a lithium atom with three electrons, the $2s$ and $2p$ orbitals have different energies.
    3. The effective nuclear charge felt by an electron in a $2p$ orbital is greater for a carbon atom than for a boron atom.

    a) 1 only   b) 2 only   c) 3 only   d) 2 and 3   e) 1, 2, and 3

    Answer: e

7.3 Electron Configurations of Atoms

13. Which of the following elements is a p-block element?

    a) C   b) Cd   c) Cs   d) Ca   e) Cf

    Answer: a

14. Which of the following elements is a d-block element?

    a) C   b) Cs   c) Cd   d) Cl   e) Cf

    Answer: c

15. Which of the following atoms is paramagnetic?

    a) Zn   b) Sr   c) Kr   d) Te   e) Ca

    Answer: d

16. Which of the following atoms is diamagnetic?

    a) Rb   b) N   c) Br   d) Cu   e) Hg

    Answer: e

17. Which element has the electron configuration $1s^2 2s^2 2p^6 3s^2 3p^2$?

    a) Mg   b) Si   c) Ge   d) Ti   e) Ga

    Answer: b

18. Which element has the electron configuration $[Ar]3d^{10}4s^2$?

    a) Cu   b) Zn   c) Ge   d) Ag   e) Cd

    Answer: b

19. What is the electron configuration for a bismuth atom?

a) $[Xe]6s^2 6p^3$ b) $[Kr]5d^{10}6s^2 6p^3$ c) $[Xe]4f^{14}5d^{10}6s^2 6p^3$
d) $[Xe]5d^{10}6s^2 6p^3$ e) $[Kr]4d^{10}4f^{14}5d^3$

Answer: c

20. Which of the following statements concerning potassium is/are CORRECT?
1. Potassium is paramagnetic.
2. Potassium is a *p*-block element.
3. Potassium has two valence shell electrons.

a) 1 only   b) 2 only   c) 3 only   d) 1 and 2   e) 2 and 3

Answer: a

21. Hund's rule states that the most stable arrangement of electrons (for a ground state electron configuration)

a) has a filled valence shell of electrons.
b) has three electrons per orbital, each with identical spins.
c) has m_ℓ values greater than or equal to +1.
d) has the maximum number of unpaired electrons, all with the same spin.
e) has two electrons per orbital, each with opposing spins.

Answer: d

22. Which element has the following ground state electron configuration?

$$\boxed{\uparrow\downarrow}\quad \boxed{\uparrow\downarrow}\quad \boxed{\uparrow\downarrow}\,\boxed{\uparrow}\,\boxed{\uparrow}$$
$$\;\;1s\qquad 2s\qquad\quad 2p$$

a) P   b) S   c) O   d) N   e) F

Answer: c

23. Which element has the following ground state electron configuration?

$$[Ar]\ \boxed{\uparrow\downarrow}\,\boxed{\uparrow\downarrow}\,\boxed{\uparrow}\,\boxed{\uparrow}\,\boxed{\uparrow}\qquad \boxed{\uparrow\downarrow}$$
$$\qquad\qquad 3d\qquad\qquad\quad 4s$$

a) Fe   b) Co   c) Ni   d) Cu   e) Zn

Answer: b

24. Which element has the following ground state electron configuration?

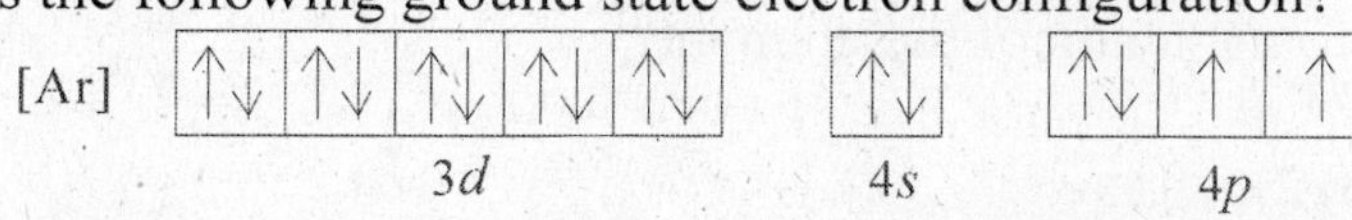

a) Se   b) As   c) S   d) Cl   e) Ge

Answer: a

25. What is the correct orbital box diagram for the ground state electron configuration of Cr?

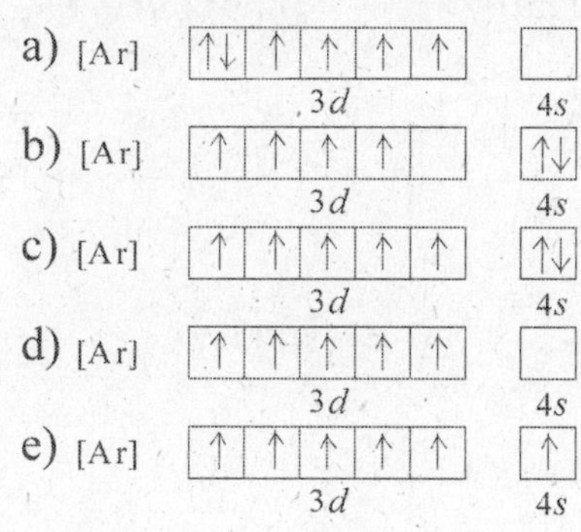

Answer: e

26. What is a possible set of quantum numbers for the unpaired electron in the orbital box diagram below?

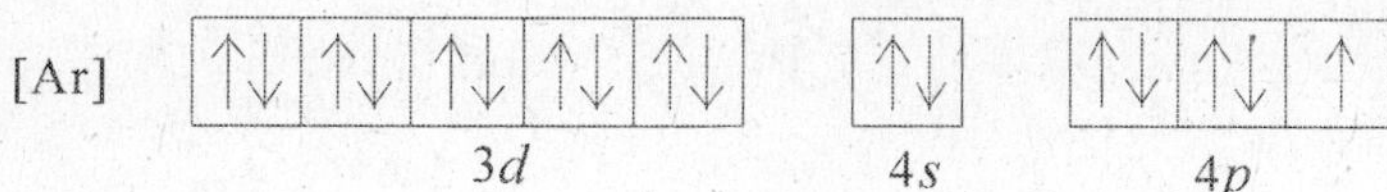

a) $n = 1$, $\ell = 1$, $m_\ell = -1$, $m_s = +1/2$ b) $n = 3$, $\ell = 2$, $m_\ell = -1$, $m_s = -1/2$

c) $n = 4$, $\ell = 2$, $m_\ell = -2$, $m_s = +1/2$ d) $n = 4$, $\ell = 0$, $m_\ell = 0$, $m_s = +1/2$

e) $n = 4$, $\ell = 1$, $m_\ell = -1$, $m_s = +1/2$

Answer: e

7.4 Electron Configurations of Ions

27. For which of the following atoms is the 2+ ion paramagnetic?

a) Sn   b) Mg   c) Zn   d) Mn   e) Ba

Answer: d

28. Which 1+ ion has the ground state electron configuration $[Kr]4d^{10}$?

a) Ru   b) Au   c) Ag   d) Tc   e) Cd

Answer: c

29. If the ground state electron configuration of an element is $[Ar]3d^{10}4s^24p^4$, what is the typical charge on the monatomic anion of the element?

a) 4+ b) 2+ c) 1- d) 2- e) 3-

Answer: d

30. What is the ground state electron configuration for Co^{3+}?

a) $[Ar]$ b) $[Ar]3d^{10}4s^2$ c) $[Ar]3d^44s^2$ d) $[Ar]3d^{10}$ e) $[Ar]3d^6$

Answer: e

31. What is the ground state electron configuration for In^+?

a) $[Kr]4d^{10}5s^2$ b) $[Kr]4d^{10}5p^2$ c) $[Kr]5s^2$
d) $[Kr]4d^{10}5s^25p^2$ e) $[Kr]4f^{14}5d^{10}5s^2$

Answer: a

32. Which of the following ions have the same ground state electron configuration: Cl^-, P^{3-}, Ca^{2+}, and Ga^+?

a) Cl^- and P^{3-} b) Cl^-, P^{3-}, and Ca^{2+} c) Ca^{2+} and Ga^+
d) P^{3-} and Ga^+ e) Cl^-, P^{3-}, Ca^{2+}, and Ga^+

Answer: b

33. What 2- ion has the following ground state electron configuration?

a) oxide ion b) nitride ion c) fluoride ion d) sulfide ion e) magnesium ion

Answer: a

34. What 2+ ion has the following ground state electron configuration?

a) Ge^{2+} b) Ni^{2+} c) Zn^{2+} d) Cu^{2+} e) None

Answer: c

35. What 2+ ion has the following ground state electron configuration?

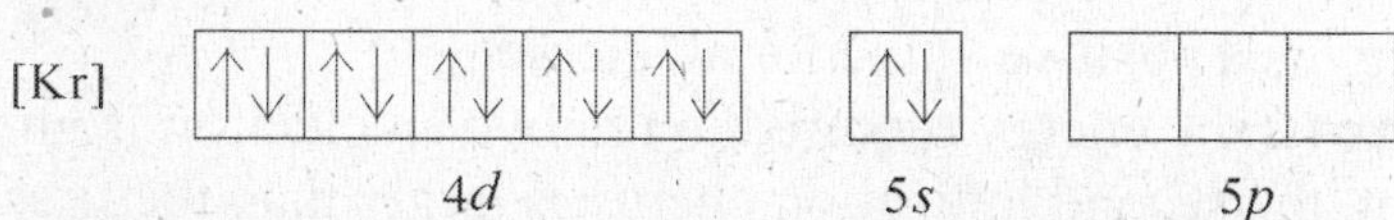

a) Cd^{2+} b) Sr^{2+} c) Zn^{2+} d) Sn^{2+} e) None

Answer: d

36. Which of the following cations has the same number of unpaired electrons as Cr^{3+}?

a) Ni^{2+} b) Fe^{2+} c) Zn^{2+} d) Mn^{2+} e) Co^{2+}

Answer: e

7.5 Atomic Properties and Periodic Trends

37. In general, atomic radii

a) decrease down a group and remain constant across a period.
b) decrease down a group and increase across a period.
c) increase down a group and increase across a period.
d) increase down a group and remain constant across a period.
e) increase down a group and decrease across a period.

Answer: e

38. Place the following atoms in order of increasing atomic radii: S, Se, Cl, and As.

a) Cl < S < Se < As   b) S < Cl < As < Se   c) S < Cl < Se < As
d) As < Se < S < Cl   e) Se < As < S < Cl

Answer: a

39. Place the following atoms in order of increasing atomic radii: K, Mg, Ca, and Rb?

a) K < Mg < Ca < Rb   b) K < Mg < Rb < Ca   c) Mg < Ca < K < Rb
d) K < Rb < Mg < Ca   e) Mg < K < Ca < Rb

Answer: c

40. Which one of the following statements is INCORRECT?

  a) Ionization energy is a positive value for all elements.
  b) Ionization energy is the energy required to remove an electron from a gaseous atom.
  c) For any element, the second ionization energy is larger than the first ionization energy.
  d) Ionization energy increases across a periodic of the periodic table.
  e) Ionization energy increases down a group of the periodic table.

Answer: e

41. Which of the following chemical equations refers to the second ionization of Ba?

  a) $Ba(s) + 2e^- \rightarrow Ba^{2-}(s)$ b) $Ba^+(g) \rightarrow Ba^{2+}(g) + e^-$
  c) $Ba(g) \rightarrow Ba^{2+}(g) + 2e^-$ d) $Ba(s) \rightarrow Ba^+(s) + e^-$
  e) $Ba^{2+}(g) + e^- \rightarrow Ba^+(g)$

Answer: b

42. For which of the following elements is the second ionization energy greatest?

  a) Mg   b) Al   c) Na   d) Sc   e) Ti

Answer: c

43. Rank Ca, Mg, and Ba in order of increasing first ionization energy.

  a) Ca < Mg < Ba   b) Ca < Ba < Mg   c) Mg < Ca < Ba
  d) Ba < Mg < Ca   e) Ba < Ca < Mg

Answer: e

44. The change in energy for the following reaction is referred to as the _________ for oxygen.
$$O(g) + e^- \rightarrow O^-(g)$$

  a) oxidation number   b) electron affinity   c) electronegativity energy
  d) first ionization energy   e) second ionization energy

Answer: b

45. Which of the following elements has the most negative electron affinity?
  a) C   b) P   c) O   d) Cl   e) F

Answer: e

46. Which elements have no affinity for electrons?

   a) transition metals   b) *s*-block elements   c) main group nonmetals
   d) noble gases   e) semiconductors

   Answer: d

47. Which group of the periodic table of elements forms only 1+ ions?

   a) group 1A   b) group 2A   c) group 1B   d) group 7A   e) group 8A

   Answer: a

48. Place the following ions in order from smallest to largest ionic radii: Se^{2-}, Sr^{2+}, Y^{3+}, and Br^-.

   a) $Se^{2-} < Sr^{2+} < Y^{3+} < Br^-$ b) $Se^{2-} < Br^- < Sr^{2+} < Y^{3+}$ c) $Br^- < Se^{2-} < Y^{3+} < Sr^{2+}$
   d) $Y^{3+} < Sr^{2+} < Br^- < Se^{2-}$ e) $Sr^{2+} < Y^{3+} < Se^{2-} < Br^-$

   Answer: d

Short Answer Questions

49. __________ rule states that the most stable arrangement of electrons is that which contains the maximum number of unpaired electrons, all with the same spin direction.

   Answer: Hund's

50. The element __________ has the following electron configuration: $[Rn]5f^4 6d^1 7s^2$.

   Answer: neptunium

51. As one moves horizontally from left to right across a period, the effective __________ charge increases, resulting in decreasing atomic radii.

   Answer: nuclear

52. The *f*-block elements are also referred to as the lanthanides and __________.

   Answer: actinides

53. Electron __________ is defined as the energy change for a process in which a gas phase atom acquires an electron.

   Answer: affinity

54. Explain why the first ionization energy for oxygen is lower than that for nitrogen.

Answer: Oxygen has a $1s^2 2s^2 2p^4$ electron configuration and nitrogen has a $1s^2 2s^2 2p^3$ electron configuration. Although the $2p$ electrons on an oxygen atom experience a greater effective nuclear charge than those on a nitrogen atom, two of oxygen's $2p$ electrons are paired in a single orbital, whereas each electron occupies its own orbital for nitrogen. The repulsion of these two electrons more than offsets oxygen's larger effective nuclear charge acting on the $2p$ electrons.

55. Explain the difference between paramagnetic and ferromagnetic.

Answer: Elements with unpaired electrons are paramagnetic. In the absence of a magnetic field, the unpaired electrons in paramagnetic elements are randomly oriented. In the presence of an external magnetic field, the unpaired electrons align with the magnetic field. For some paramagnetic elements (notably iron, cobalt, and nickel), clusters of unpaired electrons will remain aligned even in the absence of an external magnetic field. These elements are termed ferromagnetic.

Chapter 8
BONDING AND MOLECULAR STRUCTURE

CHAPTER OUTLINE

CHAPTER OBJECTIVES

- Understand the difference between ionic and covalent bonds (8.1-8.2)
- Draw Lewis electron dot structures for small molecules and ions (8.2-8.5)
- Use the valence shell electron-pair repulsion theory (VSEPR) to predict the shapes of simple molecules and ions and to understand the structures of more complex molecules (8.6)
- Use electronegativity and formal charge to predict the charge distribution in molecules and ions, to define the polarity of bonds, and to predict the polarity of molecules (8.3, 8.7)

CHAPTER 8: Bonding and Molecular Structure

- Understand the properties of covalent bonds and their influence on molecular structure (8.9)

LESSON PLAN

AP Standards

I.B.1.a	Types of binding forces: ionic, covalent, metallic, hydrogen bonding, van der Waals (including London dispersion forces
I.B.1.b	Relationships of binding forces to states, structure, and properties of matter
I.B.1.c	Polarity of bonds, electronegativities
I.B.2.a	Molecular models: Lewis structures
I.B.2.b	Valence bond: resonance
I.B.3	Geometry of molecules and ions, structural isomerism of simple organic molecules and coordination complexes; dipole moments of molecules; relation of properties to structure

Suggested Time:

Six traditional classes or 3 blocks. The entire chapter is essential for the AP material.

ASSESSMENT

Exercises:

8.1	Drawing Lewis Structures (p. 355)
8.2	Predicting Lewis Structures (p. 357)
8.3	Lewis Structures of Acids and Their Anions (p. 358)
8.4	Identifying Isoelectronic Species (.538
8.5	Calculating Formal Charge (p. 360)
8.6	Drawing Resonance Structures (p.363)
8.7	Lewis Structures in Which the Central Atom Has More Than Eight Electrons (p. 366)
8.8	Predicting Molecular Shape (p. 370)
8.9	VSEPR and Molecular Shape (p. 371)
8.10	Predicting Molecular Shape (p. 373)
8.11	Determining Molecular Shapes (p. 375)
8.12	Bond Polarity (p. 377)
8.13	Formal Charge, Bond Polarity, and Electronegativity (p. 379)
8.14	Molecular Polarity (p. 386)
8.15	Molecular Polarity (p. 386)
8.16	Bond Order and Bond Length (p. 388)
8.17	Using Bond Dissociation Enthalpies (p. 391)

Answers to Exercises p. A46-A47

Practicing Skills pp. 395-398, Answers IRM pp. 145-157

General Questions pp. 399-401, Answers IRM pp. 157-161

In the Laboratory Questions pp. 401-402, Answers IRM p. 162

Summary and Conceptual Questions pp. 402-403, Answers IRM pp. 162-163

GLOSSARY

bond angle The angle between two atoms bonded to a central atom

bond dissociation enthalpy The enthalpy change for breaking a bond in a molecule with the reactants and products in the gas phase at standard conditions

bond length The distance between the nuclei of two bonded atoms

bond pair Two electrons, shared by two atoms, that contribute to the bonding attraction between the atoms

bonding the chemical process of atoms joining by reducing the potential energy of their electrons

chemical bond An interaction between two or more atoms that holds them together by reducing the potential energy of their electrons

coordinate covalent bond Interatomic attraction resulting from the sharing of a lone pair of electrons from one atom with another atom

core electron The electrons in an atom's completed set of shells

covalent bond An interatomic attraction resulting from the sharing of electrons between atoms

dipole moment(μ) The product of the magnitude of the partial charges in a molecule and the distance by which they are separated

double bond A bond formed by sharing two pairs of electrons, one pair in a sigma bond and the other in a pi bond

electronegativity (χ) A measure of the ability of an atom in a molecule to attract electrons to itself

electron-pair geometry The geometry determined by all the bond pairs an dlone pairs in the valence shell of the central atom

formal charge The charge on an atom in a molecule or ion calculated by assuming equal sharing of the bonding electrons

free radical A neutral atom or molecule containing an unpaired electron

ionic bond The attraction between a positive and a negative ion resulting from the complete (or nearly complete) transfer of one or more electrons from one atom to another

isoelectronic species Molecules or ions that have the same number of valence electrons and comparable Lewis structures

Lewis electron dot structures (Lewis structure) A notation for the electron configuration of an atom or molecule

lone pair (nonbonding electrons) Pairs of valence electrons that do not contribute to bonding in a covalent molecule

molecular geometry The arrangement in space of the central atom and the atoms directly attached to it

non polar covalent bond A covalent bond in which there is equal sharing of the bonding electron pair

octet When forming bonds, atoms of main group elements gain, lose, or share electrons to achieve a stable configuration having eight valence electrons

order of bond The number of bonding electron pairs shared by two atoms in a molecule

partial charge The charge on an atom in a molecule or ion calculated by assuming the sharing of the bonding electrons is proportional to the electronegativity of the atom

polar covalent bond A covalent bond in which there is unequal sharing of the bonding electron pair

resonance The condition when more than one Lewis structure can be written for a molecule differing by the number of bond pairs between given pairs of atoms

resonance hybrid A compilation of resonance structures showing partial bond character

resonance structures The possible structures of a molecule for which more than one Lewis structure can be written, differing by the number of bond pairs between a given pair of atoms

single bond A bond formed by sharing one pair of electrons in a sigma bond

structure The way that atoms are arranged in space

triple bond A bond formed by sharing three pairs of electrons, one in a sigma bond and two in pi bonds

valence electron The outermost and most reactive electrons of an atom

valence shell electron pair repulsion (VSEPR) model A model for predicting the shapes of molecules in which structural electron pairs are arranged around each atom to maximize the angles between them

MEDIA MENU

The Media Menu section of this guide lists the interactive exercises, tutorials, active figures (animated, interactive figures) and additional resources available online for Chemistry Now, a unique web-based, assessment-centered personalized learning system for chemistry students. Go to www.cengagelearning.com/login

Exercises
- Screen 7.2   Effective nuclear charge
- Screen 8.6   Drawing Lewis structures
- Screen 8.11   Practice determining formal charge
- Screen 8.14   Practice predicting molecular geometry
- Screen 8.16   Practice determining polarity

Tutorials
- Screen 8.2   The correlation of the periodic table and valence electrons
- Screen 8.5   Lewis structures
- Screen 8.6   Drawing Lewis structures
- Screen 8.7   Drawing resonance structures
- Screen 8.8   Identifying electron-deficient compounds
-
-

Active Figures
- 8.5   Various geometries predicted by VSEPR
- 8.13   Polarity of triatomic molecules, AX_2
- 8.14   Polar and nonpolar molecules of the type AX_3

Additional Resources
- Screen 8.4   Factors influencing bond formation animation
- Screen 8.13   Electron-pair geometries animation
- Screen 8.13   Identifying geometries animation
- Screen 8.15   Relative electronegativity values
- Screen 8.17   Relationship between bond energy, length and order
- Screen 8.18   How reactant and product bond energies influence the energy of reactions
- Podcast Module 12   Covalent bonding and Lewis structures
- Podcast Module 13   Bond and molecular polarity

DIAGNOSTIC QUIZ QUESTIONS

Goal #1 Understand the difference between ionic and covalent bonds

1. Which of the following statements is/are CORRECT?
    1. Ionic bonds form when one or more valence electrons are transferred from one atom to another.
    2. Covalent bonds involve sharing of electrons between atoms.
    3. In most covalently bonded compounds, electrons are shared equally between the atoms.

    a) 1 only   b) 2 only   c) 3 only   d) 1 and 2   e) 1, 2, and 3

    Answer: d

2. Which one of the following compounds is ionic?

    a) $SnCl_2$ b) $SiCl_4$ c) PCl_3 d) SCl_2 e) PCl_5

    Answer: a

Goal #2 Draw Lewis electron dot structures for small molecules and ions

3. Which of the following is a correct Lewis structure for sulfite ion, $SO_3{}^{2-}$?

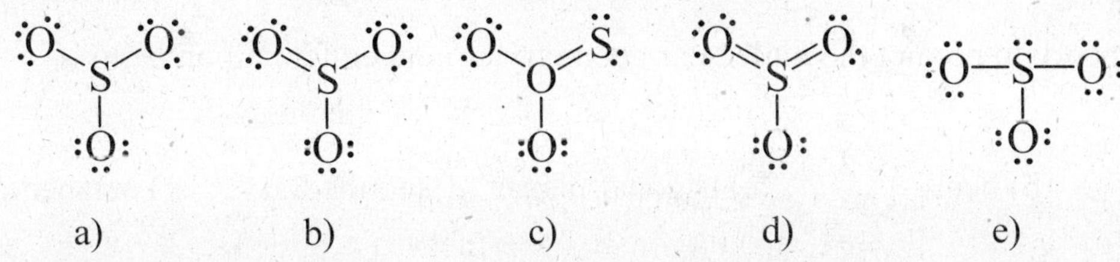

  a)   b)   c)   d)   e)

    Answer: e

4. How many lone pairs of electrons are assigned to the carbon atom in carbon monoxide?

    a) 0   b) 1   c) 2   d) 3   e) 4

    Answer: b

5. The central atom in PH_3 is surrounded by

a) three single bonds and no lone pairs of electrons.
b) three single bonds and one lone pair of electrons.
c) three single bonds and two lone pairs of electrons.
d) two single bonds, one double bond, and no lone pairs of electrons.
e) two single bonds, one double bond, and one lone pair of electrons.

Answer: b

6. Which of the following elements is able to form a molecular structure that exceeds the octet rule?

a) Ne   b) B   c) O   d) F   e) I

Answer: e

Goal #3 Use the valence shell electron-pair repulsion theory (VSEPR) to predict the shapes of simple molecules and ions and to understand the structures of more complex molecules

7. Use VSEPR theory to predict the molecular geometry of ClO_3^-.

a) trigonal-pyramidal   b) trigonal-planar   c) bent
d) T-shaped   e) linear

Answer: a

8. Use VSEPR theory to predict the molecular geometry around either carbon atom in acetylene, C_2H_2.

a) linear   b) bent   c) trigonal-planar d) tetrahedral   e) octahedral

Answer: a

9. Which of the following molecules has the smallest bond angle between any two hydrogen atoms?

a) CH_4 b) H_2O c) BH_3 d) BeH_2 e) SiH_4

Answer: b

Goal #4 Use electronegativity and formal charge to predict the charge distribution in molecules and ions, to define the polarity of bonds, and to predict the polarity of molecules

10. What is the formal charge on each atom in chloroform, $CHCl_3$?

   a) C atom = 0, H atom = 0, and each Cl atom = 0
   b) C atom = 0, H atom = +1, one Cl atom = -1, and two Cl atoms = 0
   c) C atom = +4, H = -1, and each Cl atom = -1
   d) C atom = +2, H = +1, and each Cl atom = -1
   e) C atom = -4, H = +1, and each Cl atom = +1

   Answer: a

11. Which one of the following molecules is polar?

   a) CO_2 b) SF_4 c) XeF_2 d) XeF_4 e) SO_3

   Answer: b

Goal #5 Understand the properties of covalent bonds and their influence on molecular structure

12. Predict which of the following compounds has covalent bond(s) that are the most polar.

   a) HF   b) CI_4 c) H_2S d) NBr_3 e) HI

   Answer: a

13. In molecules, as bond order increases,

   a) both bond length and bond energy increase.
   b) both bond length and bond energy decrease.
   c) bond length increases and bond energy is unchanged.
   d) bond length is unchanged and bond energy increases.
   e) bond length decreases and bond energy increases.

   Answer: e

14. Use Lewis structures to predict the bond order for a carbon-oxygen bond in carbonate ion, CO_3^{2-}.

   a) 1/2   b) 1   c) 4/3   d) 3/2   e) 2

   Answer: c

TESTBANK

<u>8.1 Chemical Bond Formation</u>

1. Which combination of atoms is most likely to produce a compound with ionic bonds?

    a) B and Cl   b) S and H   c) C and N   d) Si and I   e) Al and Br

    Answer: e

2. Which combination of atoms is most likely to produce a compound with covalent bonds?

    a) Na and Cl   b) Al and O   c) S and Br   d) Pb and F   e) K and I

    Answer: c

3. Which of the following statements is/are CORRECT?
    1. Ionic bonds form when one or more valence electrons are transferred from one atom to another.
    2. Covalent bonds involve sharing of electrons between atoms.
    3. In most covalently bonded compounds, electrons are shared equally between the atoms.

    a) 1 only   b) 2 only   c) 3 only   d) 1 and 2   e) 1, 2, and 3

    Answer: d

<u>8.2 Covalent Bonding and Lewis Structures</u>

4. Which of the following statements is/are CORRECT?
    1. All electrons in inner shells are valence electrons.
    2. For main group elements, the number of core electrons equals the group number.
    3. With the exception of He, the noble gases have eight valence electrons.

    a) 1 only   b) 2 only   c) 3 only   d) 1 and 3   e) 1, 2, and 3

    Answer: c

5. A selenium atom has ________ valence electrons.

    a) 2   b) 6   c) 16   d) 28   e) 34

    Answer: b

6. How many lone pairs of electrons are assigned to the carbon atom in carbon monoxide?

   a) 0 b) 1 c) 2 d) 3 e) 4

   Answer: b

7. How many lone pairs of electrons are assigned to the sulfur atom in H_2S?

   a) 0 b) 1 c) 2 d) 3 e) 4

   Answer: c

8. How many lone pair of electrons are assigned to each carbon atoms in C_2H_2?

   a) 0 b) 1/2 c) 1 d) 2 e) 3

   Answer: a

9. Which of the following molecules or ions are isoelectronic: SO_2, CO_2, NO_2^+, ClO_2^-?

   a) SO_2 and CO_2 b) SO_2 and NO_2^+ c) CO_2 and ClO_2^-
   d) CO_2 and NO_2^+ e) SO_2, NO_2^+, and ClO_2^-

   Answer: d

10. If two or more species have the same number of electrons, resulting in similar Lewis structures, they are said to be __________.

    a) isoelectronic b) resonant structures c) ionic
    d) neutral e) covalent

    Answer: a

11. Which of the following molecules or ions will have a Lewis structure most like that of sulfur difluoride, SF_2?

    a) NO_2^+ b) SO_2 c) O_3 d) SCN^- e) ClO_2^-

    Answer: e

12. Which of the following molecules or ions will have a Lewis structure most like that of the chlorate ion, ClO_3^-?

    a) NI_3 b) SO_3 c) CO_3^{2-} d) BF_3 e) Cl_2CO

    Answer: a

13. Which of the following is a correct Lewis structure for sulfite ion, SO_3^{2-}?

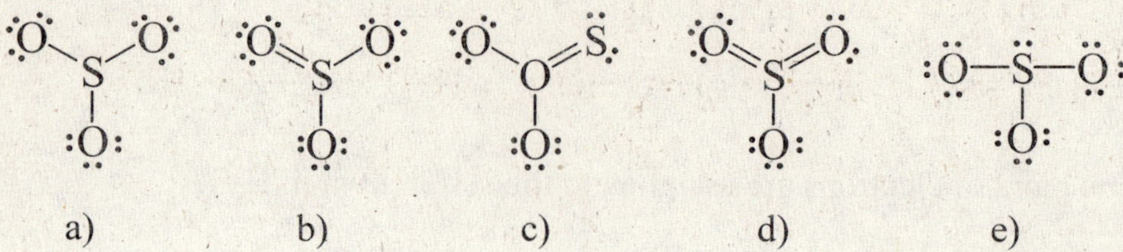

a)   b)   c)   d)   e)

Answer: e

14. Which of the following is a correct Lewis structure for nitrous acid, HNO_2?

a)   b)   c)   d)   e)

Answer: d

15. Which of the following is a correct Lewis structure for sulfate ion?

a)   b)   c)

d)   e)

Answer: a

16. Which of the following are possible Lewis structures for C_2H_6O?

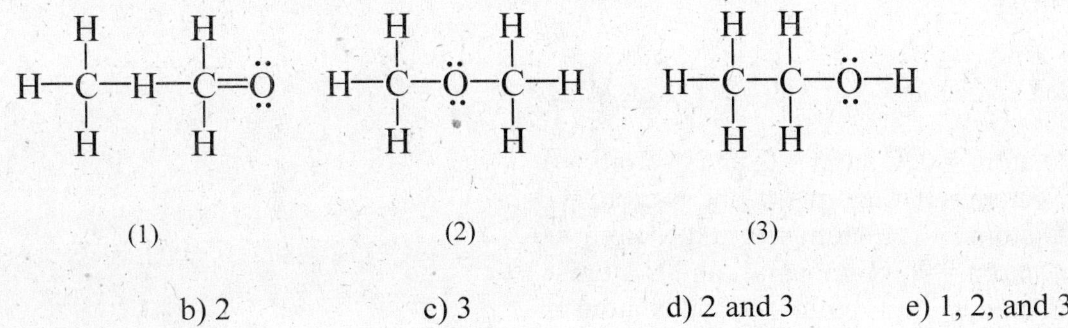

a) 1   b) 2   c) 3   d) 2 and 3   e) 1, 2, and 3

Answer: d

17. The central atom in PH_3 is surrounded by

    a) three single bonds and no lone pairs of electrons.
    b) three single bonds and one lone pair of electrons.
    c) three single bonds and two lone pairs of electrons.
    d) two single bonds, one double bond, and no lone pairs of electrons.
    e) two single bonds, one double bond, and one lone pair of electrons.

Answer: b

18. The central atom in SCl_2 is surrounded by

    a) one single bond, one double bond, and no lone pairs of electrons.
    b) one single bond, one double bond, and one lone pair of electrons.
    c) two single bonds and no lone pairs of electrons.
    d) two single bonds and two lone pairs of electrons.
    e) two double bonds, and no lone pairs of electrons.

Answer: d

8.3 Atom Formal Charges in Covalent Molecules and Ions

19. What is the formal charge on each atom in chloroform, $CHCl_3$?

    a) C atom = 0, H atom = 0, and each Cl atom = 0
    b) C atom = 0, H atom = +1, one Cl atom = -1, and two Cl atoms = 0
    c) C atom = +4, H = -1, and each Cl atom = -1
    d) C atom = +2, H = +1, and each Cl atom = -1
    e) C atom = -4, H = +1, and each Cl atom = +1

Answer: a

20. One resonance structure for OCN⁻ ion is drawn below. What is the formal charge on each atom?

$$\left[\ddot{O}=C=\ddot{N} \right]^{-}$$

a) O atom = 0, C atom = 0, and N atom = 0
b) O atom = 0, C atom = 0, and N atom = -1
c) O atom = -1, C atom = 0, and N atom = 0
d) O atom = -1, C atom = -1, and N atom = +1
e) O atom = +1, C atom = 0, and N atom = -2

Answer: b

21. One resonance structure for OCN⁻ ion is drawn below. What is the formal charge on each atom?

$$\left[\ddot{O}-C\equiv N\colon \right]^{-}$$

a) O atom = 0, C atom = 0, and N atom = 0
b) O atom = 0, C atom = 0, and N atom = -1
c) O atom = -1, C atom = 0, and N atom = 0
d) O atom = -1, C atom = -1, and N atom = +1
e) O atom = +1, C atom = 0, and N atom = -2

Answer: c

<u>8.4 Resonance</u>

22. Which of the following are resonance structures for nitrite ion, NO_2^-?

$$\left[\ddot{O}\diagdown_{N}\diagup=\ddot{O} \right]^{-} \quad \left[\ddot{O}=\diagdown_{N}\diagup\ddot{O} \right]^{-} \quad \left[\ddot{O}=\diagdown_{N}\diagup=\ddot{O} \right]^{-} \quad \left[\ddot{O}\diagdown_{N}\diagup\ddot{O} \right]^{-}$$

$$(1)(2)(3)(4)$$

a) 1 and 2   b) 2 and 4   c) 3 and 4   d) 1, 2, and 3   e) 2, 3, and 4

Answer: a

23. Which of the following are resonance structures for formate ion, HCO_2^-?

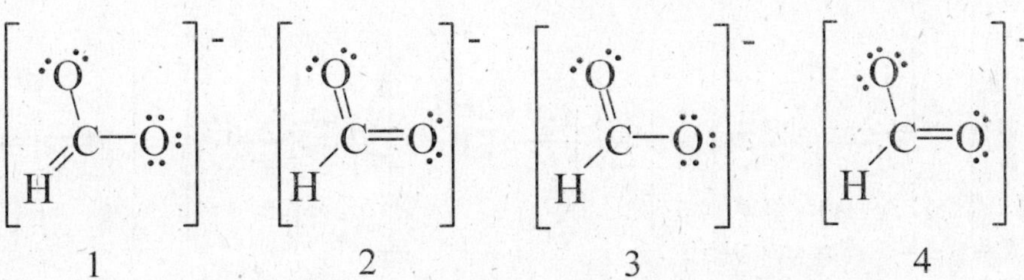

a) 1 and 2   b) 2 and 3   c) 3 and 4   d) 1, 3, and 4   e) 2, 3, and 4

Answer: c

24. Which of the following molecules or ions does not have one or more resonance structures?

a) O_3 b) SCN^- c) CO_3^{2-} d) H_2CO e) SO_2

Answer: d

8.5 Exceptions to the Octet Rule

25. Which of the following elements is able to form a molecular structure that exceeds the octet rule?

a) Ne   b) B   c) O   d) F   e) I

Answer: e

26. Which of the following elements is most likely to form a molecular structure that disobeys the octet rule?

a) B   b) C   c) N   d) O   e) F

Answer: a

27. When both of the electrons in a molecular bond originate from the same atom, the bond is called a(n)

a) sigma bond.   b) coordinate covalent bond.   c) pi bond.
d) metallic bond.   e) ionic bond.

Answer: b

28. What is the correct Lewis structure for IF_4^-?

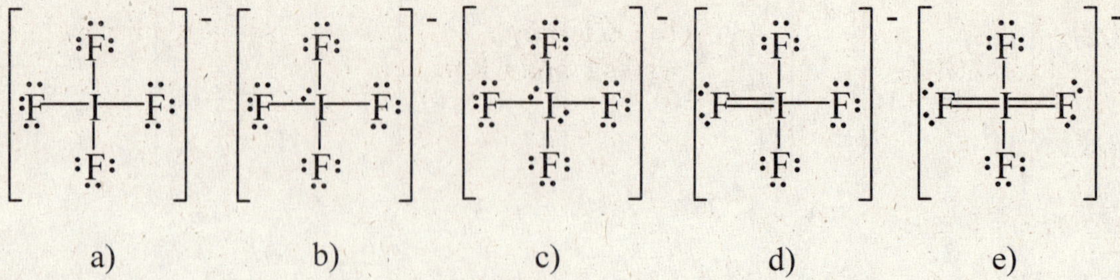

a)   b)   c)   d)   e)

Answer: c

29. What is the correct Lewis structure for IF_3?

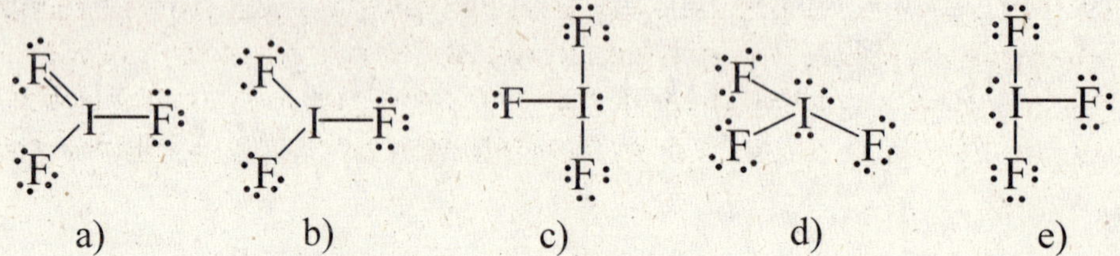

a)   b)   c)   d)   e)

Answer: e

30. The central atom in the triiodide ion, I_3^-, is surrounded by

a) two single bonds and no lone pairs of electrons.
b) two single bonds and two lone pairs of electrons.
c) two single bonds and three lone pairs of electrons.
d) one single bond, one double bond, and one lone pair of electrons.
e) two double bonds and one lone pair of electrons.

Answer: c

31. The central atom in PF_5 is surrounded by

a) five single bonds and no lone pairs of electrons.
b) five single bonds and one lone pair of electrons.
c) four single bonds, one double bond, and no lone pairs of electrons.
d) four single bonds, one double bond, and one lone pair of electrons.
e) three single bonds, two double bonds, and no lone pairs of electrons.

Answer: a

32. Which of the following species will have a Lewis structure most like that of ICl_4^+?

a) XeF_4 b) SO_4^{2-} c) PF_4^+ d) SF_4 e) IO_4^-

Answer: d

33. Which of the following molecules or ions are likely to be free radicals: N_2O, OCl^-, and ClO_2?

    a) N_2O only
    b) OCl^- only
    c) ClO_2 only
    d) N_2O and ClO_2
    e) OCl^- and ClO_2

    Answer: c

8.6 Molecular Shapes

34. Use VSEPR theory to predict the electron-pair geometry and the molecular geometry of phosphorus tribromide, PBr_3.

    a) The electron-pair geometry is linear, the molecular geometry is linear.
    b) The electron-pair geometry is trigonal-planar, the molecular geometry is trigonal-planar.
    c) The electron-pair geometry is trigonal-planar, the molecular geometry is bent.
    d) The electron-pair geometry is tetrahedral, the molecular geometry is tetrahedral.
    e) The electron-pair geometry is tetrahedral, the molecular geometry is trigonal-pyramidal.

    Answer: e

35. Use VSEPR theory to predict the electron-pair geometry and the molecular geometry of sulfur dioxide, SO_2.

    a) The electron-pair geometry is trigonal-planar, the molecular geometry is trigonal-planar.
    b) The electron-pair geometry is trigonal-planar, the molecular geometry is bent.
    c) The electron-pair geometry is tetrahedral, the molecular geometry is bent.
    d) The electron-pair geometry is tetrahedral, the molecular geometry is linear.
    e) The electron-pair geometry is trigonal-bipyramidal, the molecular geometry is linear.

    Answer: b

36. Use VSEPR theory to predict the molecular geometry of ClO_3^-.

    a) trigonal-pyramidal
    b) trigonal-planar
    c) bent
    d) T-shaped
    e) linear

    Answer: a

37. Use VSEPR theory to predict the molecular geometry around either carbon atom in acetylene, C_2H_2.

    a) linear
    b) bent
    c) trigonal-planar
    d) tetrahedral
    e) octahedral

    Answer: a

38. Use VESPR theory to predict the molecular geometry around either carbon atom in ethylene, C_2H_4.

a) linear   b) bent   c) trigonal-planar d) tetrahedral   e) octahedral

Answer: c

39. Use VSEPR theory to predict the molecular geometry of BrF_5.

a) tetrahedral   b) see-saw   c) trigonal-bipyramidal
d) square-pyramidal   e) octahedral

Answer: d

40. Use VSEPR theory to predict the molecular geometry around the nitrogen atom in nitrous acid, HNO_2.

a) trigonal-planar   b) tetrahedral   c) trigonal-pyramidal
d) T-shaped   e) bent

Answer: e

41. Which of the following molecules has the smallest bond angle between any two hydrogen atoms?

a) CH_4 b) H_2O c) BH_3 d) BeH_2 e) SiH_4

Answer: b

42. What are the approximate I-N-I bond angles in NI_3?

a) $109.5°$ b) $120°$ c) $109.5°$ and $120°$ d) $90°$ and $120°$ e) $90°$ and $180°$

Answer: a

43. What is the O-C-N bond angle in OCN^-?

a) $90°$ b) $107°$ c) $109.5°$ d) $120°$ e) $180°$

Answer: e

44. What are the O-P-O bond angles in PO_4^{3-}?

a) $90°$ b) $109.5°$ c) $120°$ d) $180°$ e) $90°$ and $180°$

Answer: b

45. What are the approximate F-Br-F bond angles in BrF_5?

    a) $90°$ and $180°$ b) $109.5°$ c) $90°$ and $120°$ d) $120°$ e) $180°$

    Answer: a

46. Place the following molecules in order from smallest to largest H-N-H bond angles: NH_4^+, NH_3, and NH_2^-.

    a) $NH_4^+ < NH_3 < NH_2^-$ b) $NH_4^+ < NH_2^- < NH_3$ c) $NH_2^- < NH_3 < NH_4^+$
    d) $NH_2^- < NH_4^+ < NH_3$ e) $NH_3 < NH_2^- < NH_4^+$

    Answer: c

8.7 Bond Polarity and Electronegativity

47. Which of the following compounds has polar covalent bonds: $NaBr$, Br_2, HBr, and CBr_4?

    a) $NaBr$ only   b) Br_2 only   c) HBr only
    d) Br_2 and HBr e) HBr and CBr_4

    Answer: e

48. Electronegativity is a measure of

    a) the ability of a substance to conduct electricity.
    b) the charge on a polyatomic cation.
    c) the charge on an polyatomic anion.
    d) the ability of an atom in a molecule to attract electrons to itself.
    e) the oxidation number of an atom in a molecule or polyatomic anion.

    Answer: d

49. Choose which central atom in the following molecules is most electronegative.

    a) PH_3 b) CH_4 c) H_2S d) H_2O e) NH_3

    Answer: d

50. Predict which of the following compounds has covalent bond(s) that are the most polar.

    a) HF b) CI_4 c) H_2S d) NBr_3 e) HI

    Answer: a

51. Which of the following statements is/are CORRECT?
   1. The atom with the lowest electronegativity is usually the central atom in a molecule.
   2. As the difference in electronegativity between two bonded atoms increases, the percent ionic character of the bond increases.
   3. Electronegativity decreases down each group of the periodic table.

   a) 1 only   b) 2 only   c) 3 only   d) 2 and 3   e) 1, 2, and 3

   Answer: e

8.8 Bond and Molecular Polarity

52. Which one of the following molecules is polar?

   a) CCl_4 b) PF_5 c) SO_2 d) F_2 e) BH_3

   Answer: c

53. Which one of the following molecules is polar?

   a) CO_2 b) SF_4 c) XeF_2 d) XeF_4 e) SO_3

   Answer: b

54. Three possible structures of $C_2H_2Cl_2$ are shown below. Which of these molecules are polar?

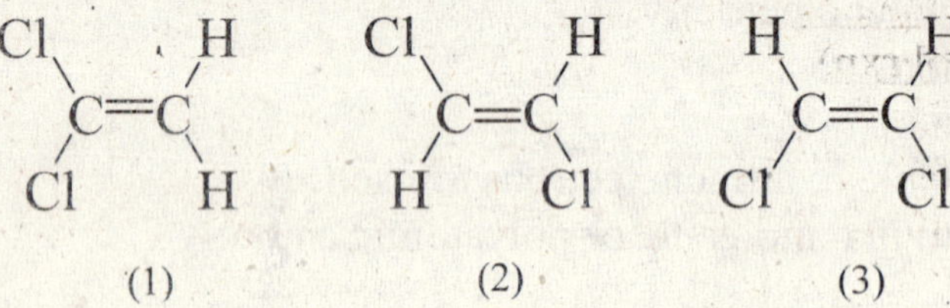

   a) 1 only   b) 2 only   c) 3 only   d) 1 and 3   e) 2 and 3

   Answer: d

8.9 Bond Properties: Order, Length, Energy

55. Linus Pauling noticed that the energy of a polar bond is often greater than expected. He attributed the greater bond energy to

   a) a coulombic attraction between atoms with partially positive and negative charges.
   b) the greater bond lengths of the heteronuclear bonds.
   c) one of the many unexplainable phenomena that scientists encounter.
   d) the ability of heteronuclear species to form double and triple bonds.
   e) the greater number of valence electrons found in heteronuclear molecules.

   Answer: a

56. In molecules, as bond order increases,

   a) both bond length and bond energy increase.
   b) both bond length and bond energy decrease.
   c) bond length increases and bond energy is unchanged.
   d) bond length is unchanged and bond energy increases.
   e) bond length decreases and bond energy increases.

   Answer: e

57. Use Lewis structures to predict the bond order for a carbon-oxygen bond in carbonate ion, CO_3^{2-}.

   a) 1/2   b) 1   c) 4/3   d) 3/2   e) 2

   Answer: c

58. Use Lewis structures to predict the bond order for a sulfur-oxygen bond in sulfur dioxide.

   a) 1/2   b) 1   c) 4/3   d) 3/2   e) 2

   Answer: d

59. Calculate the enthalpy change ($\Delta_r H$) for the reaction below,
$$Br_2(g) + 3\,F_2(g) \rightarrow 2\,BrF_3(g)$$
given the bond enthalpies of the reactants and products.

Bond	Bond Enthalpy (kJ/mol·rxn)
Br-Br	193
F-F	155
Br-F	249

   a) -836 kJ   b) -89 kJ   c) +89 kJ   d) +99 kJ   e) +836 kJ

   Answer: a

60. Use the bond enthalpies tabulated below to estimate the enthalpy of formation of $H_2O(g)$.
$$H_2(g) + \tfrac{1}{2}\,O_2(g) \rightarrow H_2O(g)$$

Bond	Bond Enthalpy (kJ/mol·rxn)
H-H	436
O-H	463
O-O	146
O=O	498

   a) -417 kJ   b) -241 kJ   c) -168 kJ   d) -8 kJ   e) +8 kJ

   Answer: b

61. When heated, azomethane decomposes into nitrogen gas and ethane gas.
$$CH_3N=NCH_3(g) \rightarrow N_2(g) + C_2H_6(g)$$

	Bond Enthalpy (kJ/mol·rxn)		Bond Enthalpy (kJ/mol·rxn)
Bond		Bond	
C-H	413	N-N	163
C-N	305	N=N	418
C-C	346	N≡N	945

Using average bond energies, calculate the enthalpy of reaction.
a) -611 kJ   b) -527 kJ   c) -429 kJ   d) -313 kJ   e) -263 kJ

Answer: e

Short Answer Questions

62. The molecular geometry of a molecule whose central atom has four single bonds and two lone pairs of electrons is __________.

Answer: square planar

63. If a molecule has a positive and a negative end, the molecule is said to have a __________ moment.

Answer: dipole

64. In PCl_5, the Cl-P-Cl bond angle between two equatorial chlorine atoms is _____ degrees.

Answer: 120

65. In benzene, C_6H_6, the six carbon atoms are arranged in a ring. Two equivalent Lewis structures can be drawn for benzene. In both structures, the carbon atoms have a trigonal planar geometry. These two equivalent structures are referred to as __________ structures.

Answer: resonance

66. One might expect PCl_3 to have a trigonal planar molecular geometry since its central atom is surrounded by three outer atoms. However, PCl_3 has a trigonal pyramidal geometry. Explain.

Answer: The shape of a molecule is dependent upon the number of electron-pairs surrounding the central atom. The phosphorus atom in PCl_3 has four electron-pairs which are distributed around the phosphorus atom in a tetrahedral geometry. Thus, the bond angles are approximately 109°, not 120°.

Chapter 9
BONDING AND MOLECULAR STRUCTURE: ORBITAL HYBRIDIZATION AND MOLECULAR ORBITALS

CHAPTER OUTLINE

CHAPTER OBJECTIVES

- Understand the difference between valence bond theory and molecular orbital theory. (9.1-9.2)
- Identify the hybridization of an atom in a molecule or ion. (9.2)
- Understand the difference between bonding and antibonding molecular orbitals and be able to write the molecular orbital configurations for simple diatomic molecules (9.3)

LESSON PLAN

AP Standards
I.B.2.b   Valence bond: hybridization of orbitals, resonance, sigma and pi bonds
I.B.3   Geometry of molecules and ions, structural isomerism of simple organic molecules and coordination complexes; dipole moments of molecules; relation of properties to structure

Suggested Time:
Three traditional classes or 2 blocks. Sections 9.1-9.2 are necessary and will take up most of the class time for this chapter. Section 9.3 can be done in one day, or assigned as reading homework. MO Theory is not part of the AP curriculum.

ASSESSMENT

Exercises:

9.1   Valence Bond Description of Bonding (p. 413)
9.2   Recognizing Hybridization (p. 416)
9.3   Bonding in Acetone (p. 419)
9.4   Triple Bonds Between Atoms (p. 420)
9.5   Bonding and Hybridization (p. 420)
9.6   Molecular Orbitals and Bond Order (p. 425)
9.7   Molecular Orbitals in Diatomic Molecules (p. 426)
9.8   Molecular Electron Configurations (p. 429)

Answers to Exercises p. A47-A48
Practicing Skills pp. 434-435, Answers IRM pp. 166-169
General Questions pp. 435-438, Answers IRM pp. 170-174
In the Laboratory Questions pp. 438-439, Answers IRM pp. 175-176
Summary and Conceptual Questions pp. 439-441, Answers IRM pp. 176-177

GLOSSARY

anti-bonding molecular orbital A molecular orbital in which the energy of the electrons is higher than that of the parent orbital

CHAPTER 9: Bonding and Molecular Structure:
Orbital Hybridization and Molecular Orbitals

bonding molecular orbital A molecular orbital which the energy of the electrons is lower than that of the parent orbital electrons

diatomic molecule Two atoms covalently bonded together in a molecule

first principle of molecular orbital theory The total number of molecular orbitals is always equalt o the total number of atomic orbitals contributed by the atoms that have combined.

fourth principle of molecular orbital theory Atomic orbitals combine to form molecular orbitals most effectively when the atomic orbitals are of similar energy.

heteronuclear diatomic molecules A molecule composed of two atoms of different elements.

homonuclear diatomic molecules A molecule composed of two atoms of the same element.

hybrid orbitals An orbital formed by mixing two or more atomic orbitals.

hybridization Process by which atomic orbitals recombine to form an equivalent set of hybrid orbitals that minimize electron-pair repulsions

isomers Two or more compounds with the same molecular formula but different arrangement of atoms

molecular orbital theory A model of bonding in which pure atomic orbitals combine to produce molecular orbitals that are delocalized over two or more atoms

molecular orbitals Process by which atomic orbitals recombine to form orbitals that are spread out, or delocalized over several atoms or an entire molecule

orbital overlap Partial occupation of the same region of space by orbitals from two atoms

pi (π) bond The second (and third, if present) bond in a multiple bond; results from sideways overlap of p atomic orbitals

second principle of molecular orbital theory Bonding molecular orbitals are lower in energy than the parent orbitals, and the antibonding orbitals are hight in energy.

sigma (σ) bond A bond formed by the overlap of orbitals head to head, and with bonding electron density concentrated along the axis of the bond

third principle of molecular orbital theory Electrons of the molecule are assigned to orbitals of successively higher energy.

valence bond theory A model of bonding in which a bond arises from the overlap of atomic orbitals on two atoms to five a bonding orbital with electrons localized between the atoms

MEDIA MENU

The Media Menu section of this guide lists the interactive exercises, tutorials, active figures (animated, interactive figures) and additional resources available online for Chemistry Now, a unique web-based, assessment-centered personalized learning system for chemistry students. Go to www.cengagelearning.com/login

Exercises
- Screen 9.3   Bond Formation
- Screen 9.4   Hybrid Orbitals
- Screen 9.7   Hybrid orbitals and sigma and pi bonding
- Screen 9.8   Isomers and multiple bonds
- Screen 9.9   Molecular orbital theory
- Screen 9.11   Molecular orbital configurations

Tutorials
- Screen 9.6   Determining hybrid orbitals
- Screen 9.5   Sigma bonding
- Screen 9.7   Hybrid orbitals and sigma and pi bonding
- Screen 9.10   Molecular electron configurations

Active Figures
- 9.1   Potential energy change during H-H bond formation from isolated hydrogen atoms
- 9.5   Hybrid orbitals for two to six electron pairs
- 9.6   Bonding in the methane molecule
- 9.10   Valence bond model of bonding in ethylene
- 9.13   Rotation around bonds

Additional Resources
- Podcast Module 14   Bonds in Valence bond Theory

DIAGNOSTIC QUIZ QUESTIONS

Goal # 1 Understand the difference between valence bond theory and molecular orbital theory.

1. Which one of the statements concerning valence bond (VB) and molecular orbital (MO) bond theories is correct?

   a) MO theory predicts that electrons are localized between pairs of atoms.
   b) In VB theory, bonding electrons are delocalized over the molecule.
   c) MO theory accurately describes bonding in O_2 and NO, VB theory does not.
   d) VB theory can describe molecular bonding in excited states.
   e) MO theory is used to accurately predict the colors of compounds.

   Answer: c

CHAPTER 9: Bonding and Molecular Structure:
Orbital Hybridization and Molecular Orbitals

Goal #2 Identify the hybridization of an atom in a molecule or ion.

2. What is the hybridization of the nitrogen atom in NCl_3?

a) sp b) sp^2 c) sp^3 d) sp^3d e) sp^3d^2

Answer: c

3. What is the hybridization of the xenon atom in XeF_2?

a) sp b) sp^2 c) sp^3 d) sp^3d e) sp^3d^2

Answer: d

4. Which of the following hybridized atoms is not possible?

a) an sp hybridized carbon atom   b) an sp^2 hybridized sulfur atom
c) an sp^3 hybridized phosphorus atom   d) an sp^3d hybridized oxygen atom
e) an sp^3d^2 hybridized xenon atom

Answer: d

Goal # 3 Understand the difference between bonding and antibonding molecular orbitals and be able to write the molecular orbital configurations for simple diatomic molecules

5. Atomic orbitals combine most effectively to form molecular orbitals when

a) electrons in the orbitals have no spins.
b) electrons in the orbitals have the same spin.
c) the atomic orbitals are hybridized.
d) the atomic orbitals have similar energies.
e) p-orbitals are half-filled.

Answer: d

6. A molecular orbital that decreases the electron density between two nuclei is said to be
________.

a) hybridized   b) bonding   c) antibonding   d) pi-bonding   e) nonpolar

Answer: c

7. According to molecular orbital theory, what is the bond order of oxygen, O_2?

a) 1   b) 3/2   c) 2   d) 5/2   e) 3

Answer: c

TESTBANK

<u>9.1 Orbitals and Theories of Chemical Bonding</u>

1. Which one of the statements concerning valence bond (VB) and molecular orbital (MO)
   bond theories is correct?

   a) MO theory predicts that electrons are localized between pairs of atoms.
   b) In VB theory, bonding electrons are delocalized over the molecule.
   c) MO theory accurately describes bonding in O_2 and NO, VB theory does not.
   d) VB theory can describe molecular bonding in excited states.
   e) MO theory is used to accurately predict the colors of compounds.

   Answer: c

<u>9.2 Valence Bond Theory</u>

2. Which of the following statements is/are CORRECT?
   1. The overlap between an s orbital and a p orbital is called a pi-bond.
   2. The overlap of two s orbitals in H_2 is called a sigma bond.
   3. HF is formed from the overlap of a hydrogen $1s$ orbital with a fluorine $2s$ orbital.

   a) 1 only   b) 2 only   c) 3 only   d) 2 and 3   e) 1, 2, and 3

   Answer: b

3. Which of the following statements concerning hybrid orbitals is/are CORRECT?
   1. The number of hybrid orbitals equals the number of atomic orbitals that are used to
      create the hybrids.
   2. When atomic orbitals are hybridized, the s orbital and at least one p orbital are always
      hybridized.
   3. To create octahedral structures, two d orbitals must be hybridized along with the s
      and all three p orbitals.

   a) 1 only   b) 2 only   c) 3 only   d) 2 and 3   e) 1, 2, and 3

   Answer: e

4. How many sigma (σ) bonds and pi (π) bonds are in ethene, C_2H_4?

   a) four σ, one π b) four σ, two π c) five σ, one π d) five σ, two π e) six σ, zero π

   Answer: c

5. How many sigma (σ) bonds and pi (π) bonds are in carbon monoxide?

   a) three σ, zero π   b) two σ, one π   c) two σ, two π   d) one σ, two π   e) zero σ, three π

   Answer: d

6. How many sigma (σ) bonds and pi (π) bonds are in the following molecule?

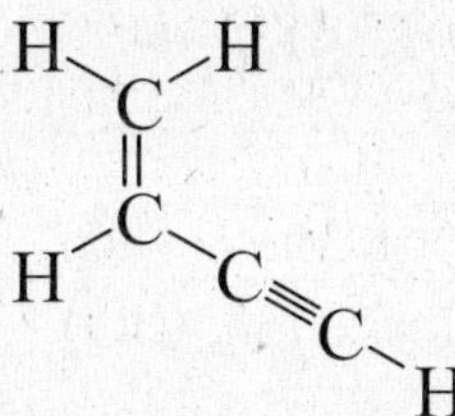

   a) seven σ and three π   b) seven σ and two π   c) five σ and five π
   d) five σ and three π   e) five σ and two π

   Answer: a

7. To form a molecule with a trigonal bipyramidal electron geometry, what set of pure atomic orbitals must be mixed?

   a) one *s* and three *p*   b) one *s*, three *p*, and one *d*   c) one *s*, three *p*, and two *d*
   d) two *s*, six *p*, and two *d*   e) two *s*, six *p*, and four *d*

   Answer: b

8. What is the maximum number of hybridized orbitals formed by a fluorine atom?

   a) 1   b) 2   c) 3   d) 4   e) 6

   Answer: d

9. What is the hybridization of either carbon atom in acetylene, C_2H_2?

   a) *sp*   b) sp^2 c) sp^3 d) sp^3d e) sp^3d^2

   Answer: a

10. What is the hybridization of the nitrogen atom in NCl_3?

    a) *sp*   b) sp^2 c) sp^3 d) sp^3d e) sp^3d^2

    Answer: c

11. What is the hybridization of the xenon atom in XeF_2?

a) sp b) sp^2 c) sp^3 d) sp^3d e) sp^3d^2

Answer: d

12. What is the hybridization of the central nitrogen atom in nitrite ion, NO_2^-?

a) sp b) sp^2 c) sp^3 d) sp^3d e) sp^3d^2

Answer: b

13. What is the hybridization of the chlorine atom in chlorite ion, ClO_2^-?

a) sp b) sp^2 c) sp^3 d) sp^3d e) sp^3d^2

Answer: c

14. What is the hybridization of the central atom in a molecule with a square-planar molecular geometry?

a) sp b) sp^2 c) sp^3 d) sp^3d e) sp^3d^2

Answer: e

15. What is the hybridization of each carbon atom in benzene, C_6H_6? Benzene contains a six-member carbon ring.

a) sp b) sp^2 c) sp^3 d) sp^3d e) sp^3d^2

Answer: b

16. For which of the following molecules does the central carbon atom have sp^2 hybridization?

a) Cl_2CO b) $CHCl_3$ c) CS_2 d) CH_2Cl_2 e) HCN

Answer: a

17. For which of the following molecules and ions does the central nitrogen atom have sp^3 hybridization?

a) NO_2^- b) HNO_3 c) $NOBr$ d) NBr_3 e) HNO_2

Answer: d

18. For which of the following molecules and ions does the central atom have *sp* hybridization: NO_2^+, O_3, and I_3^-?

    a) NO_2^+ only   b) O_3 only   c) I_3^- only   d) O_3 and I_3^- e) I_3^- and NO_2^+

    Answer: a

19. What is the molecular geometry around a central atom that is sp^3 hybridized and has two lone pairs of electrons?

    a) bent   b) linear   c) trigonal-planar
    d) trigonal-pyramidal   e) trigonal-bipyramidal

    Answer: a

20. What is the molecular geometry around a central atom that is sp^3d^2 hybridized and has one lone pair of electrons?

    a) tetrahedral   b) trigonal-bipyramidal   c) square-planar
    d) square-pyramidal   e) see-saw

    Answer: d

21. What is the molecular geometry around a central atom that is sp^2 hybridized, has three sigma bonds, and one pi bond?

    a) trigonal-planar   b) trigonal-pyramidal   c) bent
    d) T-shaped   e) tetrahedral

    Answer: a

22. What is the molecular geometry around a central atom that is sp^3d hybridized and has one lone pair of eletrons?

    a) trigonal bipyramidal   b) trigonal-pyramidal   c) see-saw
    d) tetrahedral   e) square-planar

    Answer: c

23. What is the hybridization of a central atom that has four sigma bonds and has no lone pairs of electrons?

    a) sp b) sp^2 c) sp^3 d) sp^3d e) sp^3d^2

    Answer: c

24. Upon heating, $CaCO_3$ decomposes to CaO and CO_2. What change in the hybridization of carbon occurs in this reaction?

a) sp to sp^2 b) sp^2 to sp^3 c) sp^3 to sp d) sp^2 to sp e) no change

Answer: d

25. One product of the combustion of ethane, C_2H_6, is carbon dioxide. What change in hybridization of the carbon occurs in this reaction?

a) sp^3 to sp^2 b) sp^3 to sp c) sp^2 to sp^3 d) sp^2 to sp^3d^2 e) sp^2 to sp

Answer: b

26. Nitric acid, HNO_3, dissociates in water to form nitrate ions and hydronium ions. What change in hybridization of the nitrogen atom occurs in this dissociation?

a) sp^2 to sp^3 b) sp^2 to sp c) sp^3 to sp d) sp to sp^3 e) no change

Answer: e

27. Which of the following hybridized atoms is not possible?

a) an sp hybridized carbon atom   b) an sp^2 hybridized sulfur atom
c) an sp^3 hybridized phosphorus atom   d) an sp^3d hybridized oxygen atom
e) an sp^3d^2 hybridized xenon atom

Answer: d

28. Which of the following characteristics apply to SO_2?
   1. polar bonds
   2. nonpolar molecule
   3. linear molecular shape
   4. sp hybridized

a) 1 only   b) 1 and 2   c) 3 and 4   d) 1, 2, and 3   e) 1, 2, 3, and 4

Answer: a

29. Dichloromethane, CH_2Cl_2, is a common organic solvent. Which of the following statements concerning dichloromethane is/are CORRECT?
   1. CH_2Cl_2 has two isomers. For one isomer of CH_2Cl_2, the chlorine atoms are adjacent to each other and the molecule is polar.
   2. CH_2Cl_2 has two isomers. For one isomer of CH_2Cl_2, the chlorine atoms are on opposites sides of the carbon atom and the molecule is nonpolar.
   3. The hybridization of the central carbon atom is sp^3.

a) 1 only   b) 2 only   c) 3 only   d) 1 and 2   e) 1, 2, and 3

Answer: c

30. For which of the following compounds is it possible for *cis* and *trans* isomers to exist?

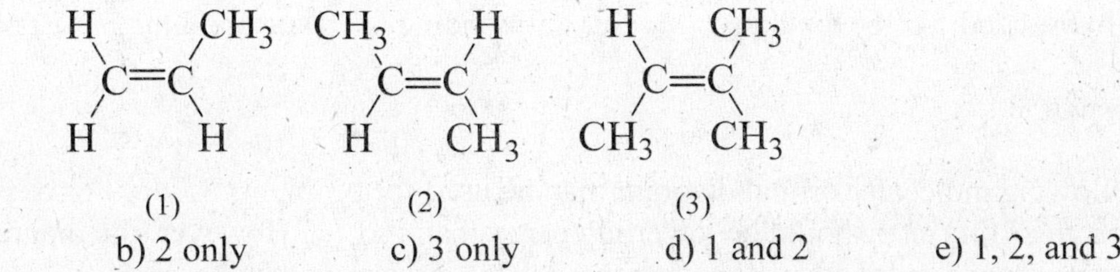

    (1)   (2)   (3)

  a) 1 only   b) 2 only   c) 3 only   d) 1 and 2   e) 1, 2, and 3

Answer: b

31. Which of the underlined atoms (C_1, C_2, N, and O) are sp^2 hybridized?

a) C_1 and C_2

  c) N and O

d) O and C_2

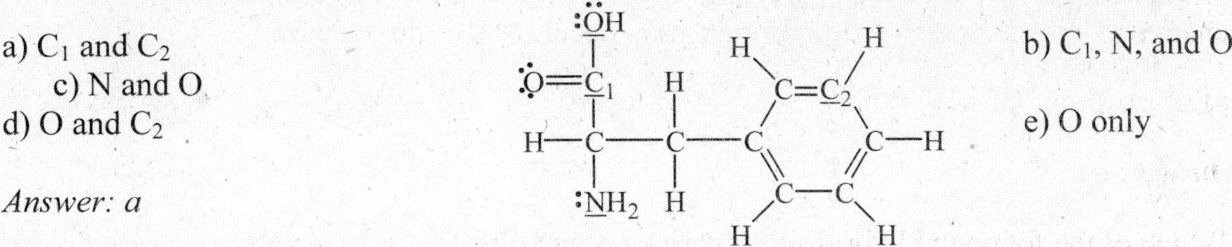

b) C_1, N, and O

e) O only

Answer: a

9.3 Molecular Orbital Theory

32. All of the following statements concerning molecular orbital (MO) theory are correct EXCEPT

a) the Pauli exclusion principle is obeyed.
b) Hund's rule is obeyed.
c) electrons are assigned to orbitals of successively higher energy.
d) a bonding molecular orbital is lower in energy than its parent atomic orbitals.
e) the combination of two atomic orbitals creates only one molecular orbital.

Answer: e

33. Atomic orbitals combine most effectively to form molecular orbitals when

a) electrons in the orbitals have no spins.
b) electrons in the orbitals have the same spin.
c) the atomic orbitals are hybridized.
d) the atomic orbitals have similar energies.
e) *p*-orbitals are half-filled.

Answer: d

34. A molecular orbital that decreases the electron density between two nuclei is said to be
__________.

a) hybridized   b) bonding   c) antibonding   d) pi-bonding   e) nonpolar

Answer: c

The following molecular orbital diagram may be used for problems 35-48. For oxygen and fluorine, the σ_{2p} orbital should be lower in energy than the π_{2p}. However, the diagram will still yield correct bond order and magnetic behavior for these molecules.

Energy

σ^*_{2p}

π^*_{2p}

σ_{2p}

π_{2p}

σ^*_{2s}

σ_{2s}

σ^*_{1s}

σ_{1s}

35. According to molecular orbital theory, which of the following species is the most likely to exist?

a) H_2^{2-} b) He_2 c) Li_2^{2-} d) Be_2 e) Be_2^{2-}

Answer: e

36. According to molecular orbital theory, which of the following species is least likely to exist?

a) Be_2 b) F_2^{2+} c) C_2^{2-} d) Li_2 e) B_2^{2-}

Answer: a

37. According to molecular orbital theory, which of the following species has the highest bond order?

a) F_2 b) F_2^{2+} c) C_2^{2-} d) Li_2 e) B_2^{2+}

Answer: c

38. According to molecular orbital theory, what is the bond order of oxygen, O_2?

   a) 1   b) 3/2   c) 2   d) 5/2   e) 3

   Answer: c

39. According to molecular orbital theory, what is the bond order of N_2^-?

   a) 1   b) 3/2   c) 2   d) 5/2   e) 3

   Answer: d

40. According to molecular orbital theory, which of the following lists ranks the oxygen species in terms of increasing bond order?

   a) $O_2^{2+} < O_2^{2-} < O_2$ b) $O_2^{2-} < O_2 < O_2^{2+}$ c) $O_2 < O_2^{2+} < O_2^{2-}$
   d) $O_2 < O_2^{2-} < O_2^{2+}$ e) $O_2^{2+} < O_2 < O_2^{2-}$

   Answer: b

41. Consider the molecules B_2, C_2, N_2 and O_2. Which two molecules have the same bond order?

   a) B_2 and C_2 b) B_2 and O_2 c) C_2 and N_2 d) C_2 and O_2 e) N_2 and O_2

   Answer: d

42. Use molecular orbital theory to predict which species is paramagnetic.

   a) N_2 b) O_2 c) F_2 d) Li_2 e) H_2

   Answer: b

43. Use molecular orbital theory to predict which ion is paramagnetic.

   a) F_2^{2+} b) O_2^{2-} c) O_2^{2+} d) N_2^{2+} e) B_2^{2-}

   Answer: a

44. What is the molecular orbital configuration of F_2?

   a) [core electrons] $(\sigma_{2s})^2 (\sigma^*_{2s})^2 (\pi_{2p})^4 (\sigma_{2p})^2 (\sigma^*_{2p})^2$
   b) [core electrons] $(\sigma_{2s})^2 (\sigma^*_{2s})^2 (\pi_{2p})^2 (\sigma_{2p})^2 (\pi^*_{2p})^2$
   c) [core electrons] $(\sigma_{2s})^2 (\sigma^*_{2s})^2 (\pi_{2p})^4 (\pi^*_{2p})^4$
   d) [core electrons] $(\sigma_{2s})^2 (\sigma^*_{2s})^2 (\pi_{2p})^4 (\sigma_{2p})^2 (\pi^*_{2p})^6$
   e) [core electrons] $(\sigma_{2s})^2 (\sigma^*_{2s})^2 (\pi_{2p})^4 (\sigma_{2p})^2 (\pi^*_{2p})^4$

   Answer: e

45. What is the molecular orbital configuration of N_2^{2+}?

a) [core electrons] $(\sigma_{2s})^2 (\sigma^*_{2s})^2 (\pi_{2p})^4 (\sigma_{2p})^2 (\pi^*_{2p})^2$
b) [core electrons] $(\sigma_{2s})^2 (\sigma^*_{2s})^2 (\pi_{2p})^4$
c) [core electrons] $(\sigma_{2s})^2 (\sigma^*_{2s})^2 (\pi_{2p})^2 (\sigma_{2p})^2$
d) [core electrons] $(\sigma_{2s})^4 (\sigma^*_{2s})^4$
e) [core electrons] $(\sigma_{2s})^2 (\sigma^*_{2s})^2 (\pi_{2p})^4 (\sigma_{2p})^2 (\pi^*_{2p})^4$

Answer: b

46. Assume that the molecular orbital energy diagram for a homonuclear diatomic molecule applies to a heteronuclear diatomic molecule. What is the molecular orbital configuration of NO?

a) [core electrons] $(\sigma_{2s})^2 (\sigma^*_{2s})^2 (\pi_{2p})^4 (\sigma_{2p})^2 (\pi^*_{2p})^1$
b) [core electrons] $(\sigma_{2s})^2 (\sigma^*_{2s})^2 (\pi_{2p})^2 (\sigma_{2p})^2 (\pi^*_{2p})^2$
c) [core electrons] $(\sigma_{2s})^2 (\sigma^*_{2s})^2 (\pi_{2p})^2 (\sigma_{2p})^3$
d) [core electrons] $(\sigma_{2s})^2 (\sigma^*_{2s})^2 (\pi_{2p})^4 (\sigma_{2p})^1$
e) [core electrons] $(\sigma_{2s})^2 (\sigma^*_{2s})^2 (\pi_{2p})^2$

Answer: a

47. Assuming that the molecular orbital energy diagram for a homonuclear diatomic molecule applies to a heteronuclear diatomic molecule, determine which of the following species has the highest bond order.

a) NO^-　　　　b) OF^-　　　　c) CN^-　　　　d) O_2　　　　e) NO

Answer: c

48. Assuming that the molecular orbital energy diagram for a homonuclear diatomic molecule applies to a heteronuclear diatomic molecule, determine which of the following species is paramagnetic.

a) NO^+　　　　b) CO　　　　c) CN^-　　　　d) OF^-　　　　e) NO

Answer: e

49. Which molecule will have the following valence molecular orbital level energy diagram?

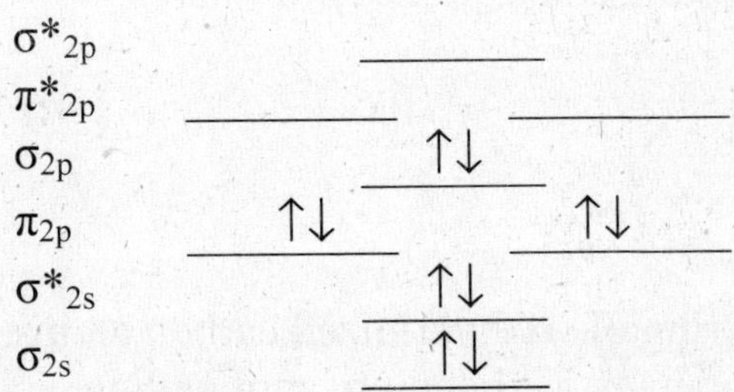

a) Li_2 b) Be_2 c) B_2 d) C_2 e) N_2

Answer: e

50. The following valence molecular orbital energy level diagram is appropriate for which one of the listed species?

a) B_2^{2-} b) C_2^{2-} c) N_2^{2-} d) O_2^{2-} e) F_2^{2-}

Answer: c

51. Which molecule will have the following valence molecular orbital energy level diagram?

a) Li_2 b) Be_2 c) B_2 d) C_2 e) F_2

Answer: d

52. In the NO_2^- ion, each atom can be viewed as sp^2 hybridized. Thus, each atom has one remaining unhybridized p orbital. How many π_{2p} molecular orbitals (including both bonding and antibonding orbitals) are formed using the unhybridized p orbitals?

a) 1   b) 3   c) 4   d) 6   e) 12

Answer: b

53. Benzene, C_6H_6, consists of a six member ring of sp^2 hybridized carbon atoms. Each carbon atom has one unhybridized p orbital. How many π_{2p} bonding, antibonding, and nonbonding molecular orbitals exist for benzene?

a) Three π_{2p} molecular orbitals exist; two bonding and one antibonding.
b) Three π_{2p} molecular orbitals exist; one bonding, one antibonding, and one nonbonding.
c) Six π_{2p} molecular orbitals exist; three bonding and three antibonding.
d) Six π_{2p} molecular orbitals exist; two bonding, two nonbonding, and two antibonding.
e) Twelve π_{2p} molecular orbitals exist; six bonding and six antibonding.

Answer: c

Short Answer Questions

54. Which theory, valence bond or molecule orbital, correctly predicts the existence of paramagnetic molecules?

Answer: molecular orbital theory

55. In valence bond theory, each sigma bond in CH_4 is formed from the overlap of a hydrogen atom's $1s$ orbital with a _____ hybridized orbital on the carbon atom.

Answer: sp^3

56. In molecular orbital theory, the bond order is defined as 1/2(the number of electrons in _________ orbitals minus the number of electrons in antibonding orbitals).

Answer: bonding

57. Triiodide ion, I_3^-, has a trigonal-bipyramidal electron-pair geometry and a linear molecular geometry. The hybridization of the central iodine atom is _________.

Answer: sp^3d

58. The hybridization of the xenon atom in $XeOF_4$ is _________.

Answer: sp^3d^2

59. Draw a Lewis structure of xenon trioxide. What is the hybridization of the xenon atom in this molecule?

Answer: The hybridization is sp^3.

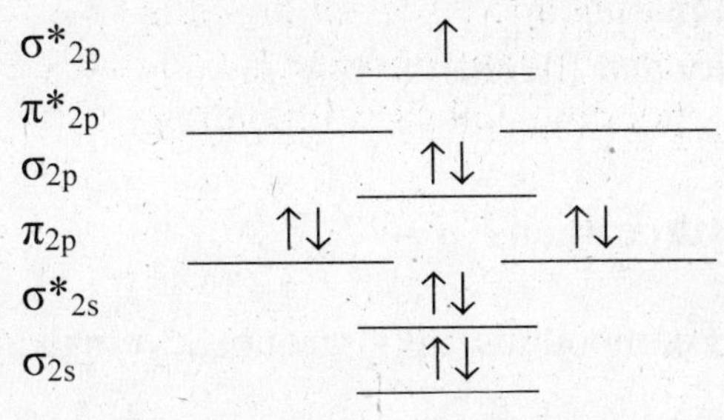

60. Draw the valence molecular orbital energy level diagram for nitrogen monoxide, NO.

Answer:

Chapter 10
CARBON: MORE THAN JUST ANOTHER ELEMENT

CHAPTER OUTLINE

CHAPTER OBJECTIVES

- Classify organic compounds based on formula and structure (10.1)
- Recognize and draw structures of structural isomers and stereoisomers for carbon compounds (10.1)
- Name and draw structures of common organic compounds (10.2-10.4)
- Know the common reactions of organic functional groups (10.2-10.4)
- Relate properties to molecular structure (10.2-10.4)
- Identify common polymers (10.5)

LESSON PLAN

AP Standards

I.B.3   Geometry of molecules and ions, structural isomerism of simple organic molecules and coordination complexes; dipole moments of molecules; relation of properties to structure

IV.3   Introduction to organic chemistry: hydrocarbons and functional groups (structure, nomenclature, chemical properties)

Suggested Time:

Three traditional classes or 2 blocks. Sections 10.3 to 10.5 should be the focus of the chapter in order to teach AP material. In sections 10.1 and 10.2, the parts that include the concepts of isomers (structural, stero, and geometric) should be included as well.

ASSESSMENT

Exercises:

- 10.1   Drawing Structural Isomers of Alkanes (p. 450)
- 10.2   Naming Alkanes (p. 452)
- 10.3   Determining Structural Isomers of Alkenes from a Formula (p. 455)
- 10.4   Reactions of Alkenes (p. 458)
- 10.5   Isomers of Aromatic Compounds (p. 460)
- 10.6   Structures of Alcohols (p. 464)
- 10.7   Aldehydes and Ketones (p. 470)
- 10.8   Aldehydes and Ketones (p. 470)
- 10.9   Esters (p. 475)
- 10.10   Esters (p. 475)
- 10.11   Functional Groups (p. 478)
- 10.12   Kevlar, a Condensation Polymer (p. 487)

Answers to Exercises p. A48-A49
Practicing Skills pp. 488-492, Answers IRM pp. 181-188
General Questions pp. 492-493, Answers IRM pp. 188-193
In the Laboratory Questions pp. 493-494, Answers IRM pp. 193-195
Summary and Conceptual Questions pp. 494-495, Answers IRM pp. 195-197

GLOSSARY

addition polymers A synthetic organic polymer formed by directly joining monomer units

addition reactions A reaction in which a molecule with the general formula X-Y adds across the carbon-carbon double bond

alcohol Any of a class of organic compounds characterized by the presence of a hydroxyl group bonded to a saturated carbon atom

aldehyde Any of a class of organic compounds characterized by the presence of a carbonyl group, in which the carbon atom is bonded to at least one hydrogen atom

alkane A class of hydrocarbons in which each carbon atom is bonded to four other atoms

alkene A class of hydrocarbons in which there is at least one carbon-carbon double bond

alkyl groups Hydrocarbon substituents

alkyne A class of hydrocarbons in which there is at least one carbon-carbon triple bond

amide An organic compound derived from a carboxylic acid and an amine

amine A derivative of ammonia in which one or more of the hydrogen atoms are replaced by organic groups

aromatic compound Any of a class of hydrocarbons characterized by the presence of a benzene ring or related structure

biodiesel An alternative to petroleum based fuels made from plant and animal oils

carbonyl compound Any of a class of organic compounds characterized by the presence of a carbonyl group

carbonyl group The functional group that characterizes aldehydes and ketones, consisting of a carbon atom doubly bonded to an oxygen atom

carboxylic acid Any of a class of organic compounds characterized by the presence of a carboxyl group

carboxylic group The functional group that consists of a carbonyl group bonded to a hydroxyl group

chiral molecule A molecule that is not superimposable on its mirror image

condensation polymers A synthetic organic polymer formed by combining monomer units in such a way that a small molecule, usually water, is split out

condensation reaction A reaction in which two monomers combine by splitting out a small molecule

coplolymer A polymer formed by combining two or more different monomers

cycloalkane Compounds constructed with tetrahedral carbon atoms joined together to form a ring

elastomer A synthetic organic polymer with a very high elasticity

enantiomers A stereoisomeric pair consisting of a chiral compound and its mirror image

ester Any of a class of organic compounds structurally related to carboxylic acids, but in which the hydrogen atom of the carboxyl group is replaced by a hydrocarbon group

esterification A reaction between a carboxylic acid and an alcohol in which a molecule of water is formed

ether Any of a class of organic compounds characterized by the presence of an oxygen atom singly bonded to two carbon atoms

fat A solid triester of a long-chain fatty acid with glycerol

fatty acid A carboxylic acid containing an unbranched chain of 10 to 20 carbon atoms

functional group A structural fragment found in all members of a class of compounds

geometric isomer Isomers in which the atoms of the molecule are arranged in different geometric relationships

hydrocarbon A compound that contains only carbon and hydrogen

hydrogenation An addition reaction in which the reagent is molecular hydrogen

hydrolysis A reaction with water in which a bond to oxygen is broken

isomer Two or more compounds with the same molecular formula but different arrangements of atoms

ketone Any of a class of organic compounds characterized by the presence of a carbonyl group, in which the carbon atom is bonded to two other carbon atoms

monomer The small units from which a polymer is constructed

oil A liquid triester of a long-chain fatty acid with glycerol

optical isomer Isomers that are nonsuperimposable mirror images of each other

peptide An amide linkage in a protein

polyamide A condensation polymer formed by elimination of water between two types of monomers, one with two carboxylic acid groups and the other with two amine groups

polyester A condensation polymer formed by elimination of water between two types of monomers, one with two carboxylic acid groups and the other with two alcohol groups

polymer A large molecule composed of many smaller repeating units, usually arranged in a chain

resonance stabilization Unusual stability due to unique delocalized pi bonding

saponification The hydrolysis of an ester

saturated compounds A hydrocarbon containing only single bonds

soap A salt produced by the hydrolysis of a fat or oil by a strong base

stereoisomer Two or more compounds with the same molecular formula and the same atom-to-atom bonding, but with different arrangements of the atoms in space

strained hydrocarbon Compounds in which an unfavorable geometry is imposed around carbon

structural isomer Two or more compounds with the same molecular formula but with different atoms bonded to each other

thermoplastics A polymer that softens but is unaltered upon heating

thermosetting plastics A polymer that degrades of decomposes on heating

unsaturated hydrocarbon A hydrocarbon containing double or triple carbon-carbon bonds

MEDIA MENU

The Media Menu section of this guide lists the interactive exercises, tutorials, active figures (animated, interactive figures) and additional resources available online for Chemistry Now, a unique web-based, assessment-centered personalized learning system for chemistry students. Go to www.cengagelearning.com/login.

Exercises
- Screen 10.6   Substitution and elimination reactions

Tutorials
- Screen 10.3   Classes of hydrocarbons
- Screen 10.5   Types of organic functional groups: structures, bonding, and chemistry

Self-Study Modules
- Screen 10.11   The polymer used in bubble gum

Active Figures
- 10.2   Optical Isomers
- 10.4   Alkanes
- 10.18   Nylon 6,6

Additional Resources
- Screen 10.4   Alkene addition reaction simulation
- Screen 10.9   Addition polymerization
- Screen 10.10   Condensation polymerization animation
- Screen 10.10   Synthesis of Nylon video
- Podcast Module 15   Hydrocarbons

DIAGNOSTIC QUIZ QUESTIONS

Goal #2 Recognize and draw structures of structural isomers and stereoisomers for carbon compounds

1. Structural isomers are compounds that have

   a) the same molar masses, but different elemental composition.
   b) have identical structures, but contain different isotopes of the same elements.
   c) two or more resonance structures.
   d) the same elemental composition, but the atoms are linked in different ways.
   e) the same physical properties, but different chemical properties.

   Answer: d

2. For which one of the following molecules do geometric isomers exist?

   a) $H_2C=CH_2$ b) $BrHC=CHCl$ c) ClH_2C-CH_2Br
   d) H_3C-CH_2Cl e) $H_2C=CCl_2$

   Answer: b

3. How many isomers, both structural and geometric, exist for C_5H_{10}?

   a) 5   b) 6   c) 7   d) 8   e) 9

   Answer: c

CHAPTER 10: Carbon: More Than Just Another Element

Goal #3 Name and draw structures of common organic compounds

4. What is the name of the following compound?

   a) 3-ethyl-5,6-dimethylheptane
   b) 2,3,5-ethylmethylethylheptane
   c) 3-ethyldecane
   d) 2,3-dimethyl-3-ethylheptane
   e) 3,5,6-trimethylheptane

$$CH_3-CH-CH-CH_2-CH-CH_2-CH_3$$

with CH_3 (top), CH_3 and C_2H_5 (bottom)

   Answer: a

5. What is the name of the following compound?

   a) *trans*-1-methyl-1-butene
   b) *cis*-1-methyl-1-butene
   c) *trans*-2-pentene
   d) *trans*-3-pentene
   e) *trans*-1-methyl-2-ethylethylene

   Answer: c

Goal #4 Know the common reactions of organic functional groups

6. Identify the product(s) of the hydrogenation of *cis*-2-hexene.

   a) 2-hydroxyhexane   b) 2,3-dihydroxyhexane   c) *trans*-2-hexane
   d) carbon dioxide and water   e) hexane

   Answer: e

7. What is the balanced chemical equation for the combustion of 2-methyl-1-butene?

   a) $2\ C_5H_{10}(g) + 15\ O_2(g) \rightarrow 10\ CO_2(g) + 10\ H_2O(\ell)$

   b) $2\ C_4H_{10}(g) + 13\ O_2(g) \rightarrow 8\ CO_2(g) + 10\ H_2O(\ell)$

   c) $C_4H_8(g) + 6\ O_2(g) \rightarrow 4\ CO_2(g) + 4\ H_2O(\ell)$

   d) $C_6H_{12}(g) + 9\ O_2(g) \rightarrow 6\ CO_2(g) + 6\ H_2O(\ell)$

   e) $C_5H_{12}(g) + 13\ O_2(g) \rightarrow 5\ CO_2(g) + 6\ H_2O(\ell)$

   Answer: a

8. What is the product of the addition of HF to 1-butene?

a) 1,1-difluorobutane   b) 1,2-difluorobutane   c) 1-fluorobutane
d) 2-fluorobutane   e) butane

Answer: d

Goal #5 Relate properties to molecular structure

9. Which of the following alkanes will have the highest boiling point?

a) hexane   b) octane   c) butane
d) pentane   e) heptane

Answer: b

10. What class of compounds is responsible for many of the distinctive odors of artificial flavors and perfumes?

a) ethers   b) aldehydes   c) esters   d) amides   e) alkenes

Answer: c

Goal #6 Identify common polymers

11. Which of the following statements concerning polyethylene is/are CORRECT?
    1. The monomer for polyethylene is CH_2CH_2.
    2. Polyethylene contains alternating single and double carbon-carbon bonds.
    3. Polyethylene is made by condensation polymerization.

a) 1 only   b) 2 only   c) 3 only   d) 1 and 2   e) 1, 2, and 3

Answer: a

12. What is the monomer of Teflon?

a) CH_2CH_2 b) CF_2CF_2 c) $CH_2C(CN)_2$ d) CH_2CHCN e) CH_2CHCH_3

Answer: b

TESTBANK

10.1 Why Carbon?

1. Structural isomers are compounds that have

    a) the same molar masses, but different elemental composition.
    b) have identical structures, but contain different isotopes of the same elements.
    c) two or more resonance structures.
    d) the same elemental composition, but the atoms are linked in different ways.
    e) the same physical properties, but different chemical properties.

    Answer: d

2. Optical isomers have

    a) nonsuperimposable mirror images.
    b) double bonded carbon atoms with either *cis* or *trans* functional groups.
    c) only sp^2 or sp hybridized central atoms.
    d) identical structures, but different physical properties (such as melting and boiling points).
    e) an odd number of electrons.

    Answer: a

3. Which one of the following statements concerning isomers is INCORRECT?

    a) Pairs of nonsuperimposable molecules are called enantiomers.
    b) Enantiomers have identical melting and boiling points.
    c) Two types of stereoisomers exist: geometric isomers and structural isomers.
    d) Optical isomers rotate polarized light in opposite directions.
    e) Structural isomers have the same composition, but the atoms are linked in different ways.

    Answer: c

4. Which one of the following molecules has a chiral center?

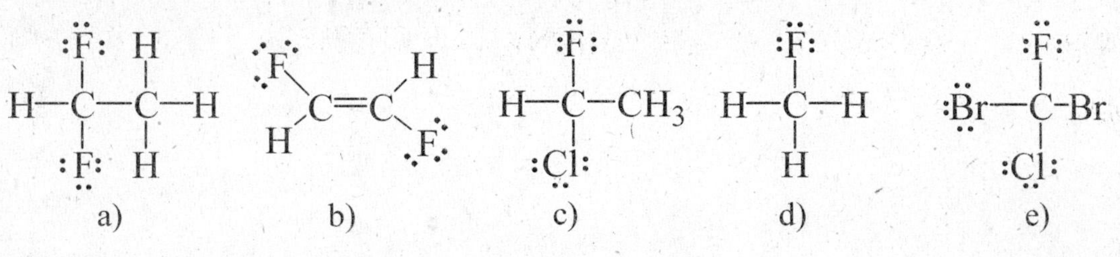

    a)   b)   c)   d)   e)

    Answer: c

5. For which one of the following molecules do geometric isomers exist?

    a) $H_2C=CH_2$ b) $BrHC=CHCl$ c) ClH_2C-CH_2Br
    d) H_3C-CH_2Cl e) $H_2C=CCl_2$

    Answer: b

10.2 Hydrocarbons

6. What is the general formula for a cycloalkane?

    a) C_nH_{2n+2} b) C_nH_{2n} c) C_nH_{n-2} d) $C_{n+2}H_n$ e) C_nH_{2n-2}

    Answer: b

7. Which of the following molecules might be an alkyne?

    a) C_3H_6 b) C_4H_{10} c) C_5H_{10} d) C_6H_{10} e) C_6H_{12}

    Answer: d

8. Which of the following molecules may be a cycloalkene?

    a) C_4H_8 b) C_5H_{12} c) C_5H_{10} d) C_6H_{12} e) C_6H_{10}

    Answer: e

9. Which of the following hydrocarbons has two π bonds?

    a) C_3H_4 b) C_4H_8 c) C_6H_{12} d) C_7H_{14} e) C_8H_{18}

    Answer: a

10. What is the molecular formula for nonane?

    a) C_6H_{12} b) C_6H_{14} c) C_8H_{18} d) C_9H_{20} e) $C_{11}H_{22}$

    Answer: d

11. What is the name of the following compound?

    a) 1,1,3-trimethylpentane
    b) 2,4-dimethyloctane
    c) 3-methyl-4-propylbutane
    d) 3,5-dimethylhexane
    e) 2,4-dimethylhexane

Answer: e

12. What is the name of the following compound?

    a) hexane
    b) cyclohexane
    c) benzene
    d) cyclohexene
    e) cyclobenzene

Answer: b

13. What is the name of the following compound?

    a) 3-ethyl-5,6-dimethylheptane
    b) 2,3,5-ethylmethylethylheptane
    c) 3-ethyldecane
    d) 2,3-dimethyl-3-ethylheptane
    e) 3,5,6-trimethylheptane

Answer: a

14. How many isomers, both structural and geometric, exist for C_5H_{10}?

    a) 5   b) 6   c) 7   d) 8   e) 9

Answer: c

15. Which one of the following hydrocarbons has *cis* and *trans* isomers?

    a) hexane   b) 1-pentene   c) 2-methylpropene
    d) 2-ethylbutane   e) 2-butene

Answer: e

16. What is the name of the following compound?

a) *trans*-1-methyl-1-butene
b) *cis*-1-methyl-1-butene
c) *trans*-2-pentene
d) *trans*-3-pentene
e) *trans*-1-methyl-2-ethylethylene

$$\begin{array}{ccc} H & & CH_2-CH_3 \\ & C=C & \\ CH_3 & & H \end{array}$$

Answer: c

17. What is the name of the following compound?

a) *trans*-acetylene
b) *trans*-1,3-pentadiene
c) *cis*-2,4-pentadiene
d) *cis*-4-methyl-1,3-butadiene
e) *cis*-4-methyl-1,3-butene

$$\begin{array}{ccc} H & & CH=CH_2 \\ & C=C & \\ CH_3 & & H \end{array}$$

Answer: b

18. What is the name of the following compound?

a) 1-methyl-2-isopropylacetylene
b) *cis*-4-methylpentyne
c) 4,4-dimethyl-2-butyne
d) 1,1-dimethyl-2-butyne
e) 4-methyl-2-pentyne

$$CH_3-C\equiv C-\underset{\underset{CH_3}{|}}{CH}-CH_3$$

Answer: e

19. What is the balanced chemical equation for the combustion of 2-methyl-1-butene?

a) $2\ C_5H_{10}(g)\ +\ 15\ O_2(g)\ \rightarrow\ 10\ CO_2(g)\ +\ 10\ H_2O(\ell)$

b) $2\ C_4H_{10}(g)\ +\ 13\ O_2(g)\ \rightarrow\ 8\ CO_2(g)\ +\ 10\ H_2O(\ell)$

c) $C_4H_8(g)\ +\ 6\ O_2(g)\ \rightarrow\ 4\ CO_2(g)\ +\ 4\ H_2O(\ell)$

d) $C_6H_{12}(g)\ +\ 9\ O_2(g)\ \rightarrow\ 6\ CO_2(g)\ +\ 6\ H_2O(\ell)$

e) $C_5H_{12}(g)\ +\ 13\ O_2(g)\ \rightarrow\ 5\ CO_2\ (g)\ +\ 6\ H_2O(\ell)$

Answer: a

20. Which one of the following hydrocarbons is aromatic?

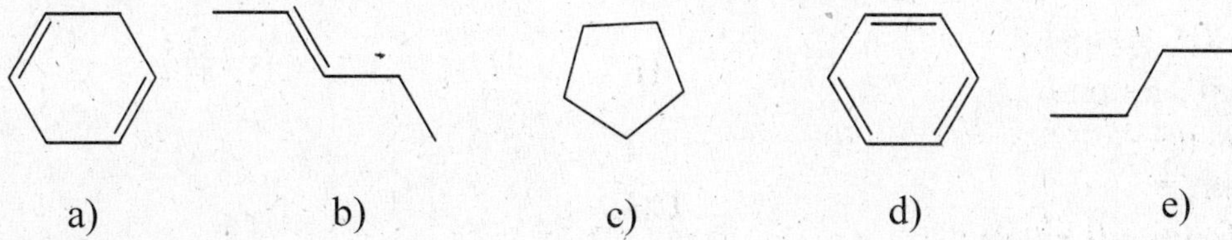

   a)   b)   c)   d)   e)

Answer: d

21. A saturated hydrocarbon is

   a) a hydrocarbon that contains oxygen.
   b) a compound in which all carbon atoms have four single bonds.
   c) a compound in which one or more carbon atoms have double or triple bonds.
   d) a hydrocarbon that is dissolved in water.
   e) a cycloalkane with five or more carbons.

Answer: b

22. What is the product of the addition of F_2 to 1-propene?

   a) 1-fluoropropene   b) 1,1-difluoropropane   c) 2,2-difluoropropane
   d) 1,2-difluoropropane   e) 2-fluoropropane

Answer: d

23. What is the product of the addition of HCl to ethylene?

   a) chloroethane   b) chloroethylene   c) 1,2-dichloroethane
   d) 1,1-dichloroethane   e) 1,1,2,2-tetrachloroethylene

Answer: a

24. What is the product of the addition of excess Br_2 to acetylene?

   a) 1,2-dibromoethane   b) ethylene   c) 1,1,2,2-tetrabromoethane
   d) ethane   e) 1,1-dibromoethylene

Answer: c

25. Identify the product(s) of the hydrogenation of *cis*-2-hexene.

   a) 2-hydroxyhexane   b) 2,3-dihydroxyhexane   c) *trans*-2-hexane
   d) carbon dioxide and water   e) hexane

Answer: e

26. How many structures (i.e., unique isomers) exist for dichlorobenzene?

a) 2   b) 3   c) 4   d) 5   e) 6

Answer: b

27. What is the product of the addition of HF to 1-butene?

a) 1,1-difluorobutane   b) 1,2-difluorobutane   c) 1-fluorobutane
d) 2-fluorobutane   e) butane

Answer: d

28. Which of the structures below has the common name *m*-dichlorobenzene (where *m*- is meta)?

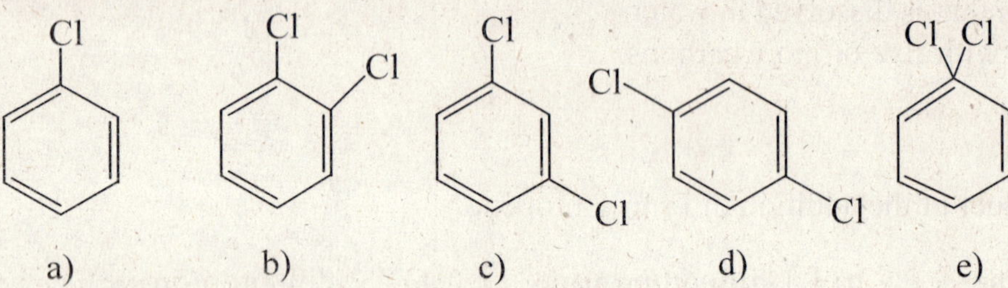

Answer: c

29. What is the name of the following benzene derivative?

a) 3-bromobenzoic acid
b) 5-bromoanaline
c) 1-acetate-3-bromobenzene
d) 5-bromobenzoic acid
e) 1,3-bromocarboxylic benzene

Answer: a

30. What is the name of the following benzene derivative?

a) dibenzene
b) 1,2-dibenzene
c) toluene
d) naphthalene
e) aniline

Answer: d

31. Which of the following statements concerning hydrocarbons is/are CORRECT?
    1. Coal is a source of many aromatic hydrocarbons.
    2. The primary components of petroleum (prior to refining) are alkenes and alkynes.
    3. Reactions of hydrocarbons with oxygen generally yield carbon dioxide and water.

    a) 1 only   b) 2 only   c) 3 only   d) 1 and 3   e) 1, 2, and 3

    Answer: d

32. Which of the following chemical equations depicts an alkylation reaction?

    a) $C_6H_6(\ell) + CH_3Br(\ell) \rightarrow C_6H_5CH_3(\ell) + HBr(g)$

    b) $CH_3CH_2OH(\ell) + 3\,O_2(g) \rightarrow 2\,CO_2(g) + 3\,H_2O(\ell)$

    c) $C_6H_{12}(\ell) \rightarrow C_6H_{10}(\ell) + H_2(g)$

    d) $C_2H_2(g) + H_2(g) \rightarrow C_2H_4(g)$

    e) $3\,C_2H_2(g) \rightarrow C_6H_6(\ell)$

    Answer: a

33. Which of the following chemical equations depicts a halogenation reaction with benzene?

    a) $C_6H_6(\ell) + 3\,Cl_2(g) \rightarrow 3\,C_2H_2Cl_2(g)$

    b) $C_6H_6(\ell) + Cl_2(g) \rightarrow C_6H_5Cl(\ell) + HCl(g)$

    c) $C_6H_6(\ell) + CH_3Cl(g) \rightarrow C_6H_5CH_3(\ell) + HCl(g)$

    d) $C_6H_6(\ell) + Cl_2(g) \rightarrow C_6H_4Cl_2(\ell) + H_2(g)$

    e) $C_6H_6(\ell) + 12\,Cl_2(g) \rightarrow 6\,CCl_4(g) + 3\,H_2(g)$

    Answer: b

10.3 Alcohols, Ethers, and Amines

34. Formulas for derivatives of hydrocarbons may be written as R-X, where R is a hydrocarbon lacking a hydrogen atom and X is a functional group. Which of the following formulas represents an ether?

    a) ROH   b) ROR'   c) RCOR'   d) RCO_2R' e) RCHO

    Answer: b

35. Which functional group does **not** contain an oxygen atom?

a) ester   b) ether   c) alcohol   d) amide   e) amine

Answer: e

36. Which functional group does **not** contain a double bond to an oxygen atom?

a) ester   b) aldehyde   c) alcohol   d) amide   e) ketone

Answer: c

37. Which of the following alcohols is likely to be least soluble in water?

a) 3-pentanol   b) 2-butanol   c) 1,2-ethanediol
d) 1,2,3-propanetriol   e) methanol

Answer: a

38. Putrescine and cadaverine are two compounds associated with decaying flesh or bad breath. These compounds belong to a class of molecules called _________.

a) ethers   b) amines   c) ketones
d) esters   e) carboxylic acids

Answer: b

39. The addition of water to ethylene has become a cheaper method than fermentation for making _________.

a) ethanol   b) antifreeze   c) diethyl ether   d) methanol   e) formic acid

Answer: a

40. Which one of the following functional groups is a base?

a) aldehyde   b) alcohol   c) carboxylic acid
d) ether   e) amine

Answer: e

10.4 Compounds with a Carbonyl Group

41. The functional group RCO_2R' is characteristic of a(n) _________.

a) ether   b) ester   c) amide   d) aldehyde   e) ketone

Answer: b

42. The carbonyl group occurs in molecules with all of the following functional groups EXCEPT
_________.

    a) esters   b) ketones   c) ethers   d) amides   e) aldehydes

    Answer: c

43. How many primary alcohols have the formula $C_5H_{12}O$?

    a) 2   b) 3   c) 4   d) 5   e) 6

    Answer: c

44. When a secondary alcohol is oxidized, the product is a(n) _________.

    a) ketone   b) ether   c) ester   d) aldehyde   e) amide

    Answer: a

45. Which of the following compounds might be used to oxidize an aldehyde to a carboxylic
acid?

    a) H_2 b) Na   c) NaH   d) $K_2Cr_2O_7$ e) $NaBH_4$

    Answer: d

46. What is the product of the reduction of propanal with sodium borohydride?

    a) propane   b) 1-propanol   c) 2-propanol   d) acetone   e) propanoic acid

    Answer: b

47. A mixture of ethanol and benzoic acid is heated in the presence of acid. What is the primary
product of the reaction?

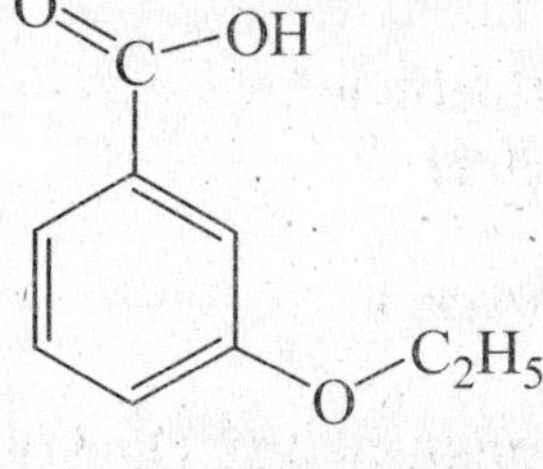

    Answer: c

48. What is the name of the product of the reaction that occurs when a mixture of ethanol and
butanoic acid is heated in the presence of acid?

   a) ethyl butylether   b) butyl ethanoate   c) 2-ethylbutanal
   d) 2-ethylbutanol   e) ethyl butanoate

Answer: e

49. The hydrolysis of an ester in the presence of NaOH produces

   a) a carboxylate salt of sodium and an alcohol.   b) an ether and an amide.
   c) an alcohol and an amine.   d) an aldehyde and an ether.
   e) an ether and a carboxylate salt of sodium.

Answer: a

50. What class of compounds is responsible for many of the distinctive odors of artificial flavors
and perfumes?

   a) ethers   b) aldehydes   c) esters   d) amides   e) alkenes

Answer: c

51. What is the classification of the molecule below?

   a) polymer   b) fatty acid   c) saccharide   d) carbohydrate   e) triglyceride

Answer: b

52. An amine reacts with a(n) __________ to form an amide.

   a) ether   b) alcohol   c) ester
   d) aldehyde   e) carboxylic acid

Answer: e

53. What is the hybridization and the bond angles around the nitrogen atom in an amide?

   a) sp hybridization, 180° bond angles
   b) sp^2 hybridization and 109° bond angles
   c) sp^2 hybridization and 120° bond angles
   d) sp^3 hybridization and 109° bond angles
   e) sp^3 hybridization and 120° bond angles

Answer: c

<u>10.5 Polymers</u>

54. Which of the following statements is INCORRECT?

a) Elastomers are materials that spring back to their original shape when stretched.
b) Thermoplastics soften and flow when they are heated and harden when they are cooled.
c) Thermosetting plastics set to a solid when heated and cannot be resoftened.
d) Polymers formed from two or more different monomers are called addition polymers.
e) A condensation reaction involves reacting two monomers and splitting out a small
   molecule, often water.

Answer: d

55. Which of the following statements concerning polyethylene is/are CORRECT?
   1. The monomer for polyethylene is CH_2CH_2.
   2. Polyethylene contains alternating single and double carbon-carbon bonds.
   3. Polyethylene is made by condensation polymerization.

a) 1 only   b) 2 only   c) 3 only   d) 1 and 2   e) 1, 2, and 3

Answer: a

56. What is the monomer of Teflon?

a) CH_2CH_2 b) CF_2CF_2 c) $CH_2C(CN)_2$ d) CH_2CHCN e) CH_2CHCH_3

Answer: b

57. Low density polyethylene (LDPE) is most likely to be used in

a) sandwich bags.   b) bottle caps.   c) rigid pipes.
d) contact lenses.   e) fabrics and apparel.

Answer: a

58 Amino acids polymerize in condensation reactions that result in the formation of an amide linkage (or peptide bond) between amino acid molecules. What dipeptide(s) may be formed in the reaction of glycine with phenylalanine?

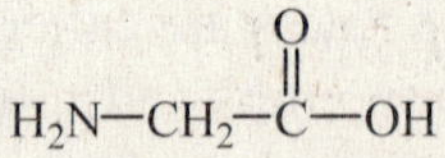

Glycine

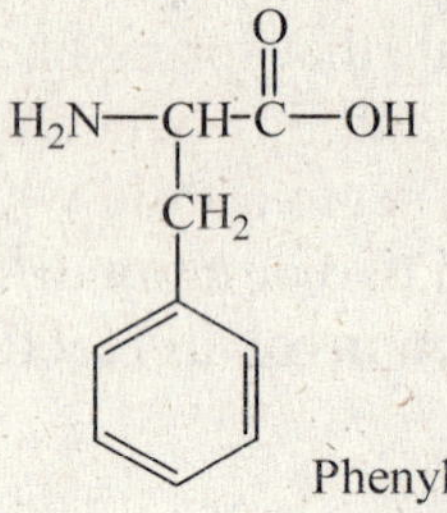

Phenylalanine

(1)

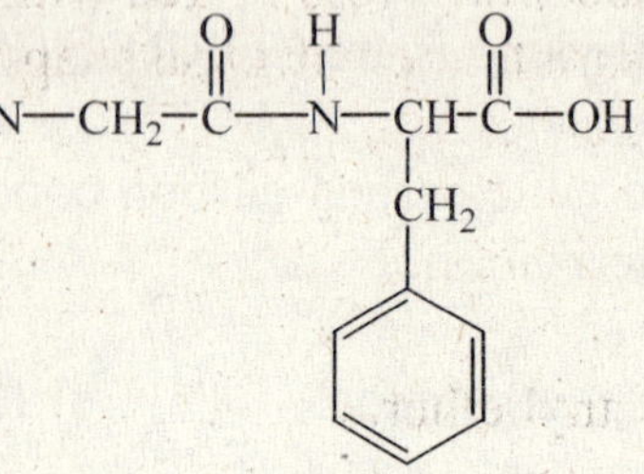

(2)

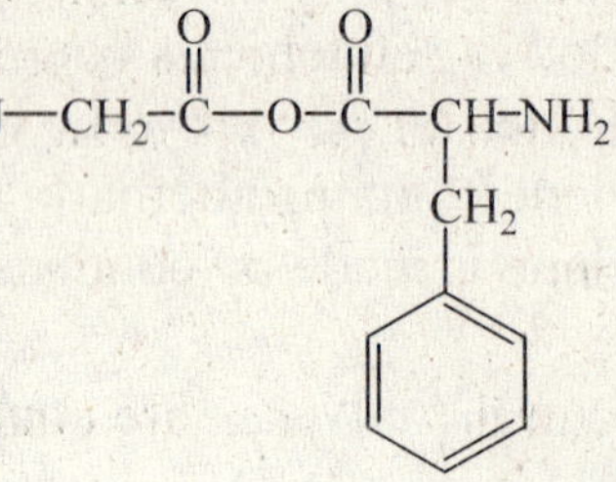

(3)

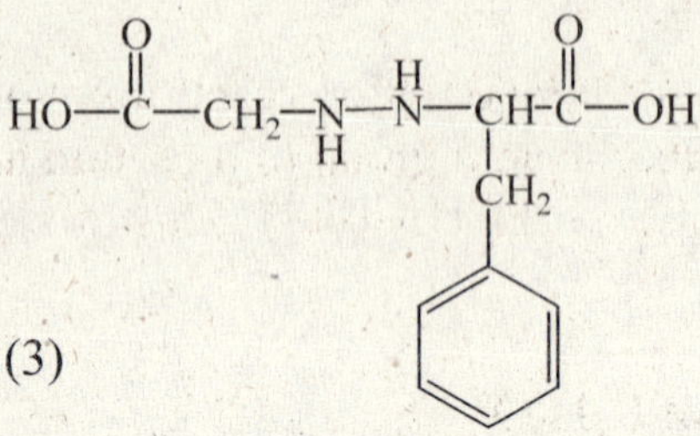

(4)

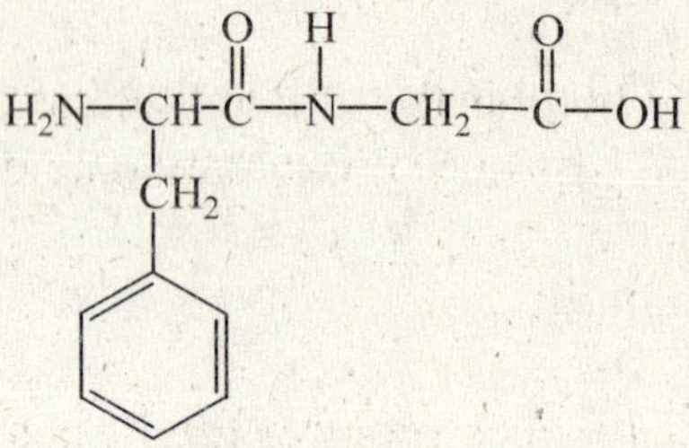

a) 1 only   b) 2 only   c) 3 only   d) 4 only   e) 1 and 4

Answer: e

59. Polyvinyl chloride is used in raincoats and pipes. It is produced by the addition reaction of chloroethylene, CH_2CHCl. What is the structure of the polymer produced in this reaction?

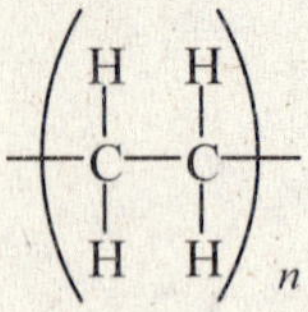

a) 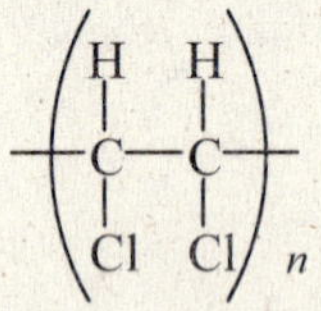   b) 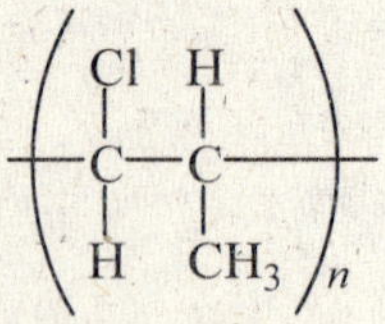   c) 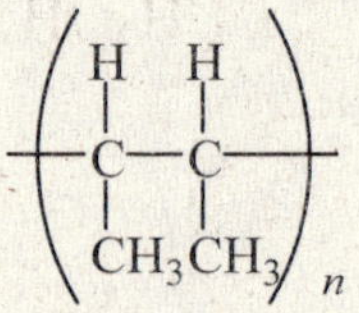   d) 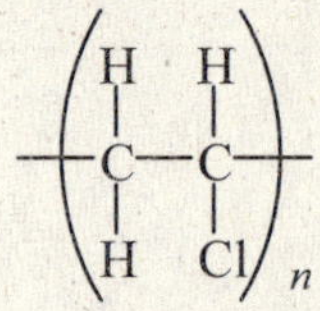   e)

Answer: e

Short Answer Questions

60. The process by which long chain hydrocarbons in petroleum are shortened is called __________.

    Answer: cracking

61. Hydrolysis of fats and oils in the presence of a strong base produces glycerol and the salts of fatty acids. The name for this reaction is __________, which means "soap making."

    Answer: saponification

62. In vulcanized rubber, the polymer chains of natural rubber are cross-linked with sulfur atoms. The polymer chains stretch when stressed, but they return to their original shape when the stress is removed. Substances that behave this way are called __________.

    Answer: elastomers

63. Examples of structural __________ are ethanol and dimethyl ether.

    Answer: isomers

64. A peptide bond is the amide linkage that is formed in a condensation reaction involving the __________ group of one amino acid with the carboxylic acid group of a second amino acid.

    Answer: amine

65. Draw Lewis structures for all possible isomers of C_3H_6O.

    Answer:

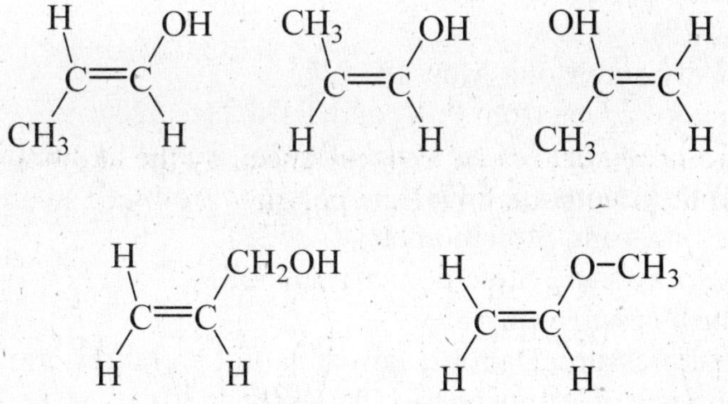

66. The molecule 2-chloro-4-methylhexane is made by addition of HCl to an alkene. Write a balanced chemical equation for this reaction.

    Answer:

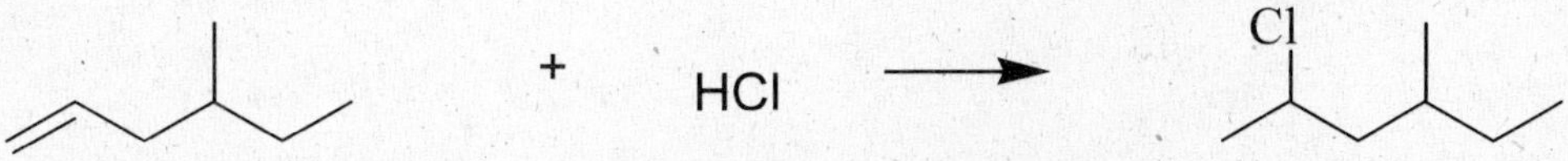

Chapter 11
GASES AND THEIR PROPERTIES

CHAPTER OUTLINE

CHAPTER 11: Gases and Their Properties

       KEY TERMS: root mean square (rms speed)
       Chemical Perspective The Earth's Atmosphere (p. 534)
       Example 11.12 Molecular Speed (p. 536)
     Kinetic-Molecular Theory and the Gas Laws (p. 537)

11.7   Diffusion and Effusion (p. 538)
       KEY TERMS: diffusion, effusion, Graham's law
       Example 11.13 Using Graham's Law of Effusion to Calculate a Molar Mass (p. 539)

11.8   Some Applications of the Gas Laws and Kinetic-Molecular Theory (p. 540)
     Separating Isotopes (p. 540)
     Deep Sea Diving (p. 540)
       Case Study You Stink! (p. 541)

11.9   Nonideal Behavior: Real Gases (p. 542)
       KEY TERMS: van der Waals equation
       Table 11.2 van der Waals Constants (p. 543)

CHAPTER OBJECTIVES

- Understand the basis of the gas laws and how to use those laws (11.1-11.2)
- Use the ideal gas law (11.3)
- Apply the gas laws to stoichiometric calculations (11.4-11.5)
- Understand kinetic molecular theory as it applied to gases, especially the distribution of molecular speeds (energies) (11.6-11.7)
- Recognize why gases do not behave like ideal gases under some conditions (11.9)

LESSON PLAN

AP Standards

II.A.1.a	Equation of state for an ideal gas
II.A.1.b	Partial Pressure
II.A.2.a	Interpretation of ideal gas laws on the basis of the kinetic-molecular theory
II.A.2.b	Avogadro's hypothesis and the mole concept
II.A.2.c	Dependence of kinetic energy of molecules on temperature
II.A.2.d	Deviations from ideal gas laws
II.C.4	Nonideal behavior (qualitative)
III.B.3	Mass volume relations with emphasis on the mole concept, including empirical formulas and limiting reagents

Suggested Time:

Four traditional classes or 3 blocks. Sections 11.3-11.8 should be the focus of the chapter in order to teach AP material. Avogadro's hypothesis is the only essential part of section 11.2. The remainder of the chapter can be skipped or assigned as reading homework.

ASSESSMENT

Exercises:

Answers to Exercises p. A49-A50
Practicing Skills pp. 546-549, Answers IRM pp. 202-209
General Questions pp. 549-551, Answers IRM pp. 209-216
In the Laboratory Questions pp. 551-552, Answers IRM pp. 216-221
Summary and Conceptual Questions pp. 552-553, Answers IRM pp. 221-223

GLOSSARY

Avogadro's hypothesis Equal volumes of gases under the same conditions of temperature and pressure have equal numbers of particles

bar A unit of pressure; 1 bar = 100 kPa

barometer An apparatus used to measure atmospheric pressure

Boyle's law The volume of a fixed amount at a given temperature is inversely proportional to the pressure exerted by the gas

Charles's law If a given quantity of gas is held at a constant pressure, its volume is directly proportional to the Kelvin temperature

compressibility The change in volume with change in pressure

Dalton's law of partial pressure The total pressure of a mixture of gases is the sum of the pressures of the components of the mixture

diffusion The gradual mixing of the molecules of two or more substances by random molecular motion

CHAPTER 11: Gases and Their Properties

effusion The movement of gas molecules through a membrane or other porous barrier by random molecular motion

gas constant (R) The proportionality constant in the ideal gas law

general (combined) gas law An equation that allows calculation of pressure, temperature, and volume when a given amount of gas undergoes a change in conditions

Graham's law The rate of effusion of a gas is inversely proportional to the square root of its molar mass.

ideal gas law A law that relates pressure, volume, number of moles, and temperature for an ideal gas

millimeters of mercury (mm Hg) A common unit of pressure, defined as the pressure that can support a 1-millimeter column of mercury; 760 mm Hg = 1 atm

mole fraction (X) The ratio of the number of moles of one substance to the total number of moles in a mixture of substances

partial pressure The pressure exerted by one gas in a mixture of gases

pascal (Pa) The SI unit of pressure 1 Pa = 1 N/m^2

pressure The force exerted on an object divided by the area over which the force is exerted

root mean square (rms) speed The square root of the average of the squares of the speeds of the molecules in a sample

standard atmospheres(atm) A unit of pressure; 1 atm = 760 mm Hg

standard molar volume The volume occupied by 1 mol of gas at standard temperature and pressure; 22.414 L

standard temperature and pressure (STP) A temperature of 0 $^{\circ}$C and a pressure of exactly 1 atm

torr A unit of pressure equivalent to one millimeter of mercury

van der Waals equation A mathematical expression that describes the behavior of non-ideal gases

MEDIA MENU

The Media Menu section of this guide lists the interactive exercises, tutorials, active figures (animated, interactive figures) and additional resources available online for Chemistry Now, a unique web-based, assessment-centered personalized learning system for chemistry students. Go to www.cengagelearning.com/login

Exercises
- Screen 11.3   The three gas laws
- Screen 11.5   Gas density
- Screen 11.12 Diffusion

Tutorials
- Screen 11.6   Using gas laws determining molar mass
- Screen 11.7   Gas laws and chemical reactions: stoichiometry
- Screen 11.8   Gas mixtures and partial pressures
- Screen 11.11 Boltzman distribution and calculation of distribution curves

Self-Study Modules
- Screen 11.9   Gases at different temperatures

Active Figures
- 11.4   An experiment to demonstrate Boyle's law
- 11.6   Charles's Law
- 11.18   Gaseous diffusion

Additional Resources
- Screen 11.4   Ideal gas law simulation
- Screen 11.10   Gas laws at the molecular level simulation
- Podcast Module 16   Kinetic-Molecular theory of Gases

DIAGNOSTIC QUIZ QUESTIONS

Goal #1 Understand the basis of the gas laws and how to use those laws

1. If the volume of a confined gas is reduced to ½ the original volume while its temperature remains constant, what change will be observed?

   a) The pressure of the gas will increase to twice its original value.
   b) The pressure of the gas will remain unchanged.
   c) The density of the gas will decrease to ½ its original value.
   d) The pressure of the gas will decrease to ½ its original value.
   e) The average velocity of the molecules will double.

   Answer: a

CHAPTER 11: Gases and Their Properties

2. A tightly sealed 4.0-L flask contains 772 mm Hg of Ar at 129 °C. The flask is cooled until the pressure is reduced 386 mm Hg. What is the temperature of the gas?

    a) -72 °C   b) 65.0 °C   c) 201 °C   d) 258 °C   e) 474 °C

    Answer: a

3. A balloon is filled with N_2 gas to a volume of 1.92 L at 37 °C. The balloon is placed in liquid nitrogen until its temperature reaches -142 °C. Assuming the pressure remains constant, what is the volume of the cooled balloon?

    a) -0.500 L   b) 0.500 L   c) 0.811 L   d) 1.43 L   e) 4.54 L

    Answer: c

4. A 0.225-L flask contains CO_2 at 22 °C and 451 mm Hg. What is the pressure of the CO_2 if the volume is increased to 0.600 L and the temperature increased to 85 °C?

    a) 43.8 mm Hg   b) 139 mm Hg   c) 205 mm Hg   d) 653mm Hg   e) 991 mm Hg

    Answer: c

Goal #2 Use the ideal gas law

5. A tightly sealed 4.0-L flask contains 772 mm Hg of Ar at 129 °C. The flask is cooled until the pressure is reduced 386 mm Hg. What is the temperature of the gas?

    a) -72 °C   b) 65.0 °C   c) 201 °C   d) 258 °C   e) 474 °C

    Answer: a

6. A balloon is filled with N_2 gas to a volume of 1.92 L at 37 °C. The balloon is placed in liquid nitrogen until its temperature reaches -142 °C. Assuming the pressure remains constant, what is the volume of the cooled balloon?

    a) -0.500 L   b) 0.500 L   c) 0.811 L   d) 1.43 L   e) 4.54 L

    Answer: c

7. A 0.225-L flask contains CO_2 at 22 °C and 451 mm Hg. What is the pressure of the CO_2 if the volume is increased to 0.600 L and the temperature increased to 85 °C?

    a) 43.8 mm Hg   b) 139 mm Hg   c) 205 mm Hg   d) 653mm Hg   e) 991 mm Hg

    Answer: c

Goal #3 Apply the gas laws to stoichiometric calculations

8. At 506 K and 2.05 atm, what mass of H_2 will react with excess N_2 to produce 18.3 L NH_3? (R = 0.08206 L·atm/mol·K)

$$N_2(g) + 3\,H_2(g) \rightarrow 2\,NH_3(g)$$

   a) 1.21 g   b) 1.36 g   c) 1.49 g   d) 1.82 g   e) 2.73 g

Answer: e

9. If 6.46 L of gaseous ethanol reacts with 32.1 L O_2, what is the maximum volume of gaseous water produced? Assume that the temperature of the reactants and products is 425 °C and the pressure remains constant at 1.00 atm.

$$CH_3CH_2OH(g) + 3\,O_2(g) \rightarrow 2\,CO_2(g) + 3\,H_2O(g)$$

   a) 6.46 L   b) 19.4 L   c) 25.6 L   d) 32.1 L   e) 38.6 L

Answer: b

Goal #4 Understand kinetic molecular theory as it applied to gases, especially the distribution of molecular speeds (energies)

10. What is the root-mean-square velocity of ammonia molecules (NH_3) at 45 °C?

   a) 21.6 m/s   b) 67.8 m/s   c) 257 m/s   d) 466 m/s   e) 682 m/s

Answer: e

11. Place the following gases in order of increasing average velocity at 300 K: CO, Ne, O_2, and N_2O.

   a) CO = Ne = O_2 = N_2O b) Ne < CO < O_2 < N_2O c) CO < O_2 < N_2O < Ne
   d) O_2 < N_2O < Ne < CO   e) N_2O < O_2 < CO < Ne.

Answer: e

Goal #5 Recognize why gases do not behave like ideal gases under some conditions (11.9)

12. Non-ideal behavior for a gas is most likely to be observed under conditions of

   a) high temperature and low pressure.   b) high temperature and high pressure.
   c) low temperature and low pressure.   d) standard temperature and pressure.
   e) low temperature and high pressure.

Answer: e

13. Which of the following statements concerning real gases is/are CORRECT?
    1. Real gases are always liquids or solids at temperatures below 273.15 K.
    2. The pressure of a real gas is higher than predicted by the ideal gas law.
    3. The molecules in a real gas are attracted to each other.

    a) 1 only b) 2 only c) 3 only d) 2 and 3 e) 1, 2 and 3

    Answer: c

TESTBANK

11.1 The Properties of Gases

1. Which of the following represents the greatest pressure?

    a) 1.0 atm b) 1.0 bar c) 1.0 mm Hg
    d) 1.0 kPa e) 1.0 psi

    Answer: a

2. Convert 221 mm Hg to kPa. (1 atm = 760 mm Hg = 101.3 kPa)

    a) 0.00287 kPa b) 0.221 kPa c) 0.291 kPa
    d) 29.5 kPa e) 1.66×10^3 kPa

    Answer: d

3. At 0.967 atm, the height of mercury in a barometer is 0.735 m. If the mercury were replaced with water, what height of water (in meters) would be supported at this pressure? The densities of Hg and H_2O are 13.5 g/cm^3 and 0.997 g/cm^3, respectively.

    a) 0.0546 m b) 0.735 m c) 0.760 m d) 9.95 m e) 13.1 m

    Answer: d

11.2 Gas Laws: The Experimental Basis

4. At constant temperature, 12.5 L of N_2 at 262 mm Hg is compressed to 4.15 L. What is the final pressure of N_2?

    a) 5.05 mm Hg b) 87.0 mm Hg c) 262 mm Hg d) 789 mm Hg e) 873 mm Hg

    Answer: d

5. If the volume of a confined gas is reduced to ½ the original volume while its temperature
    remains constant, what change will be observed?

    a) The pressure of the gas will increase to twice its original value.
    b) The pressure of the gas will remain unchanged.
    c) The density of the gas will decrease to ½ its original value.
    d) The pressure of the gas will decrease to ½ its original value.
    e) The average velocity of the molecules will double.

    Answer: a

6. The lid is tightly sealed on a rigid flask containing 4.20 L O_2 at 27 °C and 0.969 atm. If the
    flask is heated to 81 °C, what is the pressure in the flask?

    a) 0.821 atm   b) 1.14 atm   c) 1.18 atm   d) 2.91 atm   e) 3.45 atm

    Answer: b

7. A tightly sealed 4.0-L flask contains 772 mm Hg of Ar at 129 °C. The flask is cooled until
    the pressure is reduced 386 mm Hg. What is the temperature of the gas?

    a) -72 °C   b) 65.0 °C   c) 201 °C   d) 258 °C   e) 474 °C

    Answer: a

8. A balloon is filled with N_2 gas to a volume of 1.92 L at 37 °C. The balloon is placed in liquid
    nitrogen until its temperature reaches -142 °C. Assuming the pressure remains constant, what
    is the volume of the cooled balloon?

    a) -0.500 L   b) 0.500 L   c) 0.811 L   d) 1.43 L   e) 4.54 L

    Answer: c

9. A 0.225-L flask contains CO_2 at 22 °C and 451 mm Hg. What is the pressure of the CO_2 if
    the volume is increased to 0.600 L and the temperature increased to 85 °C?

    a) 43.8 mm Hg   b) 139 mm Hg   c) 205 mm Hg   d) 653mm Hg   e) 991 mm Hg

    Answer: c

10. You have a 25.0 L cylinder of helium at a pressure of 132 atm and a temperature of 19 °C.
    The He is used to fill balloons to a volume of 2.50 L at 732 mm Hg and 27 °C. How many
    balloons can be filled with He? Assume that the cylinder can provide He until its internal
    pressure reaches 1.00 atm (i.e., there are 131 atmospheres of usable He in the cylinder).

    a) 9.57×10^2 b) 1.26×10^3 c) 1.31×10^3 d) 1.36×10^3 e) 1.40×10^3

    Answer: e

11. Avogadro's hypothesis states that equal volumes of gases under the same conditions of temperature and pressure have equal _________.

a) numbers of particles   b) particle velocities   c) molar masses
d) densities   e) masses

Answer: a

12. Which of the following relationships is/are CORRECT for gases?
1. The moles of a gas is inversely proportional to its volume (at constant pressure).
2. The volume of a gas is inversely proportional to its temperature in kelvin (at constant pressure).
3. The pressure of a gas is directly proportional to its temperature in kelvin (at constant volume).

a) 1 only   b) 2 only   c) 3 only   d) 1 and 2   e) 2 and 3

Answer: c

13. What volume of Ar at 35 °C and 1.75 atm contains the same number of particles as 0.655 L of Cl_2 at 15 °C and 3.00 atm?

a) 0.481 L   b) 1.20 L   c) 1.84 L   d) 2.72 L   e) 7.49 L

Answer: b

14. At 212 mm Hg and 24 °C, a sample of gas occupies a volume of 125 mL. The gas is transferred to a 325-mL flask and the temperature is reduced to -5.0 °C. What is the pressure of the gas in the flask?

a) 17.0 mm Hg   b) 73.6 mm Hg   c) 81.5 mm Hg   d) 191 mm Hg   e) 497 mm Hg

Answer: b

15. What volume of oxygen will react with 18 mL of acetylene, assuming the gases are present at the same temperature and pressure?
$$2\ C_2H_2(g) + 5\ O_2(g) \rightarrow 4\ CO_2(g) + 2\ H_2O(g)$$

a) 7.2 mL   b) 9.0 mL   c) 18 mL   d) 36 mL   e) 45 mL

Answer: e

16. If 3.67 g of CO_2 gas is introduced into an evacuated 2.50 L flask at 65 °C, what is the pressure inside the flask? ($R = 0.08206$ L·atm/mol·K)

    a) 0.178 atm   b) 0.445 atm   c) 0.925 atm   d) 40.7 atm   e) 496 atm

    Answer: c

17. The pressure of N_2 in a 20.0 L flask is 0.603 atm at 62 °C. What mass of N_2 is in the flask? ($R = 0.08206$ L·atm/mol·K)

    a) 0.439 g   b) 6.15 g   c) 12.3 g   d) 63.9 g   e) 66.4 g

    Answer: c

18. A 48.0 L gas cylinder contains 498 g H_2 at 25 °C. What is the pressure inside the cylinder? ($R = 0.08206$ L·atm/mol·K)

    a) 0.210 atm   b) 3.27 atm   c) 10.6 atm   d) 126 atm   e) 254 atm

    Answer: d

19. What volume is occupied by 45.0 g of CH_4 at 345 mm Hg and 35 °C? ($R = 0.08206$ L·atm/mol·K)

    a) 17.8 L   b) 156 L   c) 322 L   d) 2.51×10^3 L   e) 2.78×10^3 L

    Answer: b

20. Which of the following gases has the lowest density at 25 °C and 5.0 atm?

    a) CH_4 b) O_2 c) N_2 d) Cl_2 e) C_2H_2

    Answer: a

21. Calculate the density (in g/L) of carbon monoxide gas at 21 °C and 215 mm Hg. ($R = 0.08206$ L·atm/mol·K)

    a) 0.328 g/L   b) 3.04 g/L   c) 4.60 g/L   d) 6.83 g/L   e) 249 g/L

    Answer: a

22. If the density of N_2 gas in a flask is 0.243 g/L at 18 °C, what is the pressure of nitrogen in the flask if it is heated to 36 °C? ($R = 0.08206$ L·atm/mol·K)

    a) 0.207 atm   b) 0.220 atm   c) 0.258 atm   d) 0.414 atm   e) 0.486 atm

    Answer: b

23. At what temperature does 1.00 atm of He gas have the same density as 1.00 atm of Ar gas at 273 K?

    a) 1.78 K   b) 27.4 K   c) 87.0K   d) 273 K   e) 2.73×10^3 K

    Answer: b

24. A mass of 0.645 g of an unknown gas is introduced into an evacuated 1.50 L flask. If the pressure in the flask is 0.764 atm at 96 °C, which of the following gases might be in the flask? ($R = 0.08206$ L·atm/mol·K)

    a) N_2O b) C_2H_2 c) O_2 d) HCl   e) NH_3

    Answer: e

11.4 Gas Laws and Chemical Reactions

25. Sodium azide decomposes rapidly to produce nitrogen gas.
$$2 \text{ NaN}_3(s) \rightarrow 2 \text{ Na}(s) + 3 \text{ N}_2(g)$$
    What mass of sodium azide will inflate a 56.6 L airbag for a car to a pressure of 811 mm Hg at 25 °C? ($R = 0.08206$ L·atm/mol·K)

    a) 17.6 g   b) 39.5 g   c) 107 g   d) 161 g   e) 241 g

    Answer: c

26. At 506 K and 2.05 atm, what mass of H_2 will react with excess N_2 to produce 18.3 L NH_3? ($R = 0.08206$ L·atm/mol·K)
$$\text{N}_2(g) + 3 \text{ H}_2(g) \rightarrow 2 \text{ NH}_3(g)$$

    a) 1.21 g   b) 1.36 g   c) 1.49 g   d) 1.82 g   e) 2.73 g

    Answer: e

27. What volume of O_2, measured at 82.1 °C and 713 mm Hg, will be produced by the decomposition of 5.71 g $KClO_3$? ($R = 0.08206$ L·atm/mol·K)
$$2 \text{ KClO}_3(s) \rightarrow 2 \text{ KCl}(s) + 3 \text{ O}_2(g)$$

    a) 0.335 L   b) 1.44 L   c) 2.17 L   d) 2.24 L   e) 5.55 L

    Answer: c

28. Propane, C_3H_8, reacts with excess oxygen to produce carbon dioxide gas and water. What volume of CO_2, measured at 27.2 °C and 0.918 atm is produced from the reaction of 10.1 g C_3H_8 with excess oxygen? ($R = 0.08206$ L·atm/mol·K)

a) 0.557 L   b) 0.687 L   c) 1.67 L   d) 6.15 L   e) 18.5 L

Answer: e

29. If 6.46 L of gaseous ethanol reacts with 32.1 L O_2, what is the maximum volume of gaseous water produced? Assume that the temperature of the reactants and products is 425 °C and the pressure remains constant at 1.00 atm.
$$CH_3CH_2OH(g) + 3\,O_2(g) \rightarrow 2\,CO_2(g) + 3\,H_2O(g)$$

a) 6.46 L   b) 19.4 L   c) 25.6 L   d) 32.1 L   e) 38.6 L

Answer: b

30. Ammonia gas is synthesized according to the balanced equation below.
$$N_2(g) + 3\,H_2(g) \rightarrow 2\,NH_3(g)$$
If 1.56 L N_2 react with 4.32 L H_2, what is the theoretical yield (in liters) of NH_3? Assume that the volumes of reactants and products are measured at the same temperature and pressure.

a) 2.88 L   b) 3.12 L   c) 4.32 L   d) 4.68 L   e) 5.88 L

Answer: a

31. Hydrochloric acid reacts with aluminum to produce hydrogen gas according to the reaction below.
$$6\,HCl(aq) + 2\,Al(s) \rightarrow 2\,AlCl_3(aq) + 3\,H_2(g)$$
If 650.0 mL of 2.00 M HCl is combined with 9.92 g Al, what volume of hydrogen gas can be produced? Assume the temperature and pressure of the gas are 25 °C and 0.988 atm, respectively. ($R = 0.08206$ L·atm/mol·K)

a) 1.15 L   b) 9.10 L   c) 13.7 L   d) 16.1 L   e) 32.2 L

Answer: c

32. The empirical formula of a certain hydrocarbon is CH_2. When 0.131 mole of the hydrocarbon is completely combusted with excess oxygen, 16.1 L CO_2 gas is produced at 27 °C and 1.00 atm. What is the molecular formula of the hydrocarbon? ($R = 0.08206$ L·atm/mol·K)

a) C_2H_2 b) C_2H_4 c) C_3H_6 d) C_5H_{10} e) C_6H_{12}

Answer: d

33. An unknown gaseous hydrocarbon contains 85.63% C. Its density is 1.278 g/L at 0.465 atm and 373 K. What is the molecular formula of the gas? ($R = 0.08206$ L·atm/mol·K)

a) C_2H_4 b) C_3H_6 c) C_4H_8 d) C_5H_{10} e) C_6H_{12}

Answer: e

11.5 Gas Mixtures and Partial Pressures

34. Water can be decomposed by electrolysis into hydrogen gas and oxygen gas. What mass of water must decompose to fill a 5.00 L flask to a total pressure of 2.50 atm at 298 K with a mixture hydrogen and oxygen? ($R = 0.08206$ L·atm/mol·K)

$$2 H_2O(\ell) \rightarrow 2 H_2(g) + O_2(g)$$

a) 6.14 g b) 9.21 g c) 13.8 g d) 23.5 g e) 35.3 g

Answer: a

35. Carbon monoxide reacts with oxygen to form carbon dioxide.

$$2 CO(g) + O_2(g) \rightarrow 2 CO_2(g)$$

In a 1.00 L flask, 3.40 atm of CO reacts with 2.60 atm of O_2. Assuming that the temperature remains constant, what is the final pressure in the flask?

a) 0.80 atm b) 2.60 atm c) 3.40 atm d) 4.30 atm e) 6.00 atm

Answer: d

36. A mixture of He and CO_2 is placed in a 7.50 L flask at 22 °C. The partial pressure of the He is 2.2 atm and the partial pressure of the CO_2 is 1.9 atm. What is the mole fraction of He?

a) 0.136 b) 0.463 c) 0.537 d) 0.682 e) 0.864

Answer: c

37. A 12.0 L flask at 308 K contains a mixture of Ar and N_2 with a total pressure of 1.040 atm. If the mole fraction of Ar is 0.755, what is the mass percent of Ar?

a) 35.8% b) 64.2% c) 72.6% d) 75.5% e) 81.5%

Answer: e

<u>11.6 The Kinetic-Molecular Theory of Gases</u>

38. Which of the following statements are postulates of the kinetic-molecular theory of gases?
    1. Gas particles are in constant, random motion.
    2. The distance between gas particles is large in comparison to their size.
    3. The kinetic energy of gas particles is inversely proportional to the kelvin temperature.

    a) 1 only   b) 2 only   c) 3 only   d) 1 and 2   e) 1, 2, and 3

    Answer: d

39. What is the root-mean-square velocity of ammonia molecules (NH_3) at 45 °C?

    a) 21.6 m/s   b) 67.8 m/s   c) 257 m/s   d) 466 m/s   e) 682 m/s

    Answer: e

40. Place the following gases in order of increasing average velocity at 300 K: CO, Ne, O_2, and N_2O.

    a) $CO = Ne = O_2 = N_2O$ b) $Ne < CO < O_2 < N_2O$ c) $CO < O_2 < N_2O < Ne$
    d) $O_2 < N_2O < Ne < CO$ e) $N_2O < O_2 < CO < Ne$

    Answer: b

41. The average velocity of a gas molecule is

    a) inversely proportional square root of its mass.
    b) inversely proportional to the gas constant, *R*.
    c) directly proportional to the square of its temperature in K.
    d) inversely proportional to its kinetic energy.
    e) directly proportional to the square of its temperature in °C.

    Answer: a

<u>11.7 Effusion and Diffusion</u>

42. Oxygen gas, O_2, effuses through a barrier at a rate of 0.183 mL/minute. If an unknown gas effuses through the same barrier at a rate of 0.259 mL/minute, what is the molar mass of the gas?

    a) 16.0 g/mol   b) 20.8 g/mol   c) 28.0 g/mol   d) 44.01 g/mol   e) 64.0 g/mol

    Answer: a

43. How long will it take 10.0 mL of H_2 gas to effuse through a porous barrier if it has been observed that 166 minutes are required for 10.0 mL of CH_4 gas to effuse through the same barrier?

    a) 2.62 min.   b) 20.9 min.   c) 58.9 min.   d) 468 min.   e) 1.05×10^4 min.

    Answer: c

44. If a gas effuses 2.165 times faster than Xe, what is its molar mass?

    a. 12.94 g/mol   b. 28.01 g/mol   c. 32.00 g/mol   d. 60.65 g/mol   e. 284.3 g/mol

    Answer: b

11.9 Nonideal Behavior: Real Gases

45. Non-ideal behavior for a gas is most likely to be observed under conditions of

    a) high temperature and low pressure.   b) high temperature and high pressure.
    c) low temperature and low pressure.   d) standard temperature and pressure.
    e) low temperature and high pressure.

    Answer: e

46. Which of the following statements concerning real gases is/are CORRECT?
    1. Real gases are always liquids or solids at temperatures below 273.15 K.
    2. The pressure of a real gas is higher than predicted by the ideal gas law.
    3. The molecules in a real gas are attracted to each other.

    a) 1 only   b) 2 only   c) 3 only   d) 2 and 3   e) 1, 2 and 3

    Answer: c

Short Answer Questions

47. A pressure of 1.00 atm has a metric equivalent of 1.01×10^5 _________.

    Answer: Pa (or pascals)

48. The rate of effusion of a gas is inversely proportional to the square root of its _________.

    Answer: mass (or molar mass)

49. Jacques Charles discovered that at constant pressure and number of molecules, the volume of a gas is proportional to its _________.

    Answer: temperature

50. A statement of Avogadro's hypothesis is that equal volumes of gases under the same conditions of pressure and temperature will contain equal numbers of __________.

Answer: particles (or molecules, or moles)

51. The gas constant, R, has a value of 0.08206. The units of the constant are __________.

Answer: L·atm/mol·K

52. The ideal gas law can be modified to correct for the errors arising from nonideality. The modified equation is known as the __________ equation of state for a real gas.

Answer: van der Waals

53. The pressure exerted by gas molecules inside a container is due to gas molecules colliding with the walls of the container. Why does one mole of oxygen molecules exert the same pressure inside a flask as one mole of hydrogen molecules, although the mass of an oxygen molecule is sixteen times greater than the mass of a hydrogen molecule?

Answer: At the same temperature, hydrogen molecules have a greater average velocity than oxygen molecules. The force applied to the container walls due to a collision is both a function of the mass and the velocity of the colliding molecule. In addition, hydrogen molecules collide more frequently with the walls of the container, due to their higher velocity. The combination of hydrogen's higher collision frequency and greater velocity compensates for its lower mass.

54. What are two ways in which real gases differ from ideal gases?

Answer: Gas molecules occupy a finite volume of space. Gas molecules are attracted to each other through a variety of intermolecular forces.

Chapter 12
INTERMOLECULAR FORCES AND LIQUIDS

CHAPTER OUTLINE

Vapor Pressure, enthalpy of Vaporization, and the Clausius-Clapeyron Equation (p. 575)
	KEY TERMS: Clausius-Clapeyron equation
Boiling Point (p. 576)
	KEY TERMS: normal boiling point
Critical Temperature and Pressure (p. 577)
	KEY TERMS: critical point, critical pressure, super critical fluid
	Table 12.7 Critical temperatures and Pressures for Common Compounds (p. 577)
Surface Tension, Capillary Action, and Viscosity (p. 578)
	KEY TERMS: surface tension, capillary action, adhesive forces, cohesive forces, viscosity
	Case Study The Mystery of the Disappearing Fingerprints (p. 579)

CHAPTER OBJECTIVES

- Describe intermolecular forces and their effects. (12.2-12.3)
- Understand the importance of hydrogen bonding (12.2)
- Understand the properties of liquids (12.4)

LESSON PLAN

AP Standards

II.B.1	Liquids and Solids from the kinetic-molecular viewpoint
II.B.2	Phase Diagrams of one-component systems
II.B.3	Changes of state, including critical points and triple points
III.D.3	Effect of temperature change on rates
III.E.2	Heats of vaporization and fusion

Suggested Time:
Four traditional classes or two blocks.   All sections should be covered for the AP however the sections regarding surface tension, capillary action, and viscosity could be assigned as reading or skipped.

ASSESSMENT

Exercises:
12.1	Hydration Energy (p. 558)
12.2	Hydrogen Bonding (p. 563)
12.3	Intermolecular Forces (p. 568)
12.4	Intermolecular Forces (p. 570)
12.5	Enthalpy of Vaporization (p. 573)
12.6	Vapor Pressure Curves (p. 575)
12.7	Vapor Pressure (p. 575)
12.8	Clausius-Clapeyron Equation (p. 576)
12.9	Viscosity (p. 580)

Answers to Exercises p. A50-A51

CHAPTER 12 Intermolecular Forces and Liquids

Practicing Skills pp. 581-583, Answers IRM pp. 224-226
General Questions pp. 583-584, Answers IRM pp. 226-228
In the Laboratory Questions pp. 584, Answers IRM pp. 228-229
Summary and Conceptual Questions pp. 585-587, Answers IRM pp. 229-232

GLOSSARY

adhesive force A force of attraction between molecules of two different substances

boiling point The temperature at which the vapor pressure of a liquid is equal to the external pressure on the liquid

capillary action Water level rises in a tube because the adhesive forces between water and glass and the cohesive forces within the water are balanced

Clausius-Clapeyron equation An equation that relates the natural logarithm of the vapor pressure with the reciprocal Kelvin temperature and the enthalpy of vaporization

cohesive force A force of attraction between molecules of the same substances

condensation The movement of molecules from the gas to the liquid phase

critical point The upper end of the curve of vapor pressure versus temperature

critical pressure The pressure at the critical point

dipole-dipole force The electrostatic force between two neutral molecules that have permanent dipole moments

dipole-induced dipole force The electrostatic force between two neutral molecules, one having a permanent dipole and the other having an induced dipole

dynamic equilibrium A reaction in which the forward and reverse processes are occurring

enthalpy of solvation The enthalpy change associated with the binding of solvent molecules to ions or molecules in solution

enthaply of hydration The enthalpy change associated with the hydration of ions or molecules in water

equilibrium vapor pressure The pressure of the vapor of a substance at equilibrium in contact with its liquid or solid phase in a sealed container

hydrogen bonding Attraction between a hydrogen atom and a very electronegative atom to produce an unusually strong dipole-dipole attraction

induced dipole/induced dipole force The electrostatic force between two neutral molecules, both having induced dipole forces

ion-dipole force The electrostatic force between an ion and a neutral molecule that has a permanent dipole moment

London dispersion force The only intermolecular forces that allow nonpolar molecules to interact

molar enthalpy of vaporization ($\Delta_{vap}H^{o}$) The energy required to convert one mole of a substance from a liquid to a gas

normal boiling point The boiling point when the external pressure is 1 atm

polarizability The extent to which the electron cloud of an atom or molecule can be distorted by an external electric charge

polarization The process by which the electron cloud of an atom or molecule is distorted

super critical fluid A substance at or above the critical temperature and pressure

surface tension The energy required to disrupt the surface of a liquid

vaporization Conversion of a liquid to a gas

vapor pressure The pressure of a vapor of a substance in contact with its liquid or solid phase in a sealed container

viscosity The resistance of a liquid to flow

volatility The tendency of the molecules of a substance to escape from the liquid phase into the gas phase

MEDIA MENU

The Media Menu section of this guide lists the interactive exercises, tutorials, active figures (animated, interactive figures) and additional resources available online for Chemistry Now, a unique web-based, assessment-centered personalized learning system for chemistry students. Go to www.cengagelearning.com/login

Exercises
- Screen 12.5   Intermolecular forces

Tutorials
- Screen 12.3   The importance of intermolecular forces
- Screen 12.6   Description of hydrogen bonding

CHAPTER 12 Intermolecular Forces and Liquids

- Screen 12.5   Intermolecular forces
- Screen 12.9   Three tutorials using the Clausius-Clapeyron equation

Active Figures
- 12.2   Ion-dipole interactions
- 12.6   Boiling points of some simple hydrocarbons
- 12.9   Temperature dependence of the densities of ice and water
- 12.16   Vapor pressure
- 12.17   Vapor pressure curves for diethyl ether, ethanol and water

Additional Resources
- Screen 12.2   Gases, liquids, and solids at the molecular level animation
- Screen 12.4   Ion-dipole forces animation
- Screen 12.4   Dipole-dipole forces animation
- Screen 12.7   Transformation of ice to water animation
- Screen 12.7   Table of unusual properties of water
- Screen 12.5   Induced dipole forces animation
- Screen 12.8   Vaporization process animation
- Screen 12.8   Table of enthalpy of vaporization values
- Screen 12.9   Equilibrium vapor pressure animation
- Screen 12.9   Vapor pressure curves simulation
- Screen 12.10   Boiling of water in a partial vacuum video
- Screen 12.10   Relationship of molecular composition and structure and boiling point simulations
- Podcast Module 17   Intermolecular forces involving nonpolar molecules

DIAGNOSTIC QUIZ QUESTIONS

Goal #1 Describe intermolecular forces and their effects.

1. Which of the following statements concerning the attraction of ions to polar molecules is/are CORRECT?
    1. The energy of attraction between an ion and a polar molecule is directly proportional to the square of the distance between the center of the ion and the oppositely charged pole of the dipole.
    2. The higher the ion charge, the greater the attraction between the ion and a polar molecule.
    3. The greater the magnitude of the dipole, the greater the attraction between the ion and a polar molecule.

    a) 1 only   b) 2 only   c) 3 only   d) 2 and 3   e) 1, 2, and 3

Answer: d

2. Arrange PH_3, AsH_3, and SbH_3 in order from lowest to highest boiling point.

a) $SbH_3 < AsH_3 < PH_3$ b) $PH_3 < AsH_3 < SbH_3$ c) $PH_3 < SbH_3 < AsH_3$
d) $AsH_3 < SbH_3 < PH_3$ e) $SbH_3 < PH_3 < AsH_3$

Answer: b

3. Which of the following statements concerning induced dipole/induced dipole forces is/are CORRECT?
    1. In general, induced dipole/induced dipole interactions increase as the size of a molecule increases.
    2. Induced dipole/induced dipole forces are the attractive forces in molecular solids consisting of nonpolar molecules.
    3. Induced dipole/induced dipole forces exist in both polar and nonpolar molecular solids.

a) 1 only   b) 2 only   c) 3 only   d) 1 and 2   e) 1, 2, and 3

Answer: e

4. Which intermolecular forces is/are present in liquid SO_2?
    1. induced dipole/induced dipole
    2. dipole-dipole
    3. hydrogen bonding

a) 1 only   b) 2 only   c) 3 only   d) 1 and 2   e) 1 and 3

Answer: d

Goal #2 Understand the importance of hydrogen bonding

5. Which of the following molecules is expected to form hydrogen bonds in the pure liquid or solid phase: ethylene glycol ($HOCH_2CH_2OH$), formaldehyde (H_2CO), and dimethyl ether (CH_3OCH_3)?

a) ethylene glycol only   b) formaldehyde only
c) dimethyl ether only   d) ethylene glycol and formaldehyde
e) ethylene glycol, formaldehyde and dimethyl ether

Answer: a

6. Which intermolecular force or bond is responsible for the density of $H_2O(s)$ being less than that of $H_2O(\ell)$?

a) London dispersion forces   b) hydrogen bonding   c) ionic bonding
d) covalent bonding   e) dipole/induced dipole forces

Answer: b

CHAPTER 12 Intermolecular Forces and Liquids

Goal #3 Understand the properties of liquids (12.4)

7. Equilibrium is established between a liquid and its vapor when

    a) the rate of evaporation equals the rate of condensation.
    b) equal masses exist in the liquid and gas phases.
    c) equal concentrations (in molarity) exist in the liquid and gas phases.
    d) all the liquid has evaporated.
    e) the liquid ceases to evaporate and the gas ceases to condense.

Answer: a

8. Which of the following liquids has the smallest enthalpy of vaporization?

    a) C_5H_{12} b) C_4H_{10} c) C_3H_8 d) C_2H_6 e) C_6H_{14}

Answer: d

9. Which of the following phase transitions is/are exothermic?
    1. liquid water evaporating into the gas phase.
    2. the sublimation of solid carbon dioxide into the gas phase.
    3. gaseous butane condensing into a liquid at its boiling point of -0.5 °C.

    a) 1 only   b) 2 only   c) 3 only   d) 1 and 2   e) 1, 2, and 3

Answer: c

10. Normal boiling point is defined as

    a) the temperature at which the vapor pressure of a liquid equals 1 atm.
    b) the pressure at which any liquid boils at 373.15 K.
    c) the temperature at which water always boils.
    d) the pressure of a gas when its temperature reaches 273.15 K.
    e) the temperature at which the enthalpy of vaporization equals 100.0 kJ/mol.

Answer: a

TESTBANK

<u>12.2 Intermolecular Forces Involving Polar Molecules</u>

1. Which of the following statements concerning the attraction of ions to polar molecules is/are CORRECT?
    1. The energy of attraction between an ion and a polar molecule is directly proportional to the square of the distance between the center of the ion and the oppositely charged pole of the dipole.
    2. The higher the ion charge, the greater the attraction between the ion and a polar molecule.
    3. The greater the magnitude of the dipole, the greater the attraction between the ion and a polar molecule.

    a) 1 only   b) 2 only   c) 3 only   d) 2 and 3   e) 1, 2, and 3

    Answer: d

2. Place the following cations in order from lowest to highest hydration enthalpy.

    a) $Na^+ < Rb^+ < H^+$ b) $Rb^+ < H^+ < Na^+$ c) $Rb^+ < Na^+ < H^+$
    d) $H^+ < Na^+ < Rb^+$ e) $H^+ < Rb^+ < Na^+$

    Answer: c

3. Place the following cations in order from the lowest to the highest hydration enthalpy.

    a) $K^+ < Ca^{2+} < Mg^{2+}$ b) $Mg^{2+} < Ca^{2+} < K^+$ c) $Ca^{2+} < K^+ < Mg^{2+}$
    d) $Ca^{2+} < Mg^{2+} < K^+$ e) $Mg^{2+} < K^+ < Ca^{2+}$

    Answer: a

4. Which one of the following molecules will exhibit dipole-dipole intermolecular forces as a pure liquid?

    a) C_2H_2 b) SO_3 c) CO_2 d) F_2 e) NO_2

    Answer: e

5. Hydrogen bonding is present in all of the following molecular solids EXCEPT ________ .

    a) H_2SO_4 b) NH_3 c) PH_3 d) HF e) H_2O_2

    Answer: c

6. As pure molecular solids, which of the following exhibits dipole-dipole intermolecular forces: PBr_3, SO_2, I_2, and CO_2?

a) PBr_3 only   b) PBr_3 and SO_2 c) SO_2 and CO_2
d) I_2 and CO_2 e) PBr_3, SO_2, and I_2

Answer: b

7. As pure molecular solids, which of the following exhibits dipole-dipole intermolecular forces: HCl, Cl_2, SCl_2, and CCl_4?

a) HCl only   b) HCl and Cl_2 c) HCl and SCl_2
d) Cl_2 and CCl_4 e) SCl_2 and CCl_4

Answer: c

8. Which of the following molecules is expected to form hydrogen bonds in the pure liquid or solid phase: ethylene glycol ($HOCH_2CH_2OH$), formaldehyde (H_2CO), and dimethyl ether (CH_3OCH_3)?

a) ethylene glycol only   b) formaldehyde only
c) dimethyl ether only   d) ethylene glycol and formaldehyde
e) ethylene glycol, formaldehyde and dimethyl ether

Answer: a

9. Arrange PH_3, AsH_3, and SbH_3 in order from lowest to highest boiling point.

a) $SbH_3 < AsH_3 < PH_3$ b) $PH_3 < AsH_3 < SbH_3$ c) $PH_3 < SbH_3 < AsH_3$
d) $AsH_3 < SbH_3 < PH_3$ e) $SbH_3 < PH_3 < AsH_3$

Answer: b

10. Arrange H_2O, H_2S, and H_2Se in order from lowest to highest boiling point.

a) $H_2O < H_2S < H_2Se$ b) $H_2O < H_2Se < H_2S$ c) $H_2Se < H_2O < H_2S$
d) $H_2Se < H_2S < H_2O$ e) $H_2S < H_2Se < H_2O$

Answer: e

11. Which intermolecular force or bond is responsible for the density of $H_2O(s)$ being less than that of $H_2O(\ell)$?

a) London dispersion forces   b) hydrogen bonding   c) ionic bonding
d) covalent bonding   e) dipole/induced dipole forces

Answer: b

12. Which of the following statements concerning induced dipole/induced dipole forces is/are CORRECT?
    1. In general, induced dipole/induced dipole interactions increase as the size of a molecule increases.
    2. Induced dipole/induced dipole forces are the attractive forces in molecular solids consisting of nonpolar molecules.
    3. Induced dipole/induced dipole forces exist in both polar and nonpolar molecular solids.

    a) 1 only　　　　b) 2 only　　　　c) 3 only　　　　d) 1 and 2　　　　e) 1, 2, and 3

    Answer: e

13. As pure molecular solids, which of the following exhibit only induced dipole/induced dipole forces: O_2, CH_2Cl_2, and SO_2?

    a) O_2 only　　　　　　　b) CH_2Cl_2 only　　　　　　　c) SO_2 only
    d) O_2 and CH_2Cl_2　　　　e) O_2 and SO_2

    Answer: a

14. Which of the following nonpolar molecules has the highest boiling point?

    a) F_2　　　　b) CH_4　　　　c) N_2　　　　d) O_2　　　　e) CO_2

    Answer: e

15. What intermolecular force or bond is primarily responsible for the solubility of carbon dioxide (CO_2) in water?

    a) dipole/dipole force　　　　　　　b) hydrogen bonding
    c) dipole/induced dipole force　　　　d) hydrogen bonding-dipole force
    e) ion-induced dipole force

    Answer: c

16. Which of the following statements concerning intermolecular forces is/are CORRECT?
    1. Dipole-dipole attractions occur in all molecules that contain polar bonds, regardless of whether the molecule has a dipole.
    2. Hydrogen bonding occurs in all molecules containing both oxygen and hydrogen atoms.
    3. Induced dipole/induced dipole forces exist in all molecular solids.

    a) 1 only　　　　b) 2 only　　　　c) 3 only　　　　d) 1 and 2　　　　e) 1, 2, and 3

    Answer: c

CHAPTER 12 Intermolecular Forces and Liquids

17. The boiling points of some group 7A hydrides are tabulated below.

gas	b.p. (°C)
HF	+19.7
HCl	-84.8
HBr	-66.4

Which intermolecular force or bond is responsible for the high boiling point of HF relative to HCl and HBr?

a) hydrogen bonding   b) dipole/induced dipole force
c) induced dipole/induced dipole force   d) covalent bonding
e) dipole-dipole force

Answer: a

18. Which one of the following molecules has the lowest boiling point?

a) CH_4 b) $CHCl_3$ c) CH_2Cl_2 d) CH_3Cl e) CCl_4

Answer: a

19. What intermolecular force or bond is primarily responsible for the solubility of CH_3OH in water?

a) ion-dipole force   b) dipole-dipole force   c) ionic bonding
d) covalent bonding   e) hydrogen bonding

Answer: e

20. Which intermolecular forces is/are present in liquid SO_2?
     1. induced dipole/induced dipole
     2. dipole-dipole
     3. hydrogen bonding

a) 1 only   b) 2 only   c) 3 only   d) 1 and 2   e) 1 and 3

Answer: d

21. List all the intermolecular forces present in pure acetone.

a) hydrogen bonding only
b) dipole-dipole force only
c) dipole-dipole force and induced dipole/induced dipole force
d) hydrogen bonding and induced dipole/induced dipole force
e) hydrogen bonding, dipole-dipole force, and induced dipole/induced dipole force

Answer: c

22. For which of the following pure solids is it necessary to break covalent bonds to make a liquid or gas?

a) $CO_2(s)$ b) $C(s)$ c) $C_{60}(s)$ d) $CO(s)$ e) $CCl_4(s)$

Answer: b

23. Arrange Cl_2, ICl, and Br_2 in order from lowest to highest boiling point.

a) $Cl_2 < Br_2 < ICl$ b) $Cl_2 < ICl < Br_2$ c) $ICl < Cl_2 < Br_2$
d) $Br_2 < Cl_2 < ICl$ e) $Br_2 < ICl < Cl_2$

Answer: a

12.4 Properties of Liquids

24. Equilibrium is established between a liquid and its vapor when

a) the rate of evaporation equals the rate of condensation.
b) equal masses exist in the liquid and gas phases.
c) equal concentrations (in molarity) exist in the liquid and gas phases.
d) all the liquid has evaporated.
e) the liquid ceases to evaporate and the gas ceases to condense.

Answer: a

25. Which of the following liquids has the smallest enthalpy of vaporization?

a) C_5H_{12} b) C_4H_{10} c) C_3H_8 d) C_2H_6 e) C_6H_{14}

Answer: d

26. Ammonia, NH_3, is used as a refrigerant. At its boiling point of -33 °C, the enthalpy of vaporization of ammonia is 23.3 kJ/mol. How much heat is released when 18.5 g of ammonia is condensed at -33 °C?

a) -21.5 kJ b) -25.3 kJ c) -431 kJ d) -708 kJ e) -835 kJ

Answer: b

27. At its boiling point of -34.0 °C, 6.33 kJ of heat is required to vaporize 22.0 g of chlorine (Cl_2). What is the molar enthalpy of vaporization of chlorine?

a) 0.0893 kJ/mol b) 0.286 kJ/mol c) 1.96 kJ/mol d) 20.4 kJ/mol e) 139 kJ/mol

Answer: d

28. At 65.0 °C, water has an equilibrium vapor pressure of 187.5 mm Hg. If 4.78 g H_2O is sealed in an evacuated 5.00 L flask and heated to 65.0 °C, what mass of H_2O will be found in the gas phase when liquid-vapor equilibrium is established? Assume any liquid remaining in the flask has a negligible volume. (R = 0.08206 L·atm/mol·K, 1 atm = 760 mm Hg)

    a) 0.0444 g   b) 0.801 g   c) 1.87 g   d) 4.167 g   e) 4.78 g

Answer: b

29. If 3.01 g of water is sealed in an evacuated 2.00 L flask and heated to the normal boiling point of 373 K, what is the pressure in the flask? (R = 0.08206 L·atm/mol·K)

    a) 0.391 atm   b) 0.678 atm   c) 1.00 atm   d) 2.56 atm   e) 46.1 atm

Answer: c

30. A linear relationship exists between the natural logarithm of the vapor pressure of a gas and the reciprocal of its temperature (in kelvin). What is the slope of the line?

    a) $-\dfrac{RT}{\Delta_{vap}H}$ b) $-R\times\Delta_{vap}H$ c) T d) $-\Delta_{vap}H$ e) $-\dfrac{\Delta_{vap}H}{R}$

Answer: e

31. Ethanol has an enthalpy of vaporization of 42.3 kJ/mol. The compound has a vapor pressure of 1.00 atm at 78.3 °C. At what temperature is the vapor pressure equal to 0.500 atm? (R = 8.314 J/K·mol)

    a) -96.2 °C   b) 7.19 °C   c) 36.0 °C   d) 62.2 °C   e) 281 °C

Answer: d

32. Sulfur dioxide has a vapor pressure of 462.7 mm Hg at -21.0 °C and a vapor pressure of 140.5 mm Hg at -44.0 °C. What is the enthalpy of vaporization of sulfur dioxide? (R = 8.314 J/K·mol)

    a) 0.398 kJ/mol   b) 6.33 kJ/mol   c) 14.0 kJ/mol   d) 24.9 kJ/mol   e) 39.8 kJ/mol

Answer: d

33. Mount Everest rises to a height of 8.850×10^3 m above sea level. At this height, the atmospheric pressure is 231 mm Hg. At what temperature (in °C) does water boil at the summit of Mount Everest? The vapor pressure of water at 373 K is 760.0 mm Hg. ($\Delta_{vap}H°$ for H_2O = 40.7 kJ/mole and R = 8.314 J/K·mol)

    a) 4.07 °C   b) 69.0 °C   c) 72 °C   d) 87 °C   e) 364 °C

Answer: b

34. Freon-113, $C_2Cl_3F_3$, has an enthalpy of vaporization of 27.0 kJ/mol and a normal boiling point of 48 °C. What is the vapor pressure (in atm) of Freon-113 at 5 °C? (R = 8.314 J/K·mol)

a) 0.159 atm   b) 0.209 atm   c) 0.566 atm   d)1.00 atm   e) 1.56 atm

Answer: b

35. Which of the following statements is/are CORRECT?
    1. If the intermolecular forces in a liquid decrease, the normal boiling point of the liquid decreases.
    2. If the surface area of a liquid increases, the equilibrium vapor pressure of the liquid increases.
    3. If the temperature of a liquid decreases, the equilibrium vapor pressure increases.

a) 1 only   b) 2 only   c) 3 only   d) 1 and 2   e) 1, 2, and 3

Answer: a

36. Which of the following properties of water can be attributed to hydrogen bonding?
    1. high melting point
    2. high heat of vaporization
    3. low vapor pressure
    4. high surface tension

a) 1 and 3   b) 2 and 3   c) 2, 3, and 4   d) 1, 3, and 4   e) 1, 2, 3, and 4

Answer: e

37. Which of the following phase transitions is/are exothermic?
    1. liquid water evaporating into the gas phase.
    2. the sublimation of solid carbon dioxide into the gas phase.
    3. gaseous butane condensing into a liquid at its boiling point of -0.5 °C.

a) 1 only   b) 2 only   c) 3 only   d) 1 and 2   e) 1, 2, and 3

Answer: c

38. Normal boiling point is defined as

a) the temperature at which the vapor pressure of a liquid equals 1 atm.
b) the pressure at which any liquid boils at 373.15 K.
c) the temperature at which water always boils.
d) the pressure of a gas when its temperature reaches 273.15 K.
e) the temperature at which the enthalpy of vaporization equals 100.0 kJ/mol.

Answer: a

CHAPTER 12 Intermolecular Forces and Liquids

39. Which of the following statements concerning a supercritical fluid (SCF) are correct?
    1.   An SCF has a density similar to that of a liquid.
    2.   An SCF has a viscosity similar to that of a liquid.
    3.   An SCF fills its entire container, much like a gas.

   a) 1 only   b) 2 only   c) 3 only   d) 1 and 3   e) 1, 2, and 3

   Answer: d

40. The toughness of the skin of a liquid is a measure of its _________.

   a) meniscus quality   b) adhesive forces   c) dispersion forces
   d) viscosity   e) surface tension

   Answer: e

41. Which of the following are valid reasons why vegetable oil has a greater viscosity than
    diethyl ether, CH_3OCH_3?
    1.   Unlike diethyl ether, oil molecules are not held together by hydrogen bonds.
    2.   Oil molecules have long chains that become entangled.
    3.   Intermolecular forces are greater for the larger oil molecules.

   a) 1 only   b) 2 only   c) 3 only   d) 1 and 3   e) 2 and 3

   Answer: e

Short Answer Questions

42. A cup of oil takes longer to pour from a glass than a cup of water. A liquid's resistance to
    flow is referred to as its _________.

   Answer: viscosity

43. When a glass tube with a small diameter is placed in water, the water rises in the tube. This is
    known as _________ action.

   Answer: capillary

44. _________ is a measure of the degree to which the electron cloud surrounding an atom or
    molecule can be distorted in an electric field.

   Answer: Polarizability

45. Above its critical temperature and pressure, a substance becomes a _________ fluid.

   Answer: supercritical

46. Which ion, K^+ or Ca^{2+}, is expected to have the more negative enthalpy of hydration? Why?

Answer: Ca^{2+} will have the more negative enthalpy of hydration. A calcium ion is smaller than a potassium ion. The smaller ion will bond more closely to water's dipole, allowing for a stronger ion-dipole interaction. In addition, the greater the charge on the ion, the greater the ion-dipole interaction.

47. Two important allotropes of phosphorus are white phosphorus and red phosphorus. White phosphorus, P_4, has a melting point of 44.1 °C and spontaneously reacts with oxygen. Red phosphorus, a network solid, melts at 280 °C and is stable in air. Use your knowledge of intermolecular and intramolecular bonding to explain why these two forms of the same element have such different properties.

Answer: White phosphorus is a molecular solid composed of tetrahedral P_4 molecules. The P_4 molecules are held together by London dispersion forces. Red phosphorus is a network solid; all phosphorus atoms are connected through covalent bonds. It is easier to disrupt the weaker London dispersion forces of P_4 than it is to break the covalent bonds of red phosphorus.

Chapter 13
THE CHEMISTRY OF SOLIDS

CHAPTER OUTLINE

CHAPTER OBJECTIVES

- Understand cubic unit cells (13.1)
- Relate unit cells for ionic compounds to formulas (13.2)
- Describe the properties of solids (13.4-13.5)
- Understand the nature of phase diagrams (13.5)

LESSON PLAN

AP Standards

II.B.1	Liquids and Solids from the kinetic-molecular viewpoint
II.B.2	Phase Diagrams of one-component systems
II.B.3	Changes of state, including critical points and triple points
III.D.3	Effect of temperature change on rates
III.E.2	Heats of vaporization and fusion

Suggested Time:

Three traditional classes or 2 blocks.   All sections should be covered for the AP however the sections regarding unit cell calculations may be minimized.

ASSESSMENT

Exercises:
- 13.1 Determining an Atom Radius from Lattice Dimensions (p 594)
- 13.2 The Structure of Solid Iron (p. 594)
- 13.3 Structure and Formula (p. 599)
- 13.4 Density from Cell Dimensions (p. 599)
- 13.5 Using Lattice Enthalpies (p. 602)

Answers to Exercises p. A51

Practicing Skills pp. 610-612, Answers IRM pp. 234-237

General Questions pp. 613-614, Answers IRM pp. 237-241

In the Laboratory Questions pp. 614-615, Answers IRM pp. 241-242

Summary and Conceptual Questions pp. 615, Answers IRM pp. 242-243

GLOSSARY

body-centered cubic Unit cell with a net of two atoms within the unit cell

Born-Haber cycle Application of Hess's law which enables calculations of lattice enthalpy

crystal lattice The arrangement of unit cells in the three dimensional structure of a solid

cubic unit cell Simplest of the seven crystal lattices with edges of equal length that meet at 90° angles

CHAPTER 13 The Chemistry of Solids

enthalpy of sublimation($\Delta_{sublimation} H$) The energy required to convert one mole of a substance from a solid to a gas

face-centered cubic Unit cell with a net of four atoms within the unit cell

graphene Single sheet of six-member carbon rings

lattice energy ($\Delta_{lattice} U$) The energy of formation of one mole of a solid crystalline ionic compound when ions in the gas phase combine

lattice enthalpy($\Delta_{lattice} H$) The enthalpy of formation of one mole of a solid crystalline ionic compound when ions in the gas phase combine

lattice points The corners of the cube or other geometric object that constitutes the unit cell

molecular solids A solid formed by the condensation of covalently bonded molecules

nanotubes Six-membered carbon rings arranged into sheets that fold themselves into tubes

network solid A solid composed of a network of covalently bonded atoms

octahedral holes The octahedral arrangement of spaces surrounding an ion in a face-centered cubic arrangement

phase diagram A graph showing which phases of a substance exist at various temperatures and pressures

primitive cubic Unit cell with a net of one atom within the unit cell

supercritical fluid A substance at or above the critical temperature and pressure

tetrahedral holes The tetrahedral arrangement of spaces surrounding an ion in a face-centered cubic arrangement

triple point The temperature and pressure at which the solid, liquid, and vapor phases of a substance are in equilibrium

unit cell The smallest repeating unit in a crystal lattice

MEDIA MENU

The Media Menu section of this guide lists the interactive exercises, tutorials, active figures (animated, interactive figures) and additional resources available online for Chemistry Now, a unique web-based, assessment-centered personalized learning system for chemistry students. Go to www.cengagelearning.com/login

Exercises
- Screen 13.5   Molecular solids
- Screen 13.8   Phase diagrams

Active Figures
- 13.18   Phase diagram for water

Additional Resources
- Screen 13.2   Self-study module on crystal lattices
- Screen 13.3   Ionic unit cells animation
- Screen 13.4   Lattice and lattice energy illustration
- Screen 13.6   Self-study module on network solids
- Screen 13.7   Self-study module on silicate materials
- Screen 13.8   Phase diagrams
- Podcast Module 18   Crystal Lattices and Unit Cells

DIAGNOSTIC QUIZ QUESTIONS

Goal # 1 Understand cubic unit cells

1. Which of the following statements concerning the cubic unit cell is/are CORRECT?
    1. For cubic unit cells, three cell symmetries occur: primitive cubic, face-centered cubic, and body-centered cubic.
    2. The cell edges of a cubic unit cell are all equal in length.
    3. The corner angles of a cubic cell are 90°.

    a) 1 only   b) 2 only   c) 3 only   d) 2 and 3   e) 1, 2, and 3

    Answer: e

2. Arrange the three common unit cells in order from least dense to most dense packing.

    a) face-centered cubic < body-centered cubic < primitive cubic
    b) primitive cubic < body-centered cubic < face-centered cubic
    c) primitive cubic < face-centered cubic < body-centered cubic
    d) body-centered cubic < primitive cubic < face-centered cubic
    e) body-centered cubic < face-centered cubic < primitive cubic

    Answer: b

Goal #2 Relate unit cells for ionic compounds to formulas

3. If an ionic compound with the formula MX forms a primitive cubic unit cell with the anions (X^{n-}) at the lattice points, the cations (M^{2n+}) will occupy

a) all of the tetrahedral holes in each unit cell.
b) half of the tetrahedral holes in each unit cell.
c) the cubic hole in the center of the each unit cell.
d) the center of each face in each unit cell.
e) all of the octahedral holes in each unit cell.

Answer: c

4. Which of the following compounds is expected to have the strongest ionic bonds?

a) MgO   b) KBr   c) NaI   d) BaO   e) SrS

Answer: a

5. The lattice energy of KF is -821 kJ/mol. This energy corresponds to which reaction below?

a) $K(s) + \frac{1}{2} F_2(g) \rightarrow KF(s)$ b) $K(g) + F(g) \rightarrow KF(s)$ c) $K(g) + F(g) \rightarrow KF(s)$
d) $K^+(g) + F^-(g) \rightarrow KF(s)$ e) $K^+(aq) + F^-(aq) \rightarrow KF(s)$

Answer: d

Goal #3 Describe the properties of solids

6. Which of the following is NOT a network solid?

a) elemental silicon, $Si(s)$ b) diamond, $C(s)$
c) buckminster fullerene, $C_{60}(s)$ d) silicon dioxide, $SiO_2(s)$
e) aluminum oxide, $Al_2O_3(s)$

Answer: c

7. Which of the following is/are physical properties of amorphous solids?
    1. Amorphous solids have well defined melting points.
    2. At the particulate level, amorphous solids do not have long range order.
    3. Polymeric materials never form amorphous solids.

a) 1 only   b) 2 only   c) 3 only   d) 1 and 3   e) 2 and 3

Answer: b

Goal #4 Understand the nature of phase diagrams

8. The phase diagram for CO_2 has a triple point at -56.6 °C and 5.19 atm, and a critical point at 31.0 °C and 73 atm. The solid and gas phases are in equilibrium at -78.7 °C and 1.00 atm. Which of the following statements regarding CO_2 is/are CORRECT?
    1. Sublimation occurs if the temperature of the solid phase is increased from -79.0 °C to 0.0 °C at a constant pressure of 2.5 atm.
    2. CO_2 is a supercritical fluid at 55 °C and 75 atm.
    3. Only the gas phase exists at a temperature of -56.6 °C and pressure of 1.00 atm.

    a) 1 only   b) 2 only   c) 3 only   d) 1 and 2   e) 1, 2, and 3

    Answer: e

9. Which statement concerning the phase diagram below is INCORRECT?

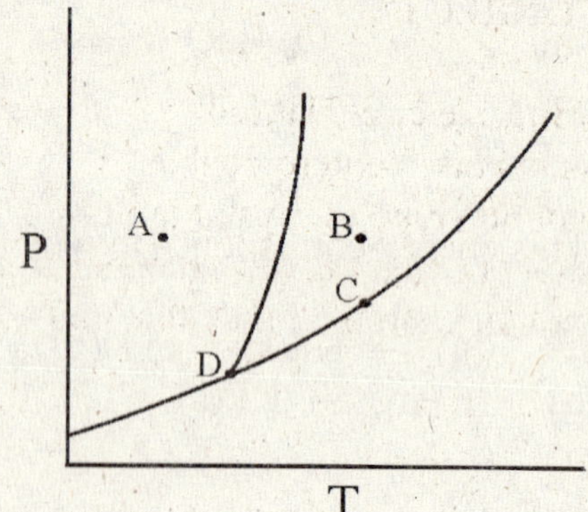

    a) Only the solid phase exists at point A.
    b) At point D, the triple point, all three phases (gas, liquid, and solid) are in equilibrium.
    c) Moving from point A to B results in a phase transition from solid to liquid.
    d) At point C, the solid and liquid phases are in equilibrium.
    e) Only the liquid phase exists at point B.

    Answer: d

TESTBANK

13.1 Crystal Lattices and Unit Cells

1. How many unit cells share an atom which is located at a corner (or lattice point) of a unit cell?

    a) 1   b) 2   c) 4   d) 6   e) 8

    Answer: e

2. An atom located on a face of a unit cell is shared equally between _________ unit cells.

    a) 2   b) 3   c) 4   d) 6   e) 8

Answer: a

3. Which of the following statements concerning the cubic unit cell is/are CORRECT?
    1. For cubic unit cells, three cell symmetries occur: primitive cubic, face-centered cubic, and body-centered cubic.
    2. The cell edges of a cubic unit cell are all equal in length.
    3. The corner angles of a cubic cell are 90°.

    a) 1 only   b) 2 only   c) 3 only   d) 2 and 3   e) 1, 2, and 3

Answer: e

4. Which of the following statements is INCORRECT?

    a) Polonium is the only metal that has a primitive cubic lattice.
    b) The lattice structure of the alkali metals is body-centered cubic.
    c) Many metals can crystallize in more than one type of crystal lattice.
    d) Metals with a face-centered cubic lattice contain a net of two metal atoms per unit cell.
    e) A hexagonal close packed structure is not an example of a cubic unit cell.

Answer: d

5. Which equation represents the number of atoms in a body-centered cubic unit cell?

    a) # atoms $= \dfrac{1}{8}(8)$ b) # atoms $= \dfrac{1}{4}(8)$ c) # atoms $= 1 + \dfrac{1}{8}(8)$

    d) # atoms $= \dfrac{1}{2}(6) + \dfrac{1}{8}(8)$ e) # atoms $= 1 + \dfrac{1}{2}(6) + \dfrac{1}{8}(8)$

Answer: c

6. Which equation represents the number of atoms in a face-centered cubic unit cell?

    a) # atoms $= \dfrac{1}{8}(8)$ b) # atoms $= \dfrac{1}{4}(8)$ c) # atoms $= 1 + \dfrac{1}{8}(8)$

    d) # atoms $= \dfrac{1}{2}(6) + \dfrac{1}{8}(8)$ e) # atoms $= 1 + \dfrac{1}{2}(6) + \dfrac{1}{8}(8)$

Answer: d

7. Arrange the three common unit cells in order from least dense to most dense packing.

a) face-centered cubic < body-centered cubic < primitive cubic
b) primitive cubic < body-centered cubic < face-centered cubic
c) primitive cubic < face-centered cubic < body-centered cubic
d) body-centered cubic < primitive cubic < face-centered cubic
e) body-centered cubic < face-centered cubic < primitive cubic

Answer: b

8. If a metal crystallizes in a face-centered cubic lattice, each metal atom has _________ "nearest neighbors."

a) 3   b) 4   c) 6   d) 8   e) 12

Answer: e

9. If a metal crystallizes in a body-centered cubic lattice, each metal atom has _________ "nearest neighbors."

a) 3   b) 4   c) 6   d) 8   e) 12

Answer: d

10. For a primitive cubic unit cell, what percentage of the space in the cell is occupied by the atoms at the corners of the cell?

a) 47%   b) 52%   c) 68%   d) 74%   e) 87%

Answer: b

11. What is the distance, in atomic radii (r), along any edge of a primitive unit cell?

a) r   b) $2 \times r$ c) $\dfrac{r}{2}$ d) $4 \times r$ e) $\dfrac{4 \times r}{\sqrt{3}}$

Answer: b

12. What is the distance, in atomic radii, across a diagonal face of a face-centered unit cell?

a) $\dfrac{4 \times r}{\sqrt{3}}$ b) $2 \times r$ c) $4 \times r$ d) $2\sqrt{2} \times r$ e) r

Answer: c

CHAPTER 13 The Chemistry of Solids

13. What is the distance, in atomic radii, along any edge of a face-centered unit cell?

   a) $\dfrac{4 \times r}{\sqrt{3}}$ b) $2 \times r$ c) $4 \times r$ d) $2\sqrt{2} \times r$ e) r

   Answer: d

14. What is the distance, in atomic radii, along any edge of a body-centered unit cell?

   a) $\dfrac{4 \times r}{\sqrt{3}}$ b) $2 \times r$ c) $4 \times r$ d) $2\sqrt{2} \times r$ e) r

   Answer: a

15. Barium crystallizes in a body-centered cubic unit cell. If the length of an edge of the unit cell is 0.503 nm, what is the atomic radius (in nm) of a barium atom?

   a) 0.178 nm   b) 0.218 nm   c) 0.237 nm   d) 0.252 nm   e) 0.436 nm

   Answer: b

16. Tungsten crystallizes in a body-centered cubic unit cell. If the radius of a tungsten atom is 0.137 nm, what is the length of an edge of the unit cell?

   a) 0.119 nm   b) 0.194 nm   c) 0.274 nm   d) 0.316 nm   e) 0.548 nm

   Answer: d

17. Polonium crystallizes in a primitive cubic unit cell. If the length of an edge of the unit cell is 336 pm, what is the radius of a polonium atom?

   a) 168 pm   b) 238 pm   c) 336 pm   d) 475 pm   e) 672 pm

   Answer: a

18. Rhodium crystallizes in a face-centered unit cell. If the length of an edge of the unit cell is 0.380 nm, what is the radius of a rhodium atom?

   a) 0.095 nm   b) 0.134 nm   c) 0.190 nm   d) 0.219 nm   e) 0.269 nm

   Answer: b

19. The radius of a lead atom is 175 pm. If lead crystallizes in a face-centered unit cell, what is the length of an edge of the unit cell?

   a) 247 pm   b) 303 pm   c) 351 pm   d) 404 pm   e) 495 pm

   Answer: e

20. Platinum (atomic mass 195.08 g/mol) crystallizes in a face-centered cubic unit cell. In addition, platinum has an atomic radius of 139 pm. What is the density (in g/cm^3) of platinum?

   a) 5.33 g/cm^3 b) 15.7 g/cm^3 c) 18.9 g/cm^3 d) 21.3 g/cm^3 e) 23.5 g/cm^3

   Answer: d

21. Iron (atomic mass 55.85 g/mol) crystallizes in a body-centered cubic unit cell. If the length of an edge of the unit cell is 287 pm, what is the density (in g/cm^3) of iron?

   a) 7.85 g/cm^3 b) 10.4 g/cm^3 c) 11.7 g/cm^3 d) 15.7 g/cm^3 e) 21.3 g/cm^3

   Answer: a

22. Silver has a face-centered cubic cell, and its density is 10.5 g/cm^3. What is the radius (in pm) of a silver atom? (The molar mass of silver is 107.9 g/mol)

   a) 91.0 pm   b) 102 pm   c) 144 pm   d) 257 pm   e) 408 pm

   Answer: c

13.2 Structures and Formulas of Ionic Compounds

23. Cesium bromide crystallizes in a primitive cubic unit cell with bromide ions at the lattice points. The cesium ions occupy cubic holes. How many chloride ions surround each cesium ion in cesium chloride?

   a) 1   b) 2   c) 4   d) 8   e) 12

   Answer: d

24. Potassium iodide crystallizes in a face-centered cubic unit cell with iodide ions occupying the lattice points and potassium ions occupying octahedral holes. How many iodide ions surround each potassium ion in KI?

   a) 6   b) 8   c) 12   d) 16   e) 20

   Answer: a

25. Which of the following statements is/are CORRECT? If an ionic compound with the formula MX forms a face-centered cubic unit cell with the anions (X^{n-}) at the lattice points, the cations (M^{n+}) may occupy
     1.   half of the tetrahedral holes in each unit cell.
     2.   all of the octahedral holes in each unit cell.
     3.   the center of each face in each unit cell.

   a) 1 only   b) 2 only   c) 3 only   d) 1 and 2   e) 1, 2, and 3

   Answer: d

26. If an ionic solid has a face-centered cubic lattice of anions (X^{n-}) and all the octahedral holes are occupied by metal cations (M^{m+}), what is the formula for the compound?

   a) M_2X b) MX   c) MX_2 d) M_2X_3 e) M_3X_2

   Answer: b

27. If an ionic compound with the formula MX_2 forms a face-centered cubic unit cell with the cations (M^{2n+}) at the lattice points, the anions (X^{n-}) will occupy

   a) all of the tetrahedral holes in each unit cell.
   b) half of the tetrahedral holes in each unit cell.
   c) all of the octahedral holes in each unit cell.
   d) the center of each face in each unit cell.
   e) the cubic hole in the center of the each unit cell.

   Answer: a

28. If an ionic compound with the formula MX forms a primitive cubic unit cell with the anions (X^{n-}) at the lattice points, the cations (M^{2n+}) will occupy

   a) all of the tetrahedral holes in each unit cell.
   b) half of the tetrahedral holes in each unit cell.
   c) the cubic hole in the center of the each unit cell.
   d) the center of each face in each unit cell.
   e) all of the octahedral holes in each unit cell.

   Answer: c

29. Calcium sulfide has a face-centered cubic unit cell with calcium ions in octahedral holes. If the radius of sulfide ion is 184 pm and the density of CaS is 2.60 g/cm³, what is the radius of the calcium ion?

   a) 56.5 pm   b) 70.8 pm   c) 101 pm   d) 112 pm   e) 174 pm

   Answer: c

30. Sodium iodide has a face-centered cubic unit cell with sodium ions in octahedral holes. The ionic radii of sodium ions and iodide ions are 98 pm and 220 pm, respectively. What is the density of NaI?

a) 2.6 g/cm^3 b) 3.9 g/cm^3 c) 4.1 g/cm^3 d) 4.8 g/cm^3 e) 6.2 g/cm^3

Answer: b

13.3 Bonding in Ionic Compounds: Lattice Energy

31. Which of the following compounds is expected to have the strongest ionic bonds?

a) RbI   b) RbF   c) NaI   d) CsBr   e) LiF

Answer: e

32. Which of the following compounds is expected to have the strongest ionic bonds?

a) MgO   b) KBr   c) NaI   d) BaO   e) SrS

Answer: a

33. The lattice energy of KF is -821 kJ/mol. This energy corresponds to which reaction below?

a) $K(s) + \frac{1}{2} F_2(g) \rightarrow KF(s)$ b) $K(g) + F(g) \rightarrow KF(s)$ c) $K(g) + F(g) \rightarrow KF(s)$
d) $K^+(g) + F^-(g) \rightarrow KF(s)$ e) $K^+(aq) + F^-(aq) \rightarrow KF(s)$

Answer: d

34. Lattice enthalpy may be calculated using the thermodynamic relationship known as

a) the Clausius-Clapeyron equation.   b) the Born-Haber cycle.
c) the dynamic equilibrium expression.   d) Avogadro's hypothesis.
e) cubic cell enthalpy of formation equation.

Answer: b

35. Calculate the lattice energy of NaBr(s), given the following thermochemical equations, where ΔIE and ΔEA are ionization energy and electron affinity, respectively.

$$\begin{array}{ll}
Na(s) \rightarrow Na(g) & \Delta_f H^\circ = +107 \text{ kJ} \\
Na(g) \rightarrow Na^+(g) + e^- & \Delta IE = +496 \text{ kJ} \\
\frac{1}{2} Br_2(g) \rightarrow Br(g) & \Delta_f H^\circ = +112 \text{ kJ} \\
Br(g) + e^- \rightarrow Br^-(g) & \Delta EA = -325 \text{ kJ} \\
Na(s) + \frac{1}{2} Br_2(g) \rightarrow NaBr(s) & \Delta_f H^\circ = -361 \text{ kJ}
\end{array}$$

a) -963 kJ   b) -751 kJ   c) −290 kJ   d) +290 kJ   e) +1403 kJ

Answer: b

36. Using the thermodynamic data below, and a value of -717 kJ/mole for the lattice enthalpy for KCl, calculate the ionization energy of K.

Enthalpy of atomization of K	+89 kJ/mol
Enthalpy of formation of solid KCl	-437 kJ/mol
Enthalpy of formation of $Cl(g)$	+121 kJ/mol
Electron affinity of $Cl(g)$	-349 kJ/mol

a) -576 kJ/mol   b) +141 kJ/mol   c) +419 kJ/mol   d) +576 kJ/mol   e) +597 kJ/mol

Answer: c

13.4 The Solid State: Other Kinds of Solid Materials

37. Which of the following is NOT a network solid?

a) elemental silicon, $Si(s)$ b) diamond, $C(s)$ c) buckminster fullerene, $C_{60}(s)$
d) silicon dioxide, $SiO_2(s)$ e) aluminum oxide, $Al_2O_3(s)$

Answer: c

38. Which of the following is/are physical properties of amorphous solids?
    1. Amorphous solids have well defined melting points.
    2. At the particulate level, amorphous solids do not have long range order.
    3. Polymeric materials never form amorphous solids.

a) 1 only   b) 2 only   c) 3 only   d) 1 and 3   e) 2 and 3

Answer: b

39. Which two of the following materials are most likely to be amorphous solids: water, nylon, glass, potassium nitrate?

a) water and glass   b) nylon and potassium nitrate   c) water and nylon
d) water and potassium nitrate   e) nylon and glass

Answer: e

13.5 Phase Changes Involving Solids

40. Which process requires the greatest endothermic change in enthalpy for water?

a) freezing   b) condensation   c) sublimation   d) melting   e) vaporization

Answer: c

41. A phase diagram of a pure compound has a triple point at 18 °C and 72 mm Hg, a normal
    melting point at 21 °C, and a normal boiling point at 87 °C. Which of the following
    statements regarding this compound is/are CORRECT?
    1. The density of the solid is greater than that of the liquid.
    2. Sublimation occurs if starting with a solid at a constant temperature of 17 °C the
        pressure is decreased until a phase change occurs.
    3. Condensation occurs if the temperature is decreased from 55 °C to 13 °C at a constant
        pressure of 1.00 atm.

    a) 1 only   b) 2 only   c) 3 only   d) 1 and 2   e) 1, 2, and 3

    Answer: d

42. The phase diagram for CO_2 has a triple point at -56.6 °C and 5.19 atm, and a critical point at
    31.0 °C and 73 atm. The solid and gas phases are in equilibrium at -78.7 °C and 1.00 atm.
    Which of the following statements regarding CO_2 is/are CORRECT?
    1. Sublimation occurs if the temperature of the solid phase is increased from -79.0 °C to
        0.0 °C at a constant pressure of 2.5 atm.
    2. CO_2 is a supercritical fluid at 55 °C and 75 atm.
    3. Only the gas phase exists at a temperature of -56.6 °C and pressure of 1.00 atm.

    a) 1 only   b) 2 only   c) 3 only   d) 1 and 2   e) 1, 2, and 3

    Answer: e

43. Which statement concerning the phase diagram below is INCORRECT?

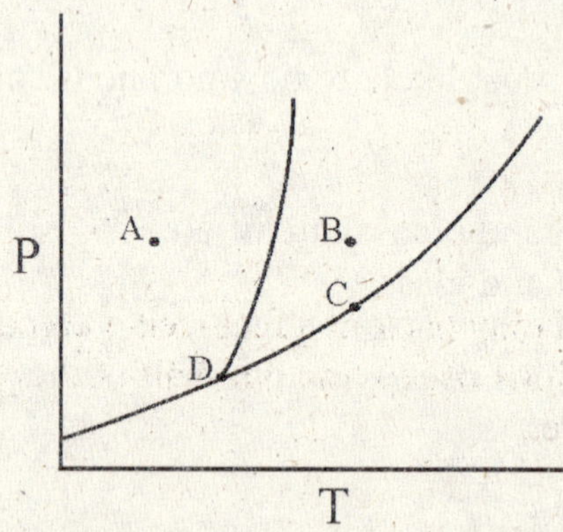

    a) Only the solid phase exists at point A.
    b) At point D, the triple point, all three phases (gas, liquid, and solid) are in equilibrium.
    c) Moving from point A to B results in a phase transition from solid to liquid.
    d) At point C, the solid and liquid phases are in equilibrium.
    e) Only the liquid phase exists at point B.

    Answer: d

44. If a pure substance begins at point B on the phase diagram below and the pressure on the substance is reduced until point C is reached, what process occurs?

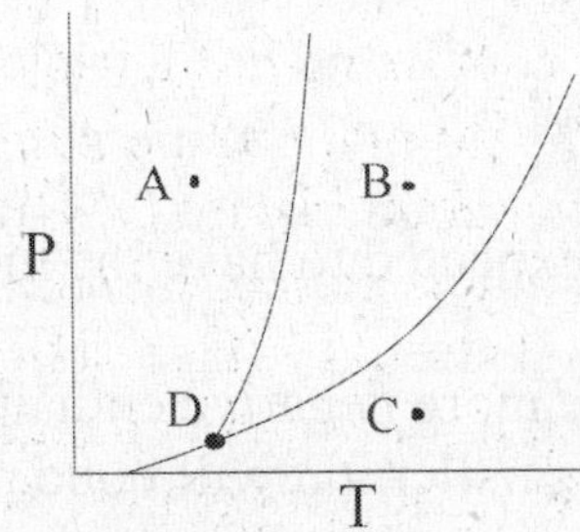

   a) fusion
   b) vaporization
   c) condensation
   d) sublimation
   e) none of the above

Answer: b

45. What process occurs when a substance is at point B on the phase diagram below, and the temperature is decreased (under constant pressure) until the substance is at point A?

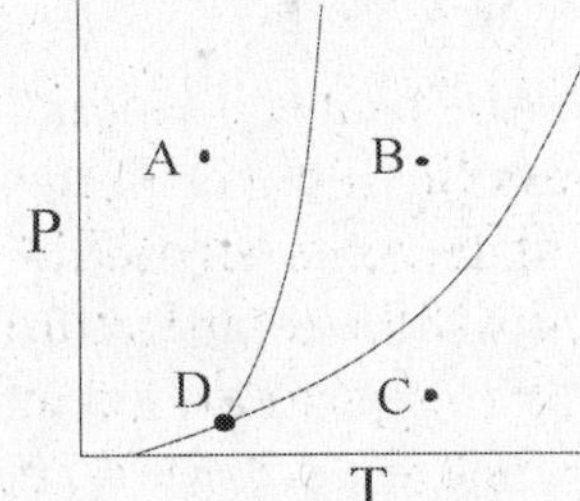

   a) condensation
   b) vaporization
   c) sublimation
   d) melting
   e) freezing

Answer: e

46. What process occurs when a substance is at point C on the phase diagram below, and the pressure is increased (under constant temperature) until the substance is at point B?

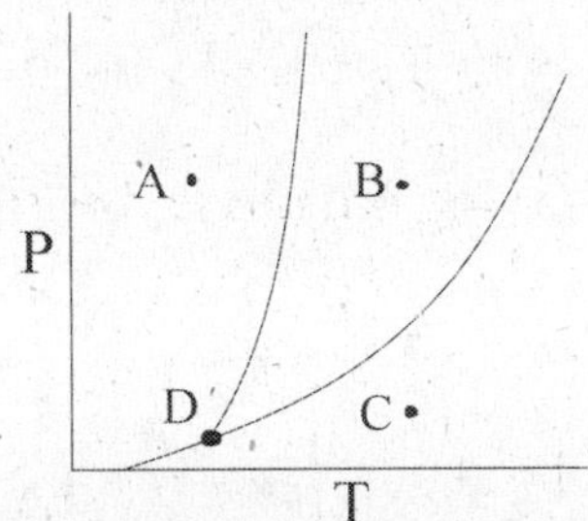

   a) condensation
   b) vaporization
   c) sublimation
   d) melting
   e) freezing

Answer: a

Short Answer Questions

47. A salt with a 1:1 ratio of anions to cations may pack in a face-centered cubic unit cell with the anions at the lattice points and the cations occupying one-half of the ________ holes. Zinc sulfide is an example of this structure.

Answer: tetrahedral

48. Sodium chloride crystallizes in a ________ cubic unit cell with chloride ions occupying the lattice points. The sodium ions occupy interstitial regions, with each cation in contact with six chloride ions.

Answer: face-centered

49. Above a substance's _________ temperature, it is not possible to compress the substance into the liquid phase. If enough pressure is applied the substance will become a supercritical fluid.

Answer: critical

50. Which ion, K^+ or Ca^{2+}, is expected to have the more negative enthalpy of hydration? Why?

Answer: Ca^{2+} will have the more negative enthalpy of hydration. A calcium ion is smaller than a potassium ion. The smaller ion will bond more closely to water's dipole, allowing for a stronger ion-dipole interaction. In addition, the higher the charge on the ion, the greater the ion-dipole interaction.

Chapter 14
SOLUTIONS AND THEIR BEHAVIOR

CHAPTER OUTLINE

Example 14.7 Determining Molar Mass from Boiling Point Elevation (p. 638)
Example 14.8 Osmotic Pressure and Molar Mass (p. 638)
Colligative Properties of Solutions Containing Ions (p. 639)
KEY TERMS: van't Hoff factor
A Closer Look Osmosis and Medicine (p. 639)
Table 14.4 Freezing Point Depressions of Some Ionic Solutions (p. 640)
Case Study Henry's Law in a Soda Bottle (p. 641)
Example 14.9 Freezing Point and Ionic Solutions (p. 641)

14.5   Colloids (p. 642)
KEY TERMS: colloidal dispersions, Tyndall effect, gel, sol
Table 14.5 Types of Colloids (p. 642)
Types of Colloids (p. 643)
KEY TERMS: hydrophobic, hydrophilic, emulsions, emulsifying agent
Surfactants (p. 645)
KEY TERMS: surfactant, detergent

CHAPTER OBJECTIVES

- Calculate and use the solution concentration units, molality, mole fraction, and weight percent (14.1-14.2)
- Understand the solution process (14.2)
- Understand and use the colligative properties of solutions (14.4)

LESSON PLAN

AP Standards

II.C.1	Types of solutions and factors affecting solubility
II.C.3	Raoult's law and colligative properties (nonvolatile solutes); osmosis
III.C.1	Le Chatelier's principle, equilibrium constants
III.C.2.b	Equilibrium constants for reactions in solution
III.D.3	Effect of temperature change on rates

Suggested Time:
Three traditional classes or two blocks. This chapter is short and can easily be done in the suggested time. Sections 14.3 and 14.3 must be covered entirely, while the remaining sections can be taught at your discretion. However, make sure you cover any topics that concern temperature change and its effect on rates.

ASSESSMENT
Exercises:
14.1   Mole Fraction, Molality, Weight Percent, and Parts per Million (p. 620)
14.2   Calculating an Enthalpy of Solution (p. 626)
14.3   Using Henry's Law (p. 627)
14.4   Using Raoult's Law (p. 631)
14.5   Boiling Point Elevation (p. 633))

CHAPTER 14 Solutions and Their Behaviors

Answers to Exercises p. A45-A52
Practicing Skills pp. 648-651, Answers IRM pp. 247-256
General Questions pp. 651-653, Answers IRM pp. 257-264
In the Laboratory Questions pp. 653, Answers IRM pp. 264-266
Summary and Conceptual Questions pp. 654-655, Answers IRM

GLOSSARY

boiling point elevation constant (K_{bp}) The proportionality constant relating molality to boiling point elevation

colligative properties The properties of a solution that depend only on the number of solute particles per solvent molecule and not on the nature of the solute or solvent

colloidal dispersions (colloids) A state of matter intermediate between a solution and a suspension, in which solute particles are large enough to scatter light, but too small to settle out

detergent A surfactant used in cleaning

emulsifying agent An additive to colloidal suspensions which enables them to appear as a homogeneous mixture, rather than separating into layers

emulsions A colloidal dispersion of one liquid in another

enthalpy of solution ($\Delta_{soln}H^o$) The amount of heat involved in the process of solution formation

freezing point depression constant (K_{fp}) The proportionality constant relating molality to freezing point depression

gel Acolloidal dispersion with a structure that prevents it from flowing

Henry's law The concentration of a gas dissolved in a liquid at a given temperature is directly proportional to the partial pressure of the gas above the liquid

hydration The process in which water molecules surround and isolate ions in the dissolving of an ionic salt

hydrophilic Water loving

hydrophobic Water fearing

ideal solution A solution that obeys Raoult's law

immiscible Liquids that do not mix to form a solution but exist in contact with each other, forming layers

Le Chatelier's principle A change in any of the factors determining an equilibrium will cause the system to adjust, to reduce, or minimize the effect of the change

miscible Liquids that mix to an appreciable extent to form a solution

molality (m) The number of moles of solute per kilogram of solvent

mole fraction (X) The ratio of the number of moles of one substance to the total number of moles in a mixture of substances

osmosis The movement of solvent molecules through a semipermeable membrane from a region of lower solute concentration to a region of higher solute concentration

osmotic pressure (Π) The pressure exerted by osmosis in a solution system at equilibrium

parts per million Grams of solute dissolved in 1.0 million grams of solution

Raoult's law The vapor pressure of the solvent is proportional to the mole fraction of the solvent in a solution

saturated A stable solution in which the maximum amount of solute has been dissolved

sol A colloidal dispersion of a solid substance in a fluid medium

solubility The concentration solute in equilibrium with undissolved solute in a saturated solution

solution A homogeneous mixture in a single phase

super-saturated solution A solution that temporarily contains more than the saturation amount of solute

surfactant A substance that changes the properties of a surface, typically in a colloidal suspension

Tyndall effect The scattering of visible light caused by particles of a colloid that are relatively large and dispersed in a solvent

unsaturated solution A solution in which the concentration of solute is less than the saturation amount

van't Hoff factor The ratio of the experimentally measured freezing point depression of a solution to the value calculated from the apparent molality

weight percent The mass of one component divided by the total mass of the mixture

CHAPTER 14 Solutions and Their Behaviors

MEDIA MENU

The Media Menu section of this guide lists the interactive exercises, tutorials, active figures (animated, interactive figures) and additional resources available online for Chemistry Now, a unique web-based, assessment-centered personalized learning system for chemistry students. Go to www.cengagelearning.com/login

Exercises
- Screen 14.2   Calculating solution concentrations in various units
- Screen 14.4   Solution energetics
- Screen 14.5   Henry's law
- Screen 14.7   Raoult's law
- Screen 14.8   Effect of a solute on the solution freezing point

Tutorials
- Screen 14.5   Henry's law
- Screen 14.6   Temperature and Le Chatelier's principle
- Screen 14.7   Raoult's law
- Screen 14.8   Effect of a solute on the solution freezing point

Active Figures
- 14.6   Solubility of non-polar iodine in polar water and nonpolar carbon tetrachloride
- 14.9   Dissolving an ionic solid in water

Additional Resources
- Screen 14.3   Visualization and problem on the solution process
- Screen 14.4   Analysis of the dissolution of KF
- Podcast Module 19   Colligative properties

DIAGNOSTIC QUIZ QUESTIONS

Goal #1 Calculate and use the solution concentration units, molality, mole fraction, and weight percent

1. If 355 g of ethanol (C_2H_5OH) is added to 645 g of water, what is the molality of the ethanol?

    a) 0.550 m b) 7.71 m c) 11.9 m d) 21.7 m e) 55.0 m

    Answer: c

2. What is the mole fraction of water in a solution that is 37.9% by weight ethylene glycol? The molar mass of ethylene glycol, $HOCH_2CH_2OH$, is 62.07 g/mol.

    a) 0.177   b) 0.379   c) 0.849   d) 0.901   e) 5.63

    Answer: c

3. What is the weight percent of acetic acid in 2.73 m CH$_3$CO$_2$H(aq)?

a) 3.95×10^{-5}%   b) 0.0455%   c) 2.73%   d) 14.1%   e) 16.4%

Answer: d

Goal #2 Understand the solution process

4. Which of the following statements is/are CORRECT?
    1. Solubility is defined as the concentration of solute in equilibrium with undissolved solute in a saturated solution.
    2. If two liquids mix to an appreciable extent to form a solution, they are miscible.
    3. If two liquids mix completely in any proportion to form a solution, the resulting solution is supersaturated.

a) 1 only   b) 2 only   c) 3 only   d) 1 and 2   e) 2 and 3

Answer: d

5. Which action(s) will increase the equilibrium concentration of a non-reacting gas in water?
    1. decreasing the temperature of the water
    2. increasing the volume of water
    3. decreasing the pressure of the gas above the liquid

a) 1 only   b) 2 only   c) 3 only   d) 1 and 3   3) 1, 2, and 3

Answer: a

Goal #3 Understand and use the colligative properties of solutions

6. Ideally, colligative properties depend only on the

a) concentration of solute particles in a solution.
b) molar masses of the solute particles in a solution.
c) density of a solution.
d) hydrated radii of the molecules or ions dissolved in a solution.
e) partial pressure of the gases above the surface of a solution.

Answer: a

CHAPTER 14 Solutions and Their Behaviors

7. Which of the following statements concerning osmosis is/are CORRECT?
   1. Osmosis involves the movement of a solvent through a semipermeable membrane.
   2. Solvents move from regions of high solute concentration to regions of lower solute concentration.
   3. Osmotic pressure is a colligative property.

   a) 1 only   b) 2 only   c) 3 only   d) 1 and 3   e) 1, 2, and 3

   Answer: d

8. What mass of ethylene glycol, when mixed with 225 g H_2O, will reduce the equilibrium vapor pressure of H_2O from 1.00 atm to 0.800 atm at 100 °C? The molar masses of water and ethylene glycol are 18.02 g/mol and 62.07 g/mol, respectively. Assume ideal behavior for the solution.

   a) 15.6 g   b) 49.9 g   c) 194 g   d) 969 g   e) 3.10×10^3 g

   Answer: c

9. Which of the following aqueous solutions should have the lowest freezing point?

   a) pure H_2O b) 1 m $MgBr_2$ c) 1 m RbI   d) 1 m NH_3 e) 1 m $C_6H_{12}O_6$

   Answer: b

TESTBANK

1. What is the definition of molality?

   a) moles of solute per kg of solvent   b) grams of solute per kg of solution
   c) grams of solute per liter of solution   d) moles of solute per liter of solvent
   e) moles of solute per liter of solution

   Answer: a

2. If 18.4 g KCl is dissolved in 2.50×10^2 g H_2O, what is the weight percent of KCl in the solution?

   a) 0.972%   b) 6.86%   c) 7.36%   d) 7.94%   e) 97.2%

   Answer: b

3. If 355 g of ethanol (C_2H_5OH) is added to 645 g of water, what is the molality of the ethanol?

   a) 0.550 m b) 7.71 m c) 11.9 m d) 21.7 m e) 55.0 m

   Answer: c

4. If 86.5 g of ethanol (C_2H_5OH) is added to 355 g of water, what is the mole fraction of ethanol?

a) 0.00425   b) 0.0870   c) 0.0953   d) 0.244   e) 5.29

Answer: b

5. What is the molality of 7.59% by weight aqueous hydrochloric acid? The molar mass of HCl is 36.46 g/mol.

a) 0.208 *m*   b) 2.08 *m*   c) 2.25 *m*   d) 2.76 *m*   e) 2.99 *m*

Answer: c

6. What is the mole fraction of calcium chloride in 4.47 *m* $CaCl_2$(aq)? The molar mass of $CaCl_2$ is 111.0 g/mol and the molar mass of water is 18.02 g/mol.

a) 0.0745   b) 0.0805   c) 0.199   d) 0.496   e) 0.860

Answer: a

7. What is the mole fraction of water in a solution that is 37.9% by weight ethylene glycol? The molar mass of ethylene glycol, $HOCH_2CH_2OH$, is 62.07 g/mol.

a) 0.177   b) 0.379   c) 0.849   d) 0.901   e) 5.63

Answer: c

8. What is the weight percent of acetic acid in 2.73 *m* CH_3CO_2H(aq)?

a) 3.95×10^{-5}%   b) 0.0455%   c) 2.73%   d) 14.1%   e) 16.4%

Answer: d

9. The weight percent of concentrated $HClO_4$(aq) is 70.5% and its density is 1.67 g/mL. What is the molarity of concentrated $HClO_4$?

a) 4.20 M   b) 7.18 M   c) 11.7 M   d) 14.2 M   e) 39.7 M

Answer: c

10. Concentrated sodium hydroxide is 19.4 M and has a density of 1.54 g/mL. What is the molality of concentrated NaOH?

a) 12.6 *m*   b) 19.8 *m*   c) 25.4 *m*   d) 29.9 *m*   e) 50.4 *m*

Answer: c

11. If the concentration of potassium carbonate in water is 210 ppm, what is the molarity of $K_2CO_3(aq)$? The molar mass of K_2CO_3 is 138.2 g/mol. Assume the density of the solution is 1.00 g/mL.

a) 1.5×10^{-3} M   b) 3.4×10^{-3} M   c) 2.9×10^{-2} M   d) 3.4×10^{-2} M   e) 0.66 M

Answer: a

12. A 25 meter by 10 meter pool of water has a depth of 2.5 meters. What mass of silver ion is present in the reservoir if the concentration of silver ion is 0.13 ppm? ($1 \text{ m}^3 = 1000$ L; assume the density of the solution is 1.00 g/mL)

a) 2.1×10^{-4} g   b) 7.5×10^{-1} g   c) 4.8 g   d) 8.8 g   e) 81 g

Answer: e

13. What concentration of potassium chloride (in ppm) is present in 7.1×10^{-5} M KCl(aq)? For very dilute aqueous solutions, you can assume the solution's density is 1.0 g/mL. The molar mass of KCl is 74.55 g/mol.

a) 0.095 ppm   b) 0.53 ppm   c) 1.1 ppm   d) 5.3 ppm   e) 11 ppm

Answer: d

14. What mass of $Zn(NO_3)_2$ must be diluted to a mass of 1.00 kg with H_2O to prepare 97 ppm $Zn^{2+}(aq)$?

a) 7.8×10^{-6} g   b) 7.8×10^{-3} g   c) 3.3×10^{-2} g   d) 1.3×10^{-1} g   e) 2.8×10^{-1} g

Answer: e

14.2 The Solution Process

15.   Which of the following statements is/are CORRECT?
1. Solubility is defined as the concentration of solute in equilibrium with undissolved solute in a saturated solution.
2. If two liquids mix to an appreciable extent to form a solution, they are miscible.
3. If two liquids mix completely in any proportion to form a solution, the resulting solution is supersaturated.

a) 1 only   b) 2 only   c) 3 only   d) 1 and 2   e) 2 and 3

Answer: d

16. Which of the following liquids will be immiscible with water: methanol (CH_3OH), carbon tetrachloride (CCl_4), benzene (C_6H_6), and/or formic acid (HCO_2H)?

a) methanol and carbon tetrachloride b) carbon tetrachloride and benzene
c) benzene and formic acid d) methanol, carbon tetrachloride, and benzene
e) carbon tetrachloride, benzene, and formic acid

Answer: b

17. Two nonpolar solvents, such as hexane and carbon tetrachloride, may be miscible even though the enthalpy of mixing of these liquids might be small. A reason that mixing occurs is that mixtures have greater disorder than pure solvents. The tendency toward disorder is a thermodynamic function called __________.

a) entropy b) enthalpy c) saturation d) adhesion e) cohesion

Answer: a

18. Which of the following statements concerning solubility is/are CORRECT?
   1. Ionic compounds, such as NaCl, are generally insoluble in nonpolar solvents.
   2. The solubility of nonpolar molecules, such as I_2, in nonpolar solvents is generally greater than their solubility in polar solvents.
   3. The solubility of polar molecules, such as sugar, in polar solvents is generally greater than their solubility in nonpolar solvents.

a) 1 only b) 2 only c) 3 only d) 1 and 3 e) 1, 2, and 3

Answer: e

19. In which solvent should Na^+ have the most negative enthalpy of solvation?

a) CCl_4 b) H_2O c) C_6H_6 d) CS_2 e) C_6H_{14}

Answer: b

20. The heat of formation of LiOH(s) is -484.9 kJ/mol and the heat of formation of LiOH(aq, 1 *m*) is -508.5 kJ/mol. Determine the heat of solution of LiOH.

a) -23.6 kJ/mol b) +23.6 kJ/mol c) -993.4 kJ/mol d) +993.4 kJ/mol e) -0.954 kJ/mol

Answer: a

CHAPTER 14 Solutions and Their Behaviors

21. The heat of formation of NaCl(s) is -411.1 kJ/mol and the heat of formation of NaCl(aq, 1 *m*) is -407.3 kJ/mol. Determine the heat of solution of NaCl. Will the solution temperature increase or decrease when NaCl is dissolved in water?

a) -3.8 kJ/mol; increase b) +3.8 kJ/mol; increase c) +3.8 kJ/mol; decrease
d) +818.4 kJ/mol; increase e) -818.4 kJ/mol; decrease

Answer: c

14.3 Factors Affecting Solubility: Pressure and Temperature

22. Which of the following statements is INCORRECT?

a) The solubility of a gas in water decreases with increasing temperature.
b) The solubility of a gas in water is proportional to the partial pressure of the gas above the water.
c) The dissolution of a gas in water is usually an exothermic process.
d) The relationship between the solubility of a gas and its partial pressure is known as Henry's law.
e) The solubility of a gas in water is inversely proportional to the molar mass of the gas.

Answer: e

23. The Henry's law constant for N_2 in water at 37 °C is 8.2×10^{-7} M/mm Hg. What is the equilibrium concentration of N_2 in water when the partial pressure of N_2 is 605 mm Hg?

a) 1.4×10^{-9} M b) 6.5×10^{-7} M c) 5.0×10^{-4} M d) 2.0×10^{3} M e) 7.4×10^{8} M

Answer: c

24. Which action(s) will increase the equilibrium concentration of a non-reacting gas in water?
   1. decreasing the temperature of the water
   2. increasing the volume of water
   3. decreasing the pressure of the gas above the liquid

a) 1 only b) 2 only c) 3 only d) 1 and 3 3) 1, 2, and 3

Answer: a

25. The Henry's law constant for O_2 in water at 25 °C is 1.66×10^{-6} M/mm Hg. What partial pressure of O_2 is necessary to achieve an equilibrium concentration of 1.5×10^{-3} M O_2?

a) 2.5×10^{-9} mm Hg b) 1.1×10^{-3} mm Hg c) 1.5×10^{-3} mm Hg
d) 9.0×10^{2} mm Hg e) 4.1×10^{8} mm Hg

Answer: d

26. Ideally, colligative properties depend only on the

    a) concentration of solute particles in a solution.
    b) molar masses of the solute particles in a solution.
    c) density of a solution.
    d) hydrated radii of the molecules or ions dissolved in a solution.
    e) partial pressure of the gases above the surface of a solution.

    Answer: a

27. What is the equilibrium partial pressure of water vapor above a mixture of 44.0 g H_2O and
    56.0 g $HOCH_2CH_2OH$ at 35 °C. The partial pressure of pure water at 35.0 °C is 42.2 mm Hg.
    Assume ideal behavior for the solution.

    a) 0.730 Hg   b) 18.7 mm Hg   c) 23.6 mm Hg   d) 30.8 mm Hg   e) 58.8 mm Hg

    Answer: d

28. The vapor pressure of pure water at 55 °C is 118 mm Hg. What is the equilibrium vapor
    pressure of water above a mixture of 77.0 g ethanol (CH_3CH_2OH, molar mass = 46.07 g/mol)
    and 32.0 g water?

    a) 2.72 mm Hg   b) 34.6 mm Hg   c) 49.0 mm Hg   d) 57.2 mm Hg   e) 60.8 mm Hg

    Answer: e

29. What mass of ethylene glycol, when mixed with 225 g H_2O, will reduce the equilibrium
    vapor pressure of H_2O from 1.00 atm to 0.800 atm at 100 °C? The molar masses of water and
    ethylene glycol are 18.02 g/mol and 62.07 g/mol, respectively. Assume ideal behavior for the
    solution.

    a) 15.6 g   b) 49.9 g   c) 194 g   d) 969 g   e) 3.10×10^3 g

    Answer: c

30.   Which of the following aqueous solutions should have the lowest freezing point?

    a) pure H_2O b) 1 m $MgBr_2$ c) 1 m RbI   d) 1 m NH_3 e) 1 m $C_6H_{12}O_6$

    Answer: b

31. Which of the following aqueous solutions should have the highest boiling point?

    a) 0.50 m NaBr   b) 0.50 m K_2SO_4 c) 0.50 m $CaBr_2$ d) 1.0 m KCl   e) 1.5 m $C_6H_{12}O_6$

    Answer: d

32. The freezing point depression constant for water is $-1.86\ °C/m$. At what temperature will a solution containing 8.27 g $CaCl_2$ and 45.0 g H_2O begin to freeze? Assume that no ion-pairing occurs between Ca^{2+} and Cl^-.

   a) -9.24 °C   b) -4.62 °C   c) -0.804 °C   d) -0.749 °C   e) +4.62 °C

   Answer: a

33. What is the freezing point of a solution containing 3.10 grams benzene (molar mass = 78.11 g/mol) dissolved in 32.0 grams paradichlorobenzene? The freezing point of pure paradichlorobenzene is 53.0 °C and the freezing point depression constant, K_{fp}, is -7.10 °C/m.

   a) 44.2 °C   b) 51.8 °C   c) 52.3 °C   d) 54.2 °C   e) 61.8 °C

   Answer: a

34. What is the molar mass of a nonpolar molecular compound if 3.42 grams dissolved in 41.8 grams benzene begins to freeze at 1.17 °C? The freezing point of pure benzene is 5.50 °C and the freezing point depression constant, K_{fp}, is -5.12 °C/m.

   a) 2.89 g/mol   b) 69.2 g/mol   c) 96.7 g/mol   d) 126 g/mol   e) 358 g/mol

   Answer: c

35. What is the boiling point of a solution containing 0.80 g caffeine, $C_8H_{10}N_4O_2$, dissolved in 13.20 g benzene? The boiling point of pure benzene is 80.1 °C and the boiling point elevation constant, K_{bp}, is 2.53 °C/m.

   a) 79.8 °C   b) 80.4 °C   c) 80.9 °C   d) 85.2 °C   e) 88.2 °C

   Answer: c

36. What mass of Na_2SO_4 must be dissolved in 75.0 grams of water to lower the freezing point by 2.50 °C? The freezing point depression constant, K_{fp}, of water is -1.86 °C/m. Assume the van't Hoff factor for Na_2SO_4 is 2.40.

   a) 0.310 g   b) 3.30 g   c) 5.97 g   d) 23.3 g   e) 44.0 g

   Answer: c

37. The freezing point depression constant of water is -1.86 °C/m. If 7.50 g NaCl is dissolved in 45.0 g H_2O, the freezing point is changed by -9.71 °C. Calculate the van't Hoff factor for NaCl.

   a) 1.00   b) 1.54   c) 1.67   d) 1.83   e) 1.91

   Answer: d

38. What concentration unit is used in the calculation of osmotic pressure for a dilute solution?

a) molality b) weight percent c) mass fraction
d) mole fraction e) molarity

Answer: e

39. Which of the following statements concerning osmosis is/are CORRECT?
  1. Osmosis involves the movement of a solvent through a semipermeable membrane.
  2. Solvents move from regions of high solute concentration to regions of lower solute concentration.
  3. Osmotic pressure is a colligative property.

a) 1 only b) 2 only c) 3 only d) 1 and 3 e) 1, 2, and 3

Answer: d

40. At 25 °C, what is the osmotic pressure of a homogeneous solution consisting of 18.0 g urea (CON_2H_4) diluted with water to 3.00 L? ($R = 0.08206$ L·atm/mol·K)

a) 0.205 atm b) 2.44 atm c) 7.33 atm d) 12.3 atm e) 14.7 atm

Answer: b

41. The osmotic pressure of blood is 7.65 atm at 37 °C. What mass of glucose $(C_6H_{12}O_6$, molar mass = 180.2 g/mol) is needed to prepare 2.25 L of solution for intravenous injection? The osmotic pressure of the glucose solution must equal the osmotic pressure of blood. ($R = 0.08206$ L·atm/mol·K)

a) 0.676 g b) 0.698 g c) 5.67 g d) 122 g e) 1.02×10^3 g

Answer: d

42. A solution is prepared by dissolving 4.78 g of an unknown nonelectrolyte in enough water to make 375 mL of solution. The osmotic pressure of the solution is 1.33 atm at 27 °C. What is the molar mass of the solute? ($R = 0.08206$ L·atm/mol·K)

a) 0.0203 g/mol b) 21.2 g/mol c) 49.4 g/mol d) 96.8 g/mol e) 236 g/mol

Answer: e

14.5 Colloids

43. All of the following are colloidal dispersions EXCEPT _________.

a) marshmallow b) butter c) milk d) whipped cream e) root beer

Answer: e

44. Clouds and fog are colloidal dispersions that contain a liquid dispersed in a gaseous medium. This type of colloid is called a(n) _________.

    a) emulsion   b) aerosol   c) foam   d) sol   e) gel

    Answer: b

45. A(n) _________ is a colloidal dispersion of a liquid in solid.

    a) gel   b) emulsion   c) aerosol   d) foam   e) sol

    Answer: a

Short Answer Questions

46. Because _________ particles are relatively large (say, 1000 nm in diameter) they scatter visible light, making the mixtures containing these particles appear cloudy. This scattering is known as the Tyndall effect.

    Answer: colloidal

47. Aqueous colloidal solutions can be classified as hydrophobic (water-fearing), or _________ (water-loving).

    Answer: hydrophilic

48. A surfactant used for cleaning is called a(n) _________.

    Answer: detergent (or soap)

49. If one of the factors determining the equilibrium of a system is changed, the system adjusts to counteract that change. This is known as _________ principle.

    Answer: Le Chatelier's

50. The following equation is known as _________ law: $P_{solvent} = X_{solvent}\, P^{\circ}_{solvent}$.

    Answer: Raoult's

51. A solution in which there is more dissolved solute than in a saturated solution is known as a(n) _________ solution.

    Answer: supersaturated

52. If an egg's shell is carefully dissolved using an acid, the egg white and yolk will remain intact inside a membrane. If the egg is then placed in distilled water, it will slowly expand until it bursts. Why?

Answer: This is an example of osmotic pressure. The membrane surrounding the egg white is permeable to solvents such as water. Since the solute concentration is greater inside the egg than outside the egg, water flows into the membrane.

53. Henry's law states that the solubility of a gas in a liquid is directly proportional to its pressure. This law holds true for gases such as nitrogen and oxygen. However, Henry's law does not hold true for hydrogen chloride gas. Why?

Answer: Henry's law is only true for gases that do not react with the solvent. Hydrogen chloride dissociates in water to form H^+ and Cl^- ions.

Chapter 15
CHEMICAL KINETICS:
THE RATES OF CHEMICAL REACTIONS

CHAPTER OUTLINE

CHAPTER OBJECTIVES

- Understand rates of reaction and the conditions affecting rates (15.1, 15.2)
- Derive the rate equation , rate constant, and reaction order from experimental data (15.3)
- Use integrated rate laws (15.4)
- Understand the collision theory of reaction rates and the role of activation energy (15.5)
- Relate reaction mechanisms and rate laws (15.5, 15.6)

LESSON PLAN

AP Standards

II.C.2	Half-lives
III.A.1	Reaction types: concepts of Arrhenius
III.D.1	Concept of rate of reaction
III.D.2	Use of experimental data and graphical analysis to determine reactant order, rate constants, and reaction rate laws
III.D.3	Effect of temperature change on rates
III.D.4	Energy of activation; the role of catalysts
III.D.5	The relationship between the rate-determining step and a mechanism

278

CHAPTER 15: Chemical Kinetics: The Rates of Chemical Reactions

Suggested Time:
Three traditional classes or 2 blocks.   This chapter is short and can easily be done in the suggested time. Sections 15.1 and 15.2 are essential for the AP material. The concepts of half-life and reaction mechanisms also need to be covered, along with any part of the chapter that concerns temperature change and its effect on rates.

ASSESSMENT

Exercises:

Answers to Exercises p. A52-A53
Practicing Skills pp. 712-716, Answers IRM pp. 267-278
General Questions pp. 716-720, Answers IRM pp. 278-285
In the Laboratory Questions pp. 720-721, Answers IRM pp. 285-285
Summary and Conceptual Questions pp. 7222-723, Answers IRM pp. 286-288

GLOSSARY

activation energy(E_a) The minimum amount of energy that must be absorbed by a system to cause it to react

Arrhenius equation A mathematical expression that relates reaction rate to the activation energy, collision frequency, molecular orientation, and temperature

bimolecular A process that involves two molecules

catalyst A substance that increases the rate of a reaction while not being consumed in the reaction

chemical kinetics The study of the rates of chemical reaction under various conditions and of reaction mechanisms

collision theory of reaction rates A theory of reaction rates that assumes that molecules must collide in order to react

elementary step A simple event in which some chemical transformation occurs; one of a sequence of events that form the reaction mechanism

enzyme A biological catalyst

frequency factor Term related to the number of collisions and the fraction with correct geometry for reacting

half-life The time required for the concentration of one of the reactants to reach half of its initial value

heterogeneous catalyst A catalyst that is in a different phase than the reaction mixture

homogeneous catalyst A catalyst that is in the same phase as the reaction mixture

initial rate The instantaneous reaction rate at the start of the reaction

integrated rate equation The rate equation derived by integrating the rate with respect to change in time

molecularity The number of particles colliding in an elementary step

order The exponent of a concentration term in the reaction's rate law

overall reaction order The sum of all the individual reactant orders in a reaction's rate law

rate constant (k) The proportionality constant in the rate equation

rate-determining step The slowest elementary step of a reaction mechanism

rate equation(rate law) The mathematical relationship between reactant concentration and reaction rate

reaction coordinate Graphical representation to show progress of a reaction

reaction intermediates A species that is produced in one step of a reaction mechanism and completely consumed in a later step

reaction mechanism(s) The sequence of events at the molecular level that control the speed and outcome of a reaction

termolecular A process that involves three molecules

transition state The arrangement of reacting molecules and atoms at the point of maximum potential energy

unimolecular A process that involves one molecule

CHAPTER 15: Chemical Kinetics: The Rates of Chemical Reactions

MEDIA MENU

The Media Menu section of this guide lists the interactive exercises, tutorials, active figures (animated, interactive figures) and additional resources available online for Chemistry Now, a unique web-based, assessment-centered personalized learning system for chemistry students. Go to www.cengagelearning.com/login

Exercises
- Screen 15.5   Determining rate equations from a study of the effect of concentration on rate
- Screen 15.14   Reaction mechanisms and the effect of catalysts
- Screen 15.14   Video exercise on catalysis
- Screen 15.12   Reaction mechanisms
- Screen 15.13   Reaction mechanisms

Tutorials
- Screen 15.5   Determining rate equations from a study of the effect of concentration on rate
- Screen 15.6   Use of the integrated first-order rate equation
- Screen 15.7   Graphical methods
- Screen 15.8   Using half-lives
- Screen 15.9   Using half-lives
- Screen 15.11   Temperature dependence of reaction rates and the Arrhenius equation
- Screen 15.14   Using half-life various catalysts

Active Figures
- 15.2   A plot of reactant concentration versus time for the decomposition of N_2O_5
- 15.6   The decomposition of azomethane, $CH_3N_2CH_3$
- 15.9   Half-life of a first-order reaction
- 15.13   Activation Energy

Additional Resources
- Screen 15.2   Visualization of ways to express reaction rates
- Screen 15.2   Self-study module on rate of reaction
- Screen 15.3   Visualization of the factors controlling rates
- Screen 15.4   Simulation of the effect of concentration on rate
- Screen 15.4   Self-study module on control of reaction rates
- Screen 15.5   Simulation on determining rate equations from a study of the effect of concentration on rate
- Screen 15.6   Self-study module on first-order rate equations
- Screen 15.9   Visualization of collision theory
- Screen 15.10   Simulation of reaction coordinate diagrams
- Screen 15.11   Simulation on the temperature dependence of reaction rates and the Arrhenius equation
- Screen 15.14   Visualization of the effect of a catalyst on activation energy
- Screen 15.14   Interview of a scientist describing using a catalyst in industry
- Podcast Module 20   Concentration-time relationships

DIAGNOSTIC QUIZ QUESTIONS

Understand rates of reaction and the conditions affecting rates (15.1, 15.2)

1. Which statement concerning relative rates of reaction is correct for the chemical equation given below?

$$2\ NOBr(g)\ \rightarrow\ 2\ NO(g)\ +\ Br_2(g)$$

 a) The rate of disappearance of NOBr is equal to the rate of appearance of Br_2.
 b) The rate of disappearance of NOBr is two times the rate of appearance of NO.
 c) The rate of disappearance of NOBr is half the rate of appearance of Br_2.
 d) The rate of appearance of NO is equal to the rate of appearance of Br_2.
 e) The rate of appearance of NO is two times the rate of appearance of Br_2.

 Answer: e

2. Which of the following factors are likely to affect the rate of a chemical reaction?
    1. the presence of a catalyst
    2. the temperature of the reactants
    3. the physical state (solid, liquid, or gas) of the reactants

 a) 1 only   b) 2 only   c) 3 only   d) 1 and 3   e) 1, 2, and 3

 Answer: e

Derive the rate equation, rate constant, and reaction order from experimental data (15.3)

3. Given the initial rate data for the reaction $A + B \rightarrow C$, determine the rate expression for the reaction.

[A], M	[B], M	$\Delta[C]/\Delta t$ (initial) M/s
0.0500	0.160	2.24×10^{-3}
0.0750	0.160	3.36×10^{-3}
0.0750	0.272	9.72×10^{-3}

 a) $\dfrac{\Delta[C]}{\Delta t} = 0.280\ M^{-1}s^{-1}[A][B]$ b) $\dfrac{\Delta[C]}{\Delta t} = 1.75\ M^{-2}s^{-1}[A][B]^2$

 c) $\dfrac{\Delta[C]}{\Delta t} = 1.75\ M^{-2}s^{-1}[A]^2[B]$ d) $\dfrac{\Delta[C]}{\Delta t} = 5.60\ M^{-2}s^{-1}[A]^2[B]$

 e) $\dfrac{\Delta[C]}{\Delta t} = 0.0303\ M^{-1}s^{-1}[B]^2$

 Answer: b

Use integrated rate laws (15.4)

4. For a reaction, $A \rightarrow B + C$, which of the following equations corresponds to the integrated expression for a first-order decomposition reaction?

    a) $[A]_t = -kt + [A]_0$

    b) $\ln \dfrac{[A]_t}{[A]_0} = -kt$

    c) $\ln[A]_t = \ln[-kt] + \ln[A]_0$

    d) $\dfrac{[A]_t}{[A]_0} = -kt$

    e) $\dfrac{1}{[A]_t} = kt + \dfrac{1}{[A]_0}$

    Answer: b

Understand the collision theory of reaction rates and the role of activation energy (15.5)

5. According to collision theory, which condition(s) must be met in order for molecules to react?
    1. The reacting molecules must collide with sufficient energy to initiate the process of breaking and forming bonds.
    2. A catalyst must be in contact with the reacting molecules for a reaction to occur.
    3. The reacting molecules must collide with an orientation that can lead to rearrangement of the atoms.

    a) 1 only   b) 2 only   c) 3 only   d) 1 and 2   e) 1 and 3

    Answer: e

Relate reaction mechanisms and rate laws (15.5, 15.6)

6. Nitrogen dioxide reacts with carbon monoxide to produce nitrogen monoxide and carbon dioxide.
$$NO_2(g) + CO(g) \rightarrow NO(g) + CO_2(g)$$
    A proposed mechanism for this reaction is
$$2\,NO_2(g) \rightleftharpoons NO_3(g) + NO(g) \qquad \text{(fast, equilibrium)}$$
$$NO_3(g) + CO(g) \rightarrow NO_2(g) + CO_2(g) \qquad \text{(slow)}$$
    What is a rate law that is consistent with the proposed mechanism?

    a) rate $= k[[NO_2]^2[CO]\,[NO]^{-1}$
    b) rate $= k[[NO_2]^2[CO]$
    c) rate $= k[NO_2][CO]$
    d) rate $= k[NO_3][CO]$
    e) rate $= k[NO_2]^2$

    Answer: a

TESTBANK

<u>15.1 Rates of Chemical Reactions</u>

1. Which statement concerning relative rates of reaction is correct for the chemical equation given below?

$$2\ NOBr(g)\ \rightarrow\ 2\ NO(g)\ +\ Br_2(g)$$

a) The rate of disappearance of NOBr is equal to the rate of appearance of Br_2.
b) The rate of disappearance of NOBr is two times the rate of appearance of NO.
c) The rate of disappearance of NOBr is half the rate of appearance of Br_2.
d) The rate of appearance of NO is equal to the rate of appearance of Br_2.
e) The rate of appearance of NO is two times the rate of appearance of Br_2.

Answer: e

2. Which relationship correctly compares the rates of the following reactants and products?

$$C_3H_8(g)\ +\ 5\ O_2(g)\ \rightarrow\ 3\ CO_2(g)\ +\ 4\ H_2O(\ell)$$

a) $\dfrac{\Delta[C_3H_8]}{\Delta t} = \dfrac{\Delta[O_2]}{\Delta t} = \dfrac{\Delta[CO_2]}{\Delta t} = \dfrac{\Delta[H_2O]}{\Delta t}$

b) $\dfrac{\Delta[C_3H_8]}{\Delta t} = \dfrac{\Delta[O_2]}{\Delta t} = -\dfrac{\Delta[CO_2]}{\Delta t} = -\dfrac{\Delta[H_2O]}{\Delta t}$

c) $-\dfrac{\Delta[C_3H_8]}{\Delta t} = -\dfrac{1}{5}\dfrac{\Delta[O_2]}{\Delta t} = \dfrac{1}{3}\dfrac{\Delta[CO_2]}{\Delta t} = \dfrac{1}{4}\dfrac{\Delta[H_2O]}{\Delta t}$

d) $\dfrac{\Delta[C_3H_8]}{\Delta t} = \dfrac{5\ \Delta[O_2]}{\Delta t} = -\dfrac{3\ \Delta[CO_2]}{\Delta t} = -\dfrac{4\ \Delta[H_2O]}{\Delta t}$

e) $\dfrac{\Delta[C_3H_8]}{\Delta t} = \dfrac{5\ \Delta[O_2]}{\Delta t} = \dfrac{3\ \Delta[CO_2]}{\Delta t} = \dfrac{4\ \Delta[H_2O]}{\Delta t}$

Answer: c

3. An aqueous solution of hydrogen peroxide decomposes to oxygen and water according to the balanced chemical equation below.

$$2\ H_2O_2(aq)\ \rightarrow\ O_2(g)\ +\ 2\ H_2O(\ell)$$

If the rate of disappearance of hydrogen peroxide is -3.8×10^{-5} M/s, what is the rate of formation of oxygen?

a) -3.8×10^{-5} M/s   b) 1.9×10^{-5} M/s   c) 3.8×10^{-5} M/s
d) 6.2×10^{-3} M/s   e) 7.6×10^{-5} M/s

Answer: b

4. For the reaction below, if the rate of appearance of Cl_2 is 0.022 M/s, what is the rate of appearance of NO?

$$2\,NOCl(g) \;\rightarrow\; 2\,NO(g) \;+\; Cl_2(g)$$

a) -0.022 M/s   b) 0.022 M/s   c) 0.033 M/s   d) 0.044 M/s   e) 0.088 M/s

Answer: d

15.2 Reaction Conditions and Rate

5. Which of the following factors are likely to affect the rate of a chemical reaction?
    1. the presence of a catalyst
    2. the temperature of the reactants
    3. the physical state (solid, liquid, or gas) of the reactants

a) 1 only   b) 2 only   c) 3 only   d) 1 and 3   e) 1, 2, and 3

Answer: e

6. What is the name given to a substance that increases the rate of a chemical reaction but is not itself consumed?

a) catalyst   b) reactant   c) intermediate   d) product   e) rate constant

Answer: a

15.3 Effect of Concentration on Reaction Rate

7. What is the overall order of the reaction below
$$2\,NO(g) \;+\; O_2(g) \;\rightarrow\; 2\,NO_2(g)$$
if it proceeds via the following rate expression?

$$-\frac{\Delta[NO]}{\Delta t} = k[NO]^2[O_2]$$

a) zero-order   b) first-order   c) second-order   d) third-order   e) fourth-order

Answer: d

8. What is the overall order of the reaction below
$$NO(g) \;+\; O_3(g) \;\rightarrow\; NO_2(g) \;+\; O_2(g)$$
if it proceeds via the following rate expression?

$$-\frac{\Delta[NO]}{\Delta t} = k[NO][O_3]$$

a) zero-order   b) first-order   c) second-order   d) third-order   e) fourth-order

Answer: c

9. Given the initial rate data for the decomposition reaction,
$$A \rightarrow B + C$$
determine the rate expression for the reaction.

[A], M	$-\Delta[A]/\Delta t$ M/s
0.0625	5.44×10^{-7}
0.0938	8.16×10^{-7}
0.125	1.09×10^{-6}

a) $\dfrac{-\Delta[A]}{\Delta t} = 8.71 \times 10^{-6} \text{ s}^{-1}[A]$

b) $\dfrac{-\Delta[A]}{\Delta t} = 1.39 \times 10^{-4} \text{ M}^{-1}\text{s}^{-1}[A]^2$

c) $\dfrac{-\Delta[A]}{\Delta t} = 6.98 \times 10^{-5} \text{ M}^{-1}\text{s}^{-1}[A]^2$

d) $\dfrac{-\Delta[A]}{\Delta t} = 5.44 \times 10^{-7} \text{ M s}^{-1}$

e) $\dfrac{-\Delta[A]}{\Delta t} = 5.44 \times 10^{-7} \text{ s}^{-1}[A]$

Answer: a

10. Given the initial rate data for the reaction $2A + B \rightarrow C$, determine the rate expression for the reaction.

[A], M	[B], M	$\Delta[C]/\Delta t$ (initial) M/s
0.180	0.250	1.36×10^{-3}
0.180	0.500	2.72×10^{-3}
0.720	0.500	1.09×10^{-2}

a) $\dfrac{\Delta[C]}{\Delta t} = 1.21 \times 10^{-1} \text{ M}^{-2}\text{s}^{-1}[A][B]^2$

b) $\dfrac{\Delta[C]}{\Delta t} = 3.02 \times 10^{-2} \text{ M}^{-2}\text{s}^{-1}[A]^2[B]$

c) $\dfrac{\Delta[C]}{\Delta t} = 1.68 \times 10^{-1} \text{ M}^{-2}\text{s}^{-1}[A]^2[B]$

d) $\dfrac{\Delta[C]}{\Delta t} = 1.36 \times 10^{-3} \text{ M}^{-1}\text{s}^{-1}[A][B]$

e) $\dfrac{\Delta[C]}{\Delta t} = 3.02 \times 10^{-2} \text{ M}^{-1}\text{s}^{-1}[A][B]$

Answer: e

11. Given the initial rate data for the reaction $A + B \rightarrow C$, determine the rate expression for the reaction.

[A], M	[B], M	$\Delta[C]/\Delta t$ (initial) M/s
0.0500	0.160	2.24×10^{-3}
0.0750	0.160	3.36×10^{-3}
0.0750	0.272	9.72×10^{-3}

a) $\dfrac{\Delta[C]}{\Delta t} = 0.280 \ M^{-1}s^{-1}[A][B]$

b) $\dfrac{\Delta[C]}{\Delta t} = 1.75 \ M^{-2}s^{-1}[A][B]^2$

c) $\dfrac{\Delta[C]}{\Delta t} = 1.75 \ M^{-2}s^{-1}[A]^2[B]$

d) $\dfrac{\Delta[C]}{\Delta t} = 5.60 \ M^{-2}s^{-1}[A]^2[B]$

e) $\dfrac{\Delta[C]}{\Delta t} = 0.0303 \ M^{-1}s^{-1}[B]^2$

Answer: b

12. Given the initial rate data for the reaction $A + B \rightarrow C$, determine the rate expression for the reaction.

[A], M	[B], M	$\Delta[C]/\Delta t$ (initial) M/s
0.0460	0.220	1.31×10^{-4}
0.0920	0.220	5.24×10^{-4}
0.0460	0.440	1.31×10^{-4}

a) $\dfrac{\Delta[C]}{\Delta t} = 1.29 \times 10^{-2} \ M^{-2}s^{-1}[A][B]$

b) $\dfrac{\Delta[C]}{\Delta t} = 2.81 \times 10^{-1} \ M^{-2}s^{-1}[A]^2[B]$

c) $\dfrac{\Delta[C]}{\Delta t} = 6.19 \times 10^{-2} \ M^{-1}s^{-1}[A]^2$

d) $\dfrac{\Delta[C]}{\Delta t} = 5.88 \times 10^{-2} \ M^{-2}s^{-1}[A][B]^2$

e) $\dfrac{\Delta[C]}{\Delta t} = 2.85 \times 10^{-3} \ s^{-1}[A]$

Answer: c

13. For the reaction $A + 2B \rightarrow C$, the rate law is

$$\frac{\Delta[C]}{\Delta t} = k[A][B] \ .$$

What are the units of the rate constant where time is measured in seconds?

a) $\dfrac{1}{M \cdot s}$

b) $\dfrac{1}{M^2 \cdot s}$

c) $\dfrac{1}{s}$

d) $\dfrac{M^2}{s}$

e) $\dfrac{M}{s}$

Answer: a

14. For the reaction A $\rightarrow$ B, the rate law is

$$\frac{\Delta[B]}{\Delta t} = k[A] \; .$$

What are the units of the rate constant where time is measured in seconds?

a) $\dfrac{1}{s}$ b) $M^2 \cdot s$ c) $\dfrac{s}{M}$ d) $\dfrac{M}{s}$ e) $\dfrac{1}{M \cdot s}$

Answer: a

15. For the reaction 2A + B $\rightarrow$ C, the rate law is

$$\frac{\Delta[C]}{\Delta t} = k[A]^2[B] \; .$$

Which of the factor(s) will affect the value of the <u>rate constant</u> for this reaction?
   1. decreasing the temperature
   2. adding a catalyst
   3. decreasing the concentration of reactant A

a) 1 only   b) 2 only   c) 3 only   d) 1 and 2   e) 2 and 3

Answer: d

16. The reaction of NO and O_2 produces NO_2.

$$2\,NO(g) + O_2(g) \rightarrow 2\,NO_2(g)$$

The reaction is second-order with respect to NO(g) and first-order with respect to O_2(g). At a given temperature, the rate constant, k, equals $7.4 \times 10^2 \, M^{-2}s^{-1}$. What is the rate of reaction when the initial concentrations of NO and O_2 are 0.030 M and 0.025 M, respectively?

a) 1.7×10^{-2} M/s   b) 0.56 M/s   c) 1.1 M/s   d) 4.9×10^5 M/s   e) 3.3×10^7 M/s

Answer: a

17. How are the exponents in a rate law determined?

a) They are equal to one for gas phase reactants and zero for solid phase reactants.
b) They are determined by experimentation.
c) They are equal to the coefficients in the overall balanced chemical equation.
d) They are equal to the reactant concentrations.
e) They are equal to one for gas phase reactions and two for aqueous reactions.

Answer: b

15.4 Concentration-Time Relationships: Integrated Rate Laws

18. Which of the following statements is correct for the first-order reaction: A → 2B?

    a) The concentration of A decreases linearly with respect to time.
    b) The concentration of A is constant with respect to time.
    c) The natural logarithm of the concentration of A decreases linearly with respect to time.
    d) The rate of reaction is constant with respect to time.
    e) The rate constant, k, of the reaction decreases linearly with respect to time.

    Answer: c

19. For a second-order decomposition reaction,
    $$2A \rightarrow B \qquad rate = k[A]^2$$
    which of the following functions can be plotted versus time to give a straight line?

    a) $[A]$ b) $\dfrac{k}{[A]^2}$ c) $\ln\dfrac{1}{[A]}$ d) $\ln[A]$ e) $\dfrac{1}{[A]}$

    Answer: e

20. A student analyzed a second-order reaction and obtained the graph below. Unfortunately, the student forgot to label the axes. What are the correct labels for the x and y axes?

    a) x axis = time, y axis = $\ln[A]$
    b) x axis = $\ln[time]$, y axis = $[A]$
    c) x axis = $\ln[time]$, y axis = $[A]$
    d) x axis = time, y axis = $1/[A]$
    e) x axis = $1/time$, y axis = $1/[A]$

    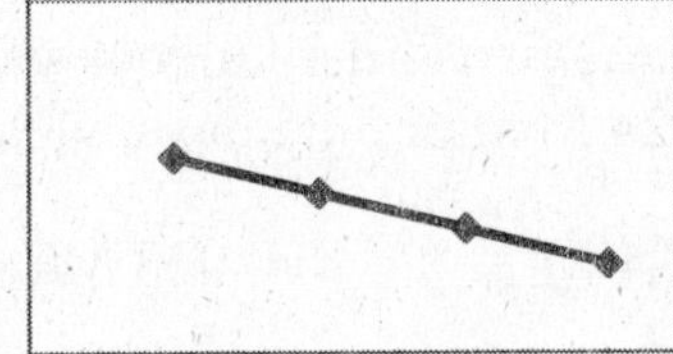

    Answer: d

21. For a reaction, A → B + C, which of the following equations corresponds to the integrated expression for a first-order decomposition reaction?

    a) $[A]_t = -kt + [A]_0$ b) $\ln\dfrac{[A]_t}{[A]_0} = -kt$ c) $\ln[A]_t = \ln[-kt] + \ln[A]_0$

    d) $\dfrac{[A]_t}{[A]_0} = -kt$ e) $\dfrac{1}{[A]_t} = kt + \dfrac{1}{[A]_0}$

    Answer: b

22. The rate constant of a first-order decomposition reaction is 0.0456 hrs^{-1}. If the initial concentration of reactant is 0.072 M, what is the concentration of reactant after 27.0 hours?

    a) 0.0033 M   b) 0.021 M   c) 0.071 M   d) 0.58 M   e) 0.25 M

    Answer: b

23. What is the half-life of a first-order reaction if the rate constant is 6.9×10^{-5} s^{-1}?

   a) 3.5×10^{-5} s   b) 0.097 s   c) 1.0×10^{1} s   d) 1.0×10^{4} s   e) 1.4×10^{4} s

   Answer: d

24. For the first-order decomposition of N_2O_5 at a high temperature, determine the rate constant if the N_2O_5 concentration decreases from 0.88 M to 0.32 M in 175 seconds.

   a) 2.08×10^{-3} s^{-1} b) 3.31×10^{-3} s^{-1} c) 5.09×10^{-2} s^{-1} d) 5.78×10^{-3} s^{-1} e) 1.73×10^{2} s^{-1}

   Answer: d

25. For a reaction, $A \rightarrow B + C$, which of the following equations corresponds to the integrated expression for a zero-order decomposition reaction?

   a) $[A]_t = -kt + [A]_0$ b) $\ln[A]_t = -kt + \ln[A]_0$ c) $\dfrac{1}{[A]_t} = kt + \dfrac{1}{[A]_0}$

   d) $\dfrac{[A]_t}{[A]_0} = -kt$ e) $\ln\dfrac{[A]_t}{[A]_0} = -kt$

   Answer: a

26. The reaction $A \rightarrow B$ follows first-order kinetics with a half-life of 14.3 days. If the concentration of A is 0.024 M after 3.25 days, what is the initial concentration of A?

   a) 0.11 M   b) 0.021 M   c) 0.028 M   d) 0.030 M   e) 0.041 M

   Answer: c

27. Hydrogen peroxide decomposes into water and oxygen in a first-order process.
$$H_2O_2(aq) \rightarrow H_2O(\ell) + 1/2\ O_2(g)$$
At 20.0 °C, the half-life for the reaction is 3.92×10^{4} seconds. If the initial concentration of hydrogen peroxide is 0.32 M, what is the concentration after 3.00 days?

   a) 0.0033 M   b) 0.0056 M   c) 0.0087 M   d) 0.048 M   e) 0.28 M

   Answer: a

28. The decomposition of formic acid follows first-order kinetics.
$$HCO_2H(g) \rightarrow CO_2(g) + H_2(g)$$
The half-life for the reaction at 550 °C is 24 seconds. How many seconds does it take for the formic acid concentration to decrease by 87.5%?

   a) 24 s   b) 36 s   c) 48 s   d) 72 s   e) 96 s

   Answer: d

29. The decomposition of phosphine, PH_3, follows first-order kinetics.

$$4\,PH_3(g) \rightarrow P_4(g) + 6\,H_2(g)$$

The half-life for the reaction at 550 °C is 81.3 seconds. What percentage of phosphine remains after 275 seconds?

a) 2.3%　　　　b) 9.6%　　　　c) 26%　　　　d) 30.%　　　　e) 74%

Answer: b

30. What is the half-life of a first-order reaction if it takes 7.1×10^{-4} seconds for the concentration to decrease from 0.460 M to 0.100 M?

a) 3.2×10^{-4} s　　b) 4.7×10^{-4} s　　c) 0.22 s　　　　d) 1.5 s　　　　e) 2.1×10^{3} s

Answer: a

31. For the second-order reaction below, the rate constant of the reaction is 3.9×10^{-5} $M^{-1}s^{-1}$. How long (in seconds) is required to decrease the concentration of A from 1.07 M to 0.24 M?

$$2A \rightarrow B \qquad rate = k[A]^2$$

a) 5.8×10^{3} s　　　　　　b) 6.5×10^{3} s　　　　　　c) 2.1×10^{4} s
d) 3.8×10^{4} s　　　　　　e) 8.3×10^{4} s

Answer: e

32. For the second-order reaction below, the initial concentration of reactant A is 0.24 M. If the rate constant for the reaction is 1.5×10^{-2} $M^{-1}s^{-1}$, what is the concentration of A after 265 seconds?

$$2A \rightarrow B + C \qquad rate = k[A]^2$$

a) 0.0036 M　　b) 0.0045 M　　c) 0.078 M　　　d) 0.12 M　　　e) 5.2 M

Answer: d

33. For the zero-order reaction below, a graph of _________ versus time will generate a straight line.

$$A \rightarrow B + C \qquad rate = k[A]^0$$

a) $[A]_t$　　　　b) $\ln\dfrac{[A]_t}{[A]_0}$　　　c) $\ln[A]_t$　　　d) $\dfrac{1}{[A]_t}$　　　e) $\dfrac{\ln(2)}{[A]_t}$

Answer: a

34. According to collision theory, which condition(s) must be met in order for molecules to react?
    1. The reacting molecules must collide with sufficient energy to initiate the process of breaking and forming bonds.
    2. A catalyst must be in contact with the reacting molecules for a reaction to occur.
    3. The reacting molecules must collide with an orientation that can lead to rearrangement of the atoms.

    a) 1 only b) 2 only c) 3 only d) 1 and 2 e) 1 and 3

    Answer: e

35. In general, as temperature increases, the rate of a chemical reaction

    a) decreases due to fewer collisions with proper molecular orientation.
    b) increases for exothermic reactions, but decreases for endothermic reactions.
    c) increases due to a greater number of effective collisions.
    d) remains unchanged.
    e) decreases due to an increase in the activation energy.

    Answer: c

36. The Arrhenius equation, $k = Ae^{-E_a/RT}$, relates the rate constant of reaction and temperature. A plot of $\ln(k)$ versus $1/T$ will yield a straight line with a slope of __________.

    a) $-E_a/R$ b) $-E_a$ c) A d) $e^{-E_a/R}$ e) $1/RT$

    Answer: a

37. Molecules must overcome a barrier called the activation energy if they are to react. The highest energy point reached during the progress of a reaction is called the __________.

    a) rate determining step b) transition state c) half-life
    d) elementary step e) intermediate state

    Answer: b

38. For a chemical reaction, the activation energy for the forward reaction is +155 kJ and the activation energy for the backward reaction is +117 kJ. What is the overall energy change for the forward reaction?

    a) -272 kJ b) -38 kJ c) +38 kJ d) +155 kJ e) +272 kJ

    Answer: c

39. Calculate the activation energy, E_a for
$$N_2O_5(g) \; \rightarrow \; 2\,NO_2(g) \; + \; \tfrac{1}{2}\,O_2(g)$$
given k (at 45.0 °C) $= 5.79 \times 10^{-4}$ s^{-1} and k (at 60.0 °C) $= 3.83 \times 10^{-3}$ s^{-1}. ($R = 8.314$ J/K·mol)

a) 0.256 kJ/mol b) 2.83 kJ/mol   c) 31.1 kJ/mol   d) 111 kJ/mol   e) 389 kJ/mol

Answer: d

40. For a given reaction, the rate constant triples when the temperature is increased from 32 °C to 71 °C. What is the activation energy for this reaction? ($R = 8.314$ J/K·mol)

a) 0.234 kJ/mol   b) 0.532 kJ/mol   c) 3.00 kJ/mol   d) 24.6 kJ/mol   e) 53.2 kJ/mol

Answer: d

41. For a given reaction, the activation energy is 21.7 kJ/mol. If the reaction rate constant is 7.4×10^{-2} M^{-1}s^{-1} at 25.0 °C, what is the reaction rate constant at -10.0 °C? ($R = 8.314$ J/K·mol)

a) 8.5×10^{-37} M^{-1}s^{-1} b) 6.7×10^{-3} M^{-1}s^{-1} c) 2.3×10^{-2} M^{-1}s^{-1}
d) 7.4×10^{-2} M^{-1}s^{-1} e) 2.4×10^{-1} M^{-1}s^{-1}

Answer: c

42. The rate constant at 366 K for a first-order reaction is 7.7×10^{-3} s^{-1} and the activation energy is 15.9 kJ/mol. What is the value of the frequency factor, A, in the Arrhenius equation? ($R = 8.314$ J/K·mol)

a) 0.0047 s^{-1} b) 0.70 s^{-1} c) 0.93 s^{-1} d) 1.1 s^{-1} e) 1.4 s^{-1}

Answer: e

43. The effect of adding a catalyst to a reaction is to

a) increase the number of collisions between reactants.
b) lower the activation energy of a reaction.
c) increase the equilibrium constant of a reaction.
d) decrease the yield of the products.
e) increase the enthalpy change of a reaction.

Answer: b

44. The elementary steps for the catalyzed decomposition of dinitrogen monoxide are shown below.

$$N_2O(g) + NO(g) \rightarrow N_2(g) + NO_2(g)$$
$$2\,NO_2(g) \rightarrow 2\,NO(g) + O_2(g)$$

Which of the following statements is/are CORRECT?
1. The overall balanced reaction is $2\,N_2O(g) \rightarrow 2\,N_2(g) + O_2(g)$.
2. NO(g) is a catalyst for the reaction.
3. $NO_2(g)$ is a reaction intermediate.

a) 1 only   b) 2 only   c) 3 only   d) 1 and 3   e) 1, 2, and 3

Answer: e

45. The mechanism of a chemical reaction is given below.

$$(CH_3)_3CCl \rightarrow (CH_3)_3C^+ + Cl^- \qquad (slow)$$
$$(CH_3)_3C^+ + OH^- \rightarrow (CH_3)_3COH \qquad (fast)$$

Which of the following statements concerning the reaction is/are CORRECT?
1. The overall balanced reaction is: $(CH_3)_3CCl + OH^- \rightarrow (CH_3)_3COH + Cl^-$
2. Chloride ion is a reaction intermediate.
3. The following rate law is consistent with the mechanism: rate $= k[(CH_3)_3CCl]$.

a) 1 only   b) 2 only   c) 3 only   d) 1 and 3   e) 1, 2, and 3

Answer: d

46. Nitrogen dioxide reacts with carbon monoxide to produce nitrogen monoxide and carbon dioxide.

$$NO_2(g) + CO(g) \rightarrow NO(g) + CO_2(g)$$

A proposed mechanism for this reaction is

$$2\,NO_2(g) \rightleftharpoons NO_3(g) + NO(g) \qquad \text{(fast, equilibrium)}$$
$$NO_3(g) + CO(g) \rightarrow NO_2(g) + CO_2(g) \qquad \text{(slow)}$$

What is a rate law that is consistent with the proposed mechanism?

a) rate $= k[[NO_2]^2[CO]\,[NO]^{-1}$
b) rate $= k[[NO_2]^2[CO]$
c) rate $= k[NO_2][CO]$
d) rate $= k[NO_3][CO]$
e) rate $= k[NO_2]^2$

Answer: a

47. For the overall reaction
$$A + 2B \rightarrow C$$
which of the following mechanisms yields the correct overall chemical equation and is consistent with the rate equation below?
$$rate = k[A][B]$$

a) $A + B \rightleftharpoons I$ (fast)   b) $A + B \rightarrow I$ (slow)
    $I + A \rightarrow C$ (slow) $I + B \rightarrow C$ (fast)

c) $2B \rightarrow I$ (slow)   d) $2B \rightleftharpoons I$ (fast)
    $A + I \rightarrow C$ (fast) $I + A \rightarrow C$ (slow)

e) $A + 2B \rightleftharpoons I$ (fast)
    $I + B \rightarrow C + B$ (slow)

Answer: b

48. For the overall reaction
$$A + 2B \rightarrow C$$
which of the following mechanisms yields the correct overall chemical equation and is consistent with the rate equation below?
$$rate = k[A]^2[B]$$

a) $A + B \rightleftharpoons I$ (fast)   b) $A + B \rightarrow I$ (slow)
    $I + A \rightarrow C$ (slow) $I + B \rightarrow C$ (fast)

c) $2B \rightarrow I$ (slow)   d) $2B \rightleftharpoons I$ (fast)
    $A + I \rightarrow C$ (fast) $I + A \rightarrow C$ (slow)

e) $A + 2B \rightleftharpoons I$ (fast)
    $I + B \rightarrow C + B$ (slow)

Answer: a

Short Answer Questions

49. If a catalyst is present in the same phase as the reactants and products, it is referred to as a __________ catalyst.

Answer: homogeneous

50. The __________ of an elementary step is defined as the number of reactant molecules that come together in the reaction.

Answer: molecularity

51. The pre-exponential, A, in the Arrhenius equation is called the __________ factor.

   Answer: frequency

52. Radioactive isotopes decay by __________-order kinetics.

   Answer: first

53. Elementary steps in a reaction mechanism often include reaction __________. These (usually) short-lived species, which are at one point produced and then later consumed, do not appear in the overall chemical reaction.

   Answer: intermediates

54. Termolecular elementary steps are rare. Why?

   Answer: A termolecular elementary step requires three reactant particles to simultaneously interact. The probability is small that three particles will effectively collide with the necessary energies and orientations.

55. Carbon dating may be used to date (once living) materials that are between 100 and 40,000 years old. The half-life of the first-order decay of carbon-14 is 5730 years. What percentage of carbon-14 remains in a sample after 40,000 years.

   Answer: 0.8%

Chapter 16
PRINCIPLES OF REACTIVITY: CHEMICAL EQUILIBRIA

CHAPTER OUTLINE

CHAPTER OBJECTIVES

- Understand the nature and characteristics of chemical equilibria (16.1)
- Understand the significance of the equilibrium constant, K, and reaction quotient, Q. (16.2)
- Understand how to use K in quantitative studies of chemical equilibria (16,2. 16.3, 16.4, 16,5, 16.6)

LESSON PLAN

AP Standards

III.C.1	Concept of dynamic equilibrium, physical and chemical; Le Chatelier's principle; Equilibrium constants
III.C.2	Quantitative treatment
III.C.2.a	Equilibrium constants for gaseous reactions : K_p, K_c
III.C.2.b	Equilibrium constants for reactions in solution
III.E.4	Relationship of change in free energy to equilibrium constants and electrode potentials
IV.1	Chemical reactivity and products of chemical reactions

Suggested Time:

Three traditional classes or 2 blocks. Sections 16.1 through 16.4 need to be addressed, as well as Le Chatelier's Principle in the beginning of 16.6. The remainder of the chapter (16.5 and most of 16.6) should be skipped or assigned as reading. Any part of the chapter that concerns AP Standard IV.1 should be included in the lesson.

ASSESSMENT

Exercises:

16.1	Writing Equilibrium Constants (p. 729)	
16.2	The Equilibrium Constant and Extent of Reaction (p. 732)	
16.3	The Reaction Quotient (p. 734)	
16.4	The Reaction Quotient (p. 734)	
16.5	Calculating an Equilibrium Constant, K (p. 736)	
16.6	Calculating Equilibrium Concentrations (p. 738)	
16.7	Calculating an Equilibrium Concentration Using an Equilibrium Constant (p. 741)	
16.8	Manipulating Equilibrium Constant Expressions (p. 743)	
16.9	Manipulating Equilibrium Constant Expressions (p. 743)	
16.10	Effect of Concentration Changes on Equilibrium (p. 746)	
16.11	Effect of Concentration and Volume Changes on Equilibria (p. 747)	
16.12	Disturbing a Chemical Equilibrium (p. 750)	

Answers to Exercises p. A53-A54

Practicing Skills pp. 752-754, Answers IRM pp. 291-296

General Questions pp. 754-757, Answers IRM pp. 298-305

In the Laboratory Questions pp. 757-758, Answers IRM p.306

Summary and Conceptual Questions p. 759, Answers IRM p.307

GLOSSARY

equilibrium A condition in which the forward and reverse reaction rates in a physical or chemical system are equal

equilibrium constant (K) The constant in the equilibrium constant expression

equilibrium constant expression A mathematical expression that relates the concentrations of the reactants and products at equilibrium at a particular temperature to a numerical constant

Haber-Bosch process The direct, catalyzed combination of gaseous nitrogen and hydrogen to product ammonia

reaction quotient(Q) The product of concentrations of products divided by the product of concentrations of the reactants, each raised to the power of its stoichiometric coefficient in the chemical equation

MEDIA MENU

The Media Menu section of this guide lists the interactive exercises, tutorials, active figures (animated, interactive figures) and additional resources available online for Chemistry Now, a unique web-based, assessment-centered personalized learning system for chemistry students. Go to www.cengagelearning.com/login.

Tutorials
- Screen 16.5   Writing Equilibrium Expressions
- Screen 16.9   The reaction quotient, Q
- Screen 16.8   Determining an equilibrium constant
- Screen 16.9   Systems at equilibrium
- Screen 16.10   Determining equilibrium concentrations
- Screen 16.13   The effect of concentration changes on an equilibrium
- Screen 16.12   The effect of temperature changes

Active Figures
- 16.2   The reaction of H_2 and I_2 reaches equilibrium
- 16.7   Addition of more reactant or product to an equilibrium system

Additional Resources
- Screen 16.2   Video of a reversible reaction
- Screen 16.3   Simulation of a chemical equilibrium
- Screen 16.4   Simulation of equilibrium and determination of the constant
- Screen 16.5   Simulation on writing equilibrium expressions
- Screen 16.6   Simulation of an equilibrium
- Screen 16.9   Simulation on the reaction quotient, Q
- Screen 16.8   Simulation on determining an equilibrium constant
- Screen 16.9   Simulation on systems at equilibrium

- Screen 16.11   Animation of Le Chatelier's principle
- Screen 16.13   Simulation on the effect of concentration changes on equilibrium
- Screen 16.14   Simulation of the effect of a volume change for a reaction involving gases
- Podcast Module 21   Using equilibrium constants in calculations

DIAGNOSTIC QUIZ QUESTIONS

Understand the nature and characteristics of chemical equilibria (16.1)

1. Which of the following statements correctly describes a chemical system that has reached equilibrium?
   1. The forward and reverse rates of the reaction are equal.
   2. No further macroscopic changes in the reaction are observed.
   3. The concentration of reactants and products remain constant.
   4. The reaction stops completely.

   a) 1 only   b) 2 only   c) 3 only
   d) 4 only   e) more than one of these

   Answer: e

Understand the significance of the equilibrium constant, K, and reaction quotient, Q.(16.2)

2. Which of the following statements is/are CORRECT?
   1. Product concentrations appear in the numerator of an equilibrium constant expression.
   2. A reaction favors the formation of products if K >> 1.
   3. A chemical system is at equilibrium if the reaction quotient (Q) equals K.

   a) 1 only   b) 2 only   c) 3 only   d) 1 and 2   e) 1, 2, and 3

3. If the reaction quotient, Q, is less than K in a gas phase reaction, then

   a) the chemical system has reached equilibrium.
   b) the temperature must be increased for the reaction to proceed in the forward direction.
   c) the reaction will proceed in the forward direction until equilibrium is established.
   d) the reaction will proceed in the backward direction until equilibrium is established.
   e) the reaction will proceed in the direction that increases the number of gas phase particles.

   Answer: c

Understand how to use K in quantitative studies of chemical equilibria (16,2. 16.3, 16.4, 16,5, 16.6)

4. An aqueous mixture of hydrocyanic acid and ammonia has initial concentrations of 0.100 M $HCN(aq)$ and 0.140 M $NH_3(aq)$. At equilibrium, the $CN^-(aq)$ concentration is 0.055 M. Calculate K for the reaction.

$$HCN(aq) + NH_3(aq) \rightleftharpoons CN^-(aq) + NH_4^+(aq)$$

a) 0.22   b) 0.79   c) 1.5   d) 3.9   e) 14

Answer: b

5. Carbonyl bromide decomposes to carbon monoxide and bromine.

$$COBr_2(g) \rightleftharpoons CO(g) + Br_2(g)$$

K_c is 0.19 at 73 °C. If an initial concentration of 0.63 M $COBr_2$ is allowed to equilibrate, what is the equilibrium concentration of $COBr_2$?

a) 0.26 M   b) 0.28 M   c) 0.35 M   d) 0.37 M   e) 0.40 M

Answer: d

TESTBANK

16.2 The Equilibrium Constant and the Reaction Quotient

1. Which of the following statements is/are CORRECT?
    1. For a chemical system, if the reaction quotient (Q) is greater than K, products must be converted to reactants to reach equilibrium.
    2. For a chemical system at equilibrium, the forward and reverse rates of reaction are equal.
    3. For a chemical system at equilibrium, the concentrations of products divided by the concentrations of reactants equals one.

a) 1 only   b) 2 only   c) 3 only   d) 1 and 2   e) 1, 2, and 3

Answer: d

2. Which of the following statements is/are CORRECT?
    1. Product concentrations appear in the numerator of an equilibrium constant expression.
    2. A reaction favors the formation of products if K >> 1.
    3. A chemical system is at equilibrium if the reaction quotient (Q) equals K.

a) 1 only   b) 2 only   c) 3 only   d) 1 and 2   e) 1, 2, and 3

Answer: e

3. Write the expression for K for the reaction below.

$$Ag^+(aq) + 2\,NH_3(aq) \rightleftharpoons Ag(NH_3)_2^+(aq)$$

a) $K = \dfrac{[Ag^+][NH_3]^2}{[Ag(NH_3)_2^+]}$

b) $K = \dfrac{[Ag(NH_3)_2^+]}{[Ag^+]}$

c) $K = \dfrac{[Ag(NH_3)_2^+]}{[Ag^+][NH_3]}$

d) $K = \dfrac{[Ag^+][NH_3]}{[Ag(NH_3)_2^+]}$

e) $K = \dfrac{[Ag(NH_3)_2^+]}{[Ag^+][NH_3]^2}$

Answer: e

4. Write the expression for K for the reaction below.

$$Mg_3(PO_4)_2(s) \rightleftharpoons 3\,Mg^{2+}(aq) + 2\,PO_4^{3-}(aq)$$

a) $K = [Mg^{2+}]^3[PO_4^{3-}]^2$

b) $K = [3\,Mg^{2+}][2\,PO_4^{3-}]$

c) $K = [Mg^{2+}][PO_4^{3-}]$

d) $K = \dfrac{[Mg_3PO_4]}{[Mg^{2+}]^3[PO_4^{3-}]^2}$

e) $K = \dfrac{[Mg^{2+}]^3[PO_4^{3-}]^2}{[Mg_3PO_4]}$

Answer: a

5. Write the expression for K for the reaction of ammonium ion with hydroxide ion.

$$NH_4^+(aq) + OH^-(aq) \rightleftharpoons NH_3(aq) + H_2O(\ell)$$

a) $K = \dfrac{[NH_4^+][OH^-]}{[NH_3][H_2O]}$

b) $K = \dfrac{[NH_3][H_2O]}{[NH_4^+][OH^-]}$

c) $K = \dfrac{[NH_3]}{[NH_4^+][OH^-]}$

d) $K = \dfrac{[NH_3]}{[NH_4^+]}$

e) $K = \dfrac{[NH_4^+][OH^-]}{[NH_3]}$

Answer: c

6. Write the expression for K_p for the reaction below.

$$4\,Ag(s) + O_2(g) \rightleftharpoons 2\,Ag_2O(s)$$

a) $K_p = \dfrac{P_{Ag}P_{O_2}}{P_{Ag_2O}}$

b) $K_p = \dfrac{P_{Ag}^4 P_{O_2}}{P_{Ag_2O}^2}$

c) $K_p = P_{Ag}^4 P_{O_2}$

d) $K_p = \dfrac{1}{P_{O_2}}$

e) $K_p = P_{O_2}$

Answer: d

7. Write the expression for K_p for the reaction below.

$$2\ NOBr(g) \rightleftharpoons 2\ NO(g) + Br_2(\ell)$$

a) $K_p = \dfrac{P_{NOBr}^2\ P_{Br_2}}{P_{NO}^2}$

b) $K_p = \dfrac{P_{NO}^2}{P_{NOBr}^2}$

c) $K_p = \dfrac{P_{NOBr}}{P_{NO}}$

d) $K_p = \dfrac{P_{NO}^2}{P_{NOBr}^2\ P_{Br_2}}$

e) $K_p = \left[Br_2\right]$

Answer: b

8. Write a balanced chemical equation which corresponds to the following equilibrium constant expression.

$$K_p = \frac{P_{N_2}^{1/2}\ P_{H_2}^{3/2}}{P_{NH_3}}$$

a) $1/2\ N_2(g) + 3/2\ H_2(g) \rightleftharpoons NH_3(g)$

b) $N_2(g) + 3\ H_2(g) \rightleftharpoons 2\ NH_3(g)$

c) $2\ NH_3(g) \rightleftharpoons N_2(g) + 3\ H_2(g)$

d) $NH_3(g) \rightleftharpoons 1/2\ N_2(g) + 3/2\ H_2(g)$

e) $2\ N_2(g) + 6\ H_2(g) \rightleftharpoons 4\ NH_3(g)$

Answer: d

9. Write a balanced chemical equation which corresponds to the following equilibrium constant expression.

$$K = [Al^{3+}]^2[CO_3^{2-}]^3$$

a) $AlCO_3^{+}(aq) \rightleftharpoons Al^{3+}(aq) + CO_3^{2-}(aq)$

b) $Al_2(CO_3)_3(s) \rightleftharpoons 2\ Al^{3+}(aq) + 3\ CO_3^{2-}(aq)$

c) $Al_3(CO_3)_2(aq) \rightleftharpoons 3\ Al^{3+}(aq) + 2\ CO_3^{2-}(aq)$

d) $Al^{3+}(aq) + CO_3^{2-}(aq) \rightleftharpoons AlCO_3^{+}(aq)$

e) $2\ Al^{3+}(aq) + 3\ CO_3^{2-}(aq) \rightleftharpoons Al_2(CO_3)_3(s)$

Answer: b

10. Write a balanced chemical equation which corresponds to the following equilibrium constant expression.

$$K = \frac{[NO_2^-][H_3O^+]}{[HNO_2]}$$

a) $HNO_2(aq) + H_2O(\ell) \rightleftharpoons NO_2^-(aq) + H_3O^+(aq)$

b) $NO_2^-(aq) + H_3O^+(aq) \rightleftharpoons HNO_2(aq) + H_2O(\ell)$

c) $NO_2^-(aq) + H_3O^+(aq) \rightleftharpoons HNO_2(aq)$

d) $H^+(aq) + OH^-(aq) \rightleftharpoons H_2O(\ell)$

e) $HNO_2(aq) \rightleftharpoons NO_2^-(aq) + H_3O^+(aq)$

Answer: a

11. For which one of the following reactions does K_p equal K_c?

a) $2\ CO(g) + O_2(g) \rightleftharpoons 2\ CO_2(g)$ b) $CH_4(g) + 2\ O_2(g) \rightleftharpoons CO_2(g) + 2\ H_2O(g)$

c) $CaCO_3(s) \rightleftharpoons CaO(s) + CO_2(g)$ d) $NH_3(g) \rightleftharpoons 3/2\ H_2(g) + 1/2\ N_2(g)$

e) $2\ O_3(g) \rightleftharpoons 3\ O_2(g)$

Answer: b

12. What is the relationship between K_p and K_c for the reaction below?

$$4\ NH_3(g) + 5\ O_2(g) \rightleftharpoons 4\ NO(g) + 6\ H_2O(g)$$

a) $K_c = \dfrac{K_p}{RT}$ b) $K_c = \dfrac{RT^2}{K_p}$ c) $K_c = (RT)K_p$

d) $K_c = \left(\dfrac{RT}{K_p}\right)^2$ e) $K_c = \left(\dfrac{K_p}{RT}\right)^2$

Answer: a

13. Ozone is formed from oxygen.

$$3\ O_2(g) \rightleftharpoons 2\ O_3(g)$$

Calculate the value of K_p, given that $K_c = 2.5 \times 10^{-29}$ at 298 K. $(R = 0.08206\ \text{L·atm/mol·K})$

a) 1.0×10^{-30} b) 2.1×10^{-30} c) 2.5×10^{-29} d) 3.3×10^{-28} e) 6.1×10^{-28}

Answer: a

14. The oxidation of sulfur dioxide produces sulfur trioxide.

$$2\,SO_3(g) \rightleftharpoons 2\,SO_2(g) + O_2(g)$$

Calculate the value of K_c, given that $K_p = 3.6 \times 10^{-3}$ at 999 K. ($R = 0.08206$ L·atm/mol·K)

a) 5.4×10^{-7} b) 4.4×10^{-5} c) 3.0×10^{-4} d) 0.044 e) 0.30

Answer: b

15. If the reaction quotient, Q, is less than K in a gas phase reaction, then

a) the chemical system has reached equilibrium.
b) the temperature must be increased for the reaction to proceed in the forward direction.
c) the reaction will proceed in the forward direction until equilibrium is established.
d) the reaction will proceed in the backward direction until equilibrium is established.
e) the reaction will proceed in the direction that increases the number of gas phase particles.

Answer: c

16. A 5.0 L flask is filled with 0.25 mol SO_3, 0.50 mol SO_2, and 1.0 mol O_2, and allowed to reach equilibrium. Assume the temperature of the mixture is chosen so that $K_c = 0.12$. Predict the effect on the concentration of SO_3 as equilibrium is achieved by using Q, the reaction quotient.

$$2\,SO_3(g) \rightleftharpoons 2\,SO_2(g) + O_2(g)$$

a) $[SO_3]$ will decrease because $Q > K$. b) $[SO_3]$ will decrease because $Q < K$.
c) $[SO_3]$ will increase because $Q < K$. d) $[SO_3]$ will increase because $Q > K$.
e) $[SO_3]$ will remain the same because $Q = K$.

Answer: d

17. Consider the reaction $A(aq) \rightleftharpoons 2\,B(aq)$ where $K_c = 4.1$ at 25 °C. If 0.50 M A(aq) and 1.5 M B(aq) are initially present in a 1.0 L flask at 25 °C, what change in concentrations (if any) will occur in time?

a) [A] will decrease and [B] will decrease. b) [A] will decrease and [B] will increase.
c) [A] will increase and [B] will decrease. d) [A] will increase and [B] will increase.
e) [A] and [B] remain unchanged.

Answer: c

18. At a high temperature, equal concentrations of 0.200 mol/L of $H_2(g)$ and $I_2(g)$ are initially present in a flask. The H_2 and I_2 react according to the balanced equation below.

$$H_2(g) + I_2(g) \rightleftharpoons 2\,HI(g)$$

When equilibrium is reached, the concentration of $H_2(g)$ has decreased to 0.046 mol/L. What is the equilibrium constant, K_c, for the reaction?

a) 3.3   b) 11   c) 15   d) 22   e) 45

Answer: e

19. An aqueous mixture of hydrocyanic acid and ammonia has initial concentrations of 0.100 M $HCN(aq)$ and 0.140 M $NH_3(aq)$. At equilibrium, the $CN^-(aq)$ concentration is 0.055 M. Calculate K for the reaction.

$$HCN(aq) + NH_3(aq) \rightleftharpoons CN^-(aq) + NH_4^+(aq)$$

a) 0.22   b) 0.79   c) 1.5   d) 3.9   e) 14

Answer: b

20. Excess $PbBr_2(s)$ is placed in water at 25°C. At equilibrium, the solution contains 0.012 M $Pb^{2+}(aq)$. What is the equilibrium constant for the reaction below?

$$PbBr_2(s) \rightleftharpoons Pb^{2+}(aq) + 2\,Br^-(aq)$$

a) 4.3×10^{-7} b) 1.7×10^{-6} c) 6.9×10^{-6} d) 1.4×10^{-4} e) 2.9×10^{-4}

Answer: c

21. At 25 °C, 0.138 mg AgBr dissolves in 10.0 L of water. What is the equilibrium constant for the reaction below?

$$AgBr(s) \rightleftharpoons Ag^+(aq) + Br^-(aq)$$

a) 5.40×10^{-13} b) 5.40×10^{-11} c) 1.90×10^{-8} d) 7.35×10^{-7} e) 1.90×10^{-6}

Answer: a

22. At a given temperature, 0.0524 mol $NO_2(g)$ is placed in a 1.00 L flask. After reaching equilibrium, the concentration of $NO_2(g)$ is 3.9×10^{-3} M. What is K_c for the reaction below?

$$2\,NO_2(g) \rightleftharpoons N_2O_4(g)$$

a) 2.9×10^{-4} b) 13   c) 1.6×10^{3} d) 3.2×10^{3} e) 3.4×10^{3}

Answer: c

23. When 0.50 mole CH_3CO_2H is dissolved in water to a volume of 1.00 L, 0.60% of the CH_3CO_2H dissociates to form $CH_3CO_2^-(aq)$. What is the equilibrium constant for the reaction?

$$CH_3CO_2H(aq) + H_2O(\ell) \rightleftharpoons CH_3CO_2^-(aq) + H_3O^+(aq)$$

a) 1.8×10^{-5} b) 3.6×10^{-5} c) 3.0×10^{-3} d) 6.0×10^{-3} e) 0.45

Answer: a

24. A sealed tube is prepared with 1.66 atm PCl_5 at 500 K. The PCl_5 decomposes until equilibrium is established.

$$PCl_5(g) \rightleftharpoons PCl_3(aq) + Cl_2(g)$$

The equilibrium pressure in the tube is 2.28 atm. Calculate K_p.

a) 0.17   b) 0.23   c) 0.37   d) 0.62   e) 4.3

Answer: c

25. Nitrosyl bromide decomposes according to the chemical equation below.

$$2\ NOBr(g) \rightleftharpoons 2\ NO(g) + Br_2(g)$$

When 0.260 atm of NOBr is sealed in a flask and allowed to reach equilibrium, 22% of the NOBr decomposes. What is the equilibrium constant, K_p, for the reaction?

a) 2.3×10^{-3} b) 4.5×10^{-3} c) 3.5×10^{-2} d) 4.8×10^{-2} e) 8.0×10^{-2}

Answer: a

16.4 Using Equilibrium Constants in Calculations

26. At 2010 K, the equilibrium constant, K_c, for the following reaction is 2.5×10^3.

$$2\ NO(g) \rightleftharpoons N_2(g) + O_2(g)$$

If the equilibrium concentrations of N_2 and O_2 are 0.28 mol/L and 0.38 mol/L at 2010 K, what is the equilibrium concentration of NO?

a) 1.8×10^{-9} M   b) 4.3×10^{-5} M   c) 3.8×10^{-3} M   d) 6.5×10^{-3} M   e) 153 M

Answer: d

27. A gaseous mixture of NO_2 and N_2O_4 is in equilibrium. If the concentration of N_2O_4 is 5.3×10^{-5} M, what is the concentration of NO_2?

$$2\ NO_2(g) \rightleftharpoons N_2O_4(g) \qquad K_c = 170$$

a) 9.7×10^{-14} M   b) 3.1×10^{-7} M   c) 5.6×10^{-4} M   d) 9.0×10^{-3} M   e) 9.5×10^{-2} M

Answer: c

28.   At 25 °C, the decomposition of dinitrogen tetraoxide
$$N_2O_4(g) \rightleftharpoons 2\,NO_2(g)$$
has an equilibrium constant (K_p) of 0.144. At equilibrium, the total pressure of the system is 0.0758 atm. What is the partial pressure of each gas?

a) 0.0745 atm $NO_2(g)$ and 0.0385 $N_2O_4(g)$ b) 0.0549 atm $NO_2(g)$ and 0.0209 $N_2O_4(g)$
c) 0.0531 atm $NO_2(g)$ and 0.0227 $N_2O_4(g)$ d) 0.0502 atm $NO_2(g)$ and 0.0256 $N_2O_4(g)$
e) 0.0381 atm $NO_2(g)$ and 0.0377 $N_2O_4(g)$

Answer: b

29. The equilibrium constant at 25 °C for the dissolution of silver iodide is 8.5×10^{-17}.
$$AgI(s) \rightleftharpoons Ag^+(aq) + I^-(aq)$$
If an excess quantity of AgI(s) is added to water and allowed to equilibrate, what is the equilibrium concentration of I^-?

a) 7.2×10^{-33} M   b) 4.3×10^{-17} M   c) 8.5×10^{-17} M   d) 6.5×10^{-9} M   e) 9.2×10^{-9} M

Answer: e

30. Hydrogen iodide can decompose into hydrogen and iodine gases.
$$2\,HI(g) \rightleftharpoons H_2(g) + I_2(g) \qquad K_c = 0.0180$$
If 0.068 mol HI(g) is sealed in a 1.0 L flask, what is the concentration of $I_2(g)$ when equilibrium is established?

a) 0.0012 M   b) 0.0072 M   c) 0.0091 M   d) 0.018 M   e) 0.035 M

Answer: b

31. If 0.98 mol PCl_5 is placed in a 1.0 L flask and allowed to reach equilibrium at a given temperature, what is the final concentration of Cl_2 in the flask?
$$PCl_5(g) \rightleftharpoons PCl_3(aq) + Cl_2(g) \qquad K_c = 0.47$$

a) 0.30 M   b) 0.40 M   c) 0.48 M   d) 0.50 M   e) 0.68 M

Answer: c

32.   The equilibrium constant, K_c, for the following reaction is 1.0×10^{-5} at 1500 K.
$$N_2(g) + O_2(g) \rightleftharpoons 2\,NO(g)$$
If 0.570 M N_2 and 0.570 M O_2 are allowed to equilibrate at 1500 K, what is the concentration of NO?

a) 5.7×10^{-6} M   b) 9.0×10^{-4} M   c) 2.4×10^{-4} M   d) 1.8×10^{-3} M   e) 2.4×10^{-3} M

Answer: d

33. At 800 K, the equilibrium constant, K_p, for the following reaction is 3.2×10^{-7}.
$$2\,H_2S(g) \rightleftharpoons 2\,H_2(g) + S_2(g)$$
A reaction vessel at 800 K initially contains 3.50 atm of H_2S. If the reaction is allowed to equilibrate, what is the equilibrium pressure of H_2?

a) 1.0×10^{-3} atm   b) 2.1×10^{-3} atm   c) 5.0×10^{-3} atm   d) 9.9×10^{-3} atm   e) 2.0×10^{-2} atm

Answer: e

34. Carbonyl bromide decomposes to carbon monoxide and bromine.
$$COBr_2(g) \rightleftharpoons CO(g) + Br_2(g)$$
K_c is 0.19 at 73 °C. If an initial concentration of 0.63 M $COBr_2$ is allowed to equilibrate, what is the equilibrium concentration of $COBr_2$?

a) 0.26 M   b) 0.28 M   c) 0.35 M   d) 0.37 M   e) 0.40 M

Answer: d

16.5 More about Balanced Equations and Equilibrium Constants

35. For the following reaction,
$$SO_2(g) + 1/2\,O_2(g) \rightleftharpoons SO_3(g)$$
the equilibrium constant, K_p, is 0.870 at 627 °C. What is the equilibrium constant, at 627 °C, for the reaction below?
$$2\,SO_3(g) \rightleftharpoons 2\,SO_2(g) + O_2(g)$$

a) -1.74   b) -.757   c) 1.32   d) 2.30   e) 5.28

Answer: c

36. The equilibrium constant (K_c) for the following reaction is 6.7×10^{-10} at 630 °C.
$$N_2(s) + O_2(g) \rightleftharpoons 2\,NO(g)$$
What is the equilibrium constant for the reaction below at the same temperature?
$$NO(g) \rightleftharpoons 1/2\,N_2(g) + 1/2\,O_2(g)$$
a) 3.9×10^4 b) 5.5×10^4 c) 7.5×10^8 d) 1.5×10^9 e) 3.0×10^9

Answer: a

37. Given the following chemical equilibria,

$$N_2(g) + O_2(g) \rightleftharpoons 2\,NO(g) \qquad\qquad K_1$$

$$4\,NH_3(g) + 5\,O_2(g) \rightleftharpoons 4\,NO(g) + 6\,H_2O(g) \qquad K_2$$

$$H_2(g) + 1/2\,O_2(g) \rightleftharpoons H_2O(g) \qquad\qquad K_3$$

Determine the method used to calculate the equilibrium constant for the reaction below.

$$N_2(g) + 3\,H_2(g) \rightleftharpoons 2\,NH_3(g) \qquad\qquad K_c$$

a) $K_c = K_1 \times K_2 \times K_3$

b) $K_c = \dfrac{K_1\,K_3^3}{\sqrt{K_2}}$

c) $K_c = K_1 + \dfrac{K_2}{2} + 3K_3$

d) $K_c = \dfrac{3(K_1 \times K_2)}{2K_3}$

e) $K_c = \dfrac{K_1 \times \sqrt{K_2}}{K_3^3}$

Answer: b

38. Determine the equilibrium constant for the following reaction

$$Ca(OH)_2(s) + 2\,H^+(aq) \rightleftharpoons Ca^{2+}(aq) + 2\,H_2O(\ell)$$

given the chemical reactions below.

$$Ca(OH)_2(s) \rightleftharpoons Ca^{2+}(aq) + 2\,OH^-(aq) \qquad\qquad K = 6.5 \times 10^{-6}$$

$$H_2O(\ell) \rightleftharpoons H^+(aq) + OH^-(aq) \qquad\qquad K = 1.0 \times 10^{-14}$$

a) 6.5×10^{-20} b) 1.5×10^{-9} c) 1.3×10^{9} d) 1.5×10^{19} e) 6.5×10^{22}

Answer: e

39. Given the following equilibria,

$$PbBr_2(s) \rightleftharpoons Pb^{2+}(aq) + 2\,Br^-(aq) \qquad\qquad K_1 = 6.6 \times 10^{-6}$$

$$Pb(OH)_2(s) \rightleftharpoons Pb^{2+}(aq) + 2\,OH^-(aq) \qquad\qquad K_2 = 1.4 \times 10^{-15}$$

determine the equilibrium constant, K_c, for the following reaction.

$$PbBr_2(s) + 2\,OH^-(aq) \rightleftharpoons Pb(OH)_2(s) + 2\,Br^-(aq)$$

a) 9.2×10^{-21} b) 2.1×10^{-10} c) 6.6×10^{-6} d) 4.7×10^{9} e) 1.1×10^{20}

Answer: d

40. Given the following equilibria,

$$Ni(OH)_2(s) \rightleftharpoons Ni^{2+}(aq) + 2\,OH^-(aq) \qquad K_1 = 5.6 \times 10^{-16}$$

$$Ni^{2+}(aq) + 6\,NH_3(aq) \rightleftharpoons Ni(NH_3)_6^{2+}(aq) \qquad K_2 = 5.6 \times 10^8$$

determine the equilibrium constant, K_c, for the following reaction.

$$Ni(OH)_2(s) + 6\,NH_3(aq) \rightleftharpoons Ni(NH_3)_6^{2+}(aq) + 2\,OH^-(aq)$$

a) 1.0×10^{-24} b) 3.1×10^{-7} c) 3.2×10^6 d) 5.6×10^8 e) 1.0×10^{24}

Answer: b

41. Assume that the following chemical reaction is at equilibrium.

$$2\,ICl(g) \rightleftharpoons I_2(g) + Cl_2(g) \qquad \Delta H° = +26.9 \text{ kJ}$$

At 25 °C, $K_p = 2.0 \times 10^5$. If the temperature is decreased to 5 °C, which statement applies?

a) K_p will decrease and the reaction will proceed in the backward direction.
b) K_p will decrease and the reaction will proceed in the forward direction.
c) K_p will remain unchanged and the reaction will proceed in the forward direction.
d) K_p will remain unchanged and the reaction will proceed in the backward direction.
e) K_p will increase and the reaction will proceed in the forward direction.

Answer: a

42. Assume that the following <u>endothermic</u> chemical reaction is at equilibrium.

$$C(s) + H_2O(g) \rightleftharpoons H_2(g) + CO(g)$$

Which of the following statements is/are CORRECT?
   1. Increasing the concentration of $H_2O(g)$ will cause the reaction to proceed in the forward direction, increasing the equilibrium concentration of $CO(g)$.
   2. Increasing the temperature will cause the reaction to proceed in the forward direction, increasing the equilibrium concentration of $CO(g)$.
   3. Increasing the amount of $C(s)$ will cause the reaction to proceed in the forward direction, increasing the equilibrium concentration of $CO(g)$.

a) 1 only   b) 2 only   c) 3 only   d) 1 and 2   e) 1, 2, and 3

Answer: d

43. The thermochemical equation for the formation of ammonia from elemental nitrogen and hydrogen is as follows.

$$N_2(g) + 3\,H_2(g) \rightleftharpoons 2\,NH_3(g) \qquad \Delta H = -92.2 \text{ kJ}$$

Given a system that is initially at equilibrium, which of the following actions cause the reaction to proceed to the left?

a) adding $N_2(g)$ b) removing $NH_3(g)$ c) adding a catalyst
d) decreasing the temperature   e) removing $H_2(g)$

Answer: e

44. In which of the following equilibrium systems will an increase in the pressure have no effect on the concentrations of products and reactants?

a) $H_2(g) + F_2(g) \rightleftharpoons 2\,HF(g)$

b) $N_2(g) + 3\,H_2(g) \rightleftharpoons 2\,NH_3(g)$

c) $CaCO_3(s) \rightleftharpoons CaO(s) + CO_2(g)$

d) $2\,NOBr(g) \rightleftharpoons 2\,NO(g) + Br_2(g)$

e) $2\,H_2O(g) + O_2(g) \rightleftharpoons 2\,H_2O_2(g)$

Answer: a

45. A flask contains the following chemical system at equilibrium.
$$Ca(OH)_2(s) \rightleftharpoons Ca^{2+}(aq) + 2\,OH^-(aq)$$
Addition of which of the following substances will decrease the solubility of $Ca(OH)_2(s)$ in water?
1.   aqueous hydrochloric acid
2.   aqueous sodium hydroxide
3.   solid calcium hydroxide

a) 1 only   b) 2 only   c) 3 only   d) 1 and 3   e) 1, 2, and 3

Answer: b

Short Answer Questions

46. The symbol Q is called the __________ .

Answer: reaction quotient

47. In 1913, the Haber-Bosch process was patented. The product of the Haber-Bosch process is __________ .

Answer: ammonia (or NH_3)

48. If a stress is applied to an equilibrium system, the system will respond is such a way as to relieve that stress. This is a statement of __________ principle.

Answer: Le Chatelier's

49. When 1.0 mole of acetic acid is diluted with water to a volume of 1.0 L at 25 °C, 0.42% of the acetic acid ionizes to form acetate ion and hydronium ion.
$$CH_3CO_2H(aq) + H_2O(\ell) \rightleftharpoons CH_3CO_2^-(aq) + H_3O^+(aq)$$
What percentage of the acid ionizes when 0.75 mole of acetic acid is diluted with water to 1.0 L at 25 °C?

Answer: 0.49%

50. The standard enthalpy of formation of ammonia is -46.1 kJ/mol.

$$1/2\ N_2(g)\ +\ 3/2\ H_2(g)\ \rightleftharpoons\ NH_3(g)$$

Commercially, the reaction is carried out at high temperatures. Using your knowledge of kinetics and equilibrium, explain an advantage and a disadvantage of synthesizing ammonia at high temperatures.

Answer: The rate of formation of ammonia is negligible at room temperature. The advantage of high temperatures is that reaction rates increase. The disadvantage is that the equilibrium constant for this (exothermic) reaction decreases as the temperature of the reaction increases.

Chapter 17
THE CHEMISTRY OF ACIDS AND BASES

CHAPTER OUTLINE

CHAPTER 17: The Chemistry of Acids and Bases

CHAPTER OBJECTIVES

- Use the Brønsted-Lowry and Lewis theories of acids and bases. (17.1, 17.2, 17.3)
- Apply the principles of chemical equilibrium to acids and bases in aqueous solution (17.4, 17.5, 17.7)
- Predict the outcome of reactions of acids and bases (17.6, 17.7)
- Understand the influence of structure and bonding on acid-base properties (17.9, 17.10)

LESSON PLAN

AP Standards

III.A.1   Acid-base reactions; concepts of Arrhenius, Brønsted-Lowry and Lewis; coordination complexes; amphoterism

III.A.3.c   Electrochemistry: prediction of the direction of redox reactions

III.C.2.a   Equilibrium constants for gaseous reactions : K_p, K_c

Suggested Time:

Four traditional classes or three blocks.   Sections 17.4 and 17.9 need to be covered in detail. The Brønsted Lowry theory (17.2), predicting direction of reactions (17.5), and determining constants for acids and bases (17.7) must also be included in the lessons. The remainder of the chapter can be skipped or assigned as reading homework. This chapter, along with 18, will require the students to practice many examples and problems in order to successfully grasp the material.

ASSESSMENT

Exercises:

GLOSSARY

acid ionization constant (K_a) The equilibrium constant for the ionization of an acid in aqueous solution

amphiprotic A substance that can behave as either a Brønsted acid or a Brønsted base

amphoteric A substance, such as a metal hydroxide, that can behave as either an acid or base

autoionization of water Interaction of two water molecules to produce a hydronium ion and a hydroxide ion by proton transfer

autoionization constant for water (K_w) The equilibrium constant for the autoionization of water

base ionization constant (K_b) The equilibrium constant for the ionization of a base in aqueous solution

complex ion An ion in which a metal ion is bonded to one or more molecules or ion to define a structural unit

conjugate acid-base pair A pair of compounds or ions that differ by the presence of one H^+ unit

coordination complexes An ion in which a metal ion is bonded to one or more molecules or ion to define a structural unit

Lewis acid A substance that can accept a pair of electrons to form a new bond

Lewis base A substance that can donate a pair of electrons to form a new bond

monoprotic acids A Brønsted acid that can donate one proton

oxoacids Acids that contain an atom bonded to one or more oxygen atoms

pH The negative of the base-10 logarithm of the hydrogen ion concentration

pOH The negative of the base-10 logarithm of the hydroxide ion concentration

polyprotic acids A Brønsted acid that can donate more than one proton

polyprotic bases A Brønsted base that can accept more than one proton

MEDIA MENU

The Media Menu section of this guide lists the interactive exercises, tutorials, active figures (animated, interactive figures) and additional resources available online for Chemistry Now, a unique web-based, assessment-centered personalized learning system for chemistry students. Go to www.cengagelearning.com/login.

Tutorials
- Screen 17.2   Acids, Bases, and their conjugates
- Screen 17.4   Using pH and pOH
- Screen 17.5   The pH of solutions of acids and bases
- Screen 17.8   Determining K_a and K_b values
- Screen 17.9   Estimating the pH of weak acid solutions
- Screen 17.10   Estimating the pH following an acid-base reaction

- Screen 17.12   Lewis acids and bases
- Screen 17.14   Neutral Lewis Acids

Exercises
- Screen 17.2   Acids, Bases, and their conjugates
- Screen 17.4   Using pH and pOH

Active Figures
- 17.1   pH and pOH

Additional Resources
- Screen 17.3   Simulation of the effect of temperature on K_w
- Screen 17.4   Simulation on using pH and pOH
- Screen 17.6   Table of K_a and K_b values
- Screen 17.11   Simulation on showing the pH of a number of cation/anion combinations
- Screen 17.7   Predicting the direction of acid-base reactions
- Screen 17.11   Simulation of acid-base properties of salts
- Screen 17.13   Description of cationic Lewis acids
- Screen 17.15   Self-study module on molecular interpretation of acid-base behavior
- Podcast Module 22   Calculations with equilibrium constants

DIAGNOSTIC QUIZ QUESTIONS

Use the Brønsted-Lowry and Lewis theories of acids and bases. (17.1, 17.2, 17.3)

1. According to the Brønsted-Lowry definition, a base

    a) is a weak electrolyte.
    b) increases the OH^- concentration in an aqueous solution.
    c) is an electron-pair donor.
    d) increases the pH of a solution.
    e) is a proton acceptor.

    Answer: e

2. In the following reaction,
    $$HS^-(aq) + H_2O(\ell) \rightleftharpoons S^{2-}(aq) + H_3O^+(aq)$$

    a) H_3O^+ is an acid and HS^- is its conjugate base.
    b) HS^- is an acid and S^{2-} is its conjugate base.
    c) HS^- is an acid and H_2O is its conjugate base.
    d) H_2O is an acid and S^{2-} is its conjugate base.
    e) H_3O^+ is an acid and S^{2-} is its conjugate base.

    Answer: b

Apply the principles of chemical equilibrium to acids and bases in aqueous solution (17.4, 17.5, 17.7)

3. Which of the following chemical equations corresponds to the acid ionization constant, K_a, for ammonium ion (NH_4^+)?

a) $NH_3(aq) + H_3O^+(aq) \rightleftharpoons NH_4^+(aq) + H_2O(\ell)$

b) $NH_4^+(aq) + H_2O(\ell) \rightleftharpoons NH_3(aq) + H_3O^+(aq)$

c) $NH_4^+(aq) + OH^-(aq) \rightleftharpoons NH_3(aq) + H_2O(\ell)$

d) $NH_4^+(aq) + H_3O^+(aq) \rightleftharpoons NH_5^+(aq) + H_2O(\ell)$

e) $NH_3(aq) + H_2O(\ell) \rightleftharpoons NH_4^+(aq) + OH^-(aq)$

Answer: b

4. If 0.10 M aqueous solutions are prepared of each of the following acids, which produces the solution with the lowest pH?

a) benzoic acid, $K_a = 6.3 \times 10^{-5}$ b) acetic acid, $K_a = 1.8 \times 10^{-5}$
c) hydrocyanic acid, $K_a = 4.0 \times 10^{-10}$ d) hydrogen sulfite ion, $K_a = 6.2 \times 10^{-8}$
e) formic acid, $K_a = 1.8 \times 10^{-4}$

Answer: e

Predict the outcome of reactions of acids and bases (17.6, 17.7)

5. What is the equilibrium constant for the following reaction,

$$HCN(aq) + NO_2^-(aq) \rightleftharpoons CN^-(aq) + HNO_2(aq)$$

and does the reaction favor the formation of reactants or products? The acid dissociation constant, K_a, of HCN is 4.0×10^{-10} and the acid dissociation constant of HNO_2 is 4.5×10^{-4}.

a) $K = 1.00$. The reaction favors neither the formation of reactants nor products.
b) $K = 8.9 \times 10^{-7}$. The reaction favors the formation of products.
c) $K = 8.9 \times 10^{-7}$. The reaction favors the formation of reactants.
d) $K = 1.1 \times 10^{6}$. The reaction favors the formation of products.
e) $K = 1.1 \times 10^{6}$. The reaction favors the formation of reactants.

Answer: c

Understand the influence of structure and bonding on acid-base properties (17.9, 17.10)

6. Which of the following molecules or ions is the strongest acid?

a) $CH_3CO_2^-$ b) CH_3CO_2H c) CFH_2CO_2H d) CF_2HCO_2H e) CF_3CO_2H

Answer: e

TEST BANK

<u>17.1 Acids and Bases: A Review</u>

1. According to the Arrhenius definition, a base is substance that when dissolved in water increases the concentration of

    a) hydroxide ion.   b) hydronium ion.   c) protons.
    d) electrolytes.   e) cations.

    Answer: a

2. According to the Brønsted-Lowry definition, an acid

    a) increases the H_3O^+ concentration in an aqueous solution.
    b) is a strong electrolyte.
    c) is a proton acceptor.
    d) increases the pH of a solution.
    e) is a proton donor.

    Answer: e

3. According to the Brønsted-Lowry definition, a base

    a) is a weak electrolyte.   b) increases the OH^- concentration in an aqueous solution.
    c) is an electron-pair donor.   d) increases the pH of a solution.
    e) is a proton acceptor.

    Answer: e

<u>17.2 The Brønsted-Lowry concepts of Acids and Bases</u>

4. Which of the following substances is never a Brønsted-Lowry acid in an aqueous solution?

    a) hydrogen fluoride, $HF(g)$ b) sodium phosphate, $Na_3PO_4(s)$
    c) ammonium chloride, $NH_4Cl(s)$ d) hydrogen bromide, $HBr(g)$
    e) sodium bicarbonate, $NaHCO_3(s)$

    Answer: b

5. Which of the following substances is never a Brønsted-Lowry base in an aqueous solution?

    a) potassium hydroxide, $NaOH(g)$ b) sodium dihydrogen phosphate, $NaH_2PO_4(s)$
    c) sodium phosphate, $Na_3PO_4(s)$ d) ammonium chloride, $NH_4Cl(g)$
    e) sodium bicarbonate, $NaHCO_3(s)$

    Answer: d

6. Which equation depicts hydrocyanic acid, HCN, behaving as a Brønsted-Lowry acid in water?

a) $HCN(aq) + OH^-(aq) \rightleftharpoons OCN^-(aq) + H_2(g)$

b) $HCN(aq) + H_3O^+(aq) \rightleftharpoons H_2CN^+(aq) + H_2O(\ell)$

c) $HCN(aq) + H_2O(\ell) \rightleftharpoons CN^-(aq) + H_3O^+(aq)$

d) $CN^-(aq) + H_3O^+(aq) \rightleftharpoons HCN(aq) + H_2O(\ell)$

e) $CN^-(aq) + H_2O(\ell) \rightleftharpoons OCN^-(aq) + H_2(s)$

Answer: c

7. Which equation depicts dihydrogen phosphate ion behaving as a Brønsted-Lowry base in water?

a) $H_2PO_4^-(aq) + H_2O(\ell) \rightleftharpoons H_3PO_4(aq) + OH^-(aq)$

b) $H_2PO_4^-(aq) + OH^-(aq) \rightleftharpoons HPO_4^{2-}(aq) + H_2O(\ell)$

c) $H_2PO_4^-(aq) + H_2O(\ell) \rightleftharpoons HPO_4^{2-}(aq) + H_3O^+(aq)$

d) $H_2PO_4^-(aq) + O^{2-}(aq) \rightleftharpoons PO_4^{3-}(aq) + H_2O(\ell)$

e) $H_2PO_4^-(aq) + H_2O(\ell) \rightleftharpoons 2\,H_2O(\ell) + PO_3(s)$

Answer: a

8. In the following reaction,

$$HS^-(aq) + H_2O(\ell) \rightleftharpoons S^{2-}(aq) + H_3O^+(aq)$$

a) H_3O^+ is an acid and HS^- is its conjugate base.　　b) HS^- is an acid and S^{2-} is its conjugate base.
c) HS^- is an acid and H_2O is its conjugate base.　　d) H_2O is an acid and S^{2-} is its conjugate base.
e) H_3O^+ is an acid and S^{2-} is its conjugate base.

Answer: b

9. In the following reaction
$$F^-(aq) + H_2O(\ell) \rightleftharpoons HF(aq) + OH^-(aq)$$

a) F^- is and acid and HF is its conjugate base.　　b) F^- is a base and HF is its conjugate acid.
c) F^- is a base and H_2O is its conjugate acid　　d) F^- is an acid and OH^- is its conjugate base.
e) HF is an acid and OH^- is its conjugate base.

Answer: b

10. The conjugate base of H_2O is __________.

a) H_3O^+ b) OH^+ c) O^{2-} d) H^+ e) OH^-

Answer: e

11. What is the conjugate base of $[Fe(H_2O)_6]^{3+}(aq)$?

a) H_3O^+ b) $[Fe(H_2O)_6]^{2+}$ c) $[Fe(H_2O)_5H_3O]^{4+}$
d) $[Fe(H_2O)_5OH]^{2+}$ e) $[Fe(H_2O)_5]^{3+}$

Answer: d

12. The conjugate acid of HPO_4^{2-} is __________.

a) OH^- b) PO_4^{3-} c) $H_2PO_4^-$ d) H_2O e) H_3O^+

Answer: c

13. Molecules or ions that can alternately behave as either a Brønsted-Lowry acid or base are called

a) polyanions.   b) hydronium ions.   c) polyprotic acids or bases.
d) conjugate acids or bases.   e) amphiprotic.

Answer: e

14. Which is NOT an amphiprotic species in water?

a) $CH_3CO_2^-$ b) HSO_4^- c) HPO_4^{2-} d) HS^- e) HCO_3^-

Answer: a

15. Which of the following molecules or ions is amphiprotic in water?

a) NH_4^+ b) HF c) CH_3CO_2H d) $HC_2O_4^-$ e) CN^-

Answer: d

17.3 Water and the pH Scale

16. At 20 °C, the water ionization constant, K_w, is 6.8×10^{-15}. What is the H_3O^+ concentration in neutral water at this temperature?

a) 4.6×10^{-29} M   b) 3.4×10^{-15} M   c) 6.8×10^{-15} M   d) 8.2×10^{-8} M   e) 1.0×10^{-7} M

Answer: d

17. A solution is prepared by diluting 0.055 mol HCl with water to a volume of 2.5 L. What is the pH of the solution?

a) 0.0220   b) 1.26   c) 1.66   d) 2.90   e) 3.82

Answer: c

18. What is the H_3O^+ concentration in 5.8×10^{-4} M KOH(aq) at 25 °C? ($K_w = 1.0 \times 10^{-14}$)

a) 5.8×10^{-18} M   b) 1.7×10^{-11} M   c) 5.8×10^{-4} M   d) 11 M   e) 5.8×10^{10} M

Answer: b

19. Which one of the following aqueous solutions will have a pH of 0.00 at 25 °C? ($K_w = 1.0 \times 10^{-14}$)

a) 0.10 M HNO_3 b) 1.0 M NaOH   c) 1.0×10^1 M HBr
d) 1.0 M KCl   e) 1.0 M HCl

Answer: e

20. What is the pH of 2.75 M HCl(aq) at 25 °C? ($K_w = 1.0 \times 10^{-14}$)

a) -0.439   b) 0.439   c) 1.012   d) 2.75   e) 3.636

Answer: a

21. What is the pH of 4.4×10^{-4} M KOH(aq) at 25 °C? ($K_w = 1.0 \times 10^{-14}$)

a) -3.36   b) 3.36   c) 6.27   d) 7.73   e) 10.64

Answer: e

22. What is the H_3O^+ concentration of an aqueous solution with a pH of 8.77?

a) 1.7×10^{-9} M   b) 5.9×10^{-6} M   c) 1.6×10^{-4} M   d) 5.23 M   e) 5.9×10^8 M

Answer: a

23. What is the OH^- concentration of an aqueous solution with a pH of 4.45? ($K_w = 1.0 \times 10^{-14}$)

a) 2.8×10^{-10} M   b) 3.5×10^{-5} M   c) 7.1×10^{-5} M   d) 9.55 M   e) 2.8×10^4 M

Answer: a

24. What is the OH$^-$ concentration of an aqueous solution with a pH of 13.38? ($K_w = 1.0 \times 10^{-14}$)

   a) 4.17×10^{-14} M   b) 6.47×10^{-9} M   c) 1.55×10^{-6} M   d) 0.24 M   e) 0.62 M

   Answer: d

25. An aqueous solution with a pH of 12.00 is diluted from 1.0 L to 4.0 L. What is the pH of the diluted solution?

   a) 3.00   b) 6.78   c) 7.22   d) 11.40   e) 19.93

   Answer: d

26. An aqueous solution with a pH of 2.60 is diluted from 1.0 L to 2.5 L. What is the pH of the diluted solution?

   a) 1.04   b) 2.20   c) 3.00   d) 3.22   e) 6.50

   Answer: c

<u>17.4 Equilibrium Constants for Acids and Bases</u>

27. Which of the following chemical equations corresponds to the acid ionization constant, K_a, for ammonium ion (NH$_4^+$)?

   a) $NH_3(aq) + H_3O^+(aq) \rightleftharpoons NH_4^+(aq) + H_2O(\ell)$

   b) $NH_4^+(aq) + H_2O(\ell) \rightleftharpoons NH_3(aq) + H_3O^+(aq)$

   c) $NH_4^+(aq) + OH^-(aq) \rightleftharpoons NH_3(aq) + H_2O(\ell)$

   d) $NH_4^+(aq) + H_3O^+(aq) \rightleftharpoons NH_5^+(aq) + H_2O(\ell)$

   e) $NH_3(aq) + H_2O(\ell) \rightleftharpoons NH_4^+(aq) + OH^-(aq)$

   Answer: b

28. Which of the following chemical equations corresponds to the base ionization constant, K_b, for hydrogen phosphate ion (HPO$_4^{2-}$)?

   a) $HPO_4^{2-}(aq) + H_3PO_4(aq) \rightleftharpoons 2\ H_2PO_4^-(aq)$

   b) $HPO_4^{2-}(aq) + H_2O(\ell) \rightleftharpoons PO_4^{3-}(aq) + H_3O^+(aq)$

   c) $HPO_4^{2-}(aq) + OH^-(aq) \rightleftharpoons PO_4^{3-}(aq) + H_2O(\ell)$

   d) $HPO_4^{2-}(aq) + H_3O^+(aq) \rightleftharpoons H_2PO_4^-(aq) + H_2O(\ell)$

   e) $HPO_4^{2-}(aq) + H_2O(\ell) \rightleftharpoons H_2PO_4^-(aq) + OH^-(aq)$

   Answer: e

29. If 0.10 M aqueous solutions are prepared of each of the following acids, which produces the solution with the lowest pH?

a) benzoic acid, $K_a = 6.3 \times 10^{-5}$ b) acetic acid, $K_a = 1.8 \times 10^{-5}$
c) hydrocyanic acid, $K_a = 4.0 \times 10^{-10}$ d) hydrogen sulfite ion, $K_a = 6.2 \times 10^{-8}$
e) formic acid, $K_a = 1.8 \times 10^{-4}$

Answer: e

30. Which of the following weak acids has the strongest conjugate base in an aqueous solution?

a) acetic acid, $K_a = 1.8 \times 10^{-5}$ b) hydrocyanic acid, $K_a = 4.0 \times 10^{-10}$
c) hydrogen sulfite ion, $K_a = 6.2 \times 10^{-8}$ d) nitrous acid, $K_a = 4.5 \times 10^{-4}$
e) phosphoric acid, $K_a = 7.5 \times 10^{-3}$

Answer: b

31. Ammonium ion has a pK_a value of 9.25. What is the value of pK_b for ammonia?

a) 5.6×10^{-10} b) 1.8×10^{-5} c) 0.108   d) 0.75   e) 4.75

Answer: e

32. At 25 °C, all of the following ionic compounds produce a basic aqueous solution, except __________.

a) $KClO_4$ b) Na_2CO_3 c) $NaNO_2$ d) KCN e) $NaCH_3CO_2$

Answer: a

33. At 25 °C, all of the following ions produce an acidic solution, except __________.

a) NH_4^+ b) HSO_3^- c) HPO_4^{2-} d) $[Fe(H_2O)_6]^{3+}$ e) $[Al(H_2O)_6]^{3+}$

Answer: c

34. Which ionic compound forms a pH-neutral aqueous solution at 25 °C?

a) Na_2CO_3 b) KNO_3 c) LiF d) NH_4Cl e) K_2S

Answer: b

35. What is the equilibrium constant for the following reaction,

$$HCN(aq) + NO_2^-(aq) \rightleftharpoons CN^-(aq) + HNO_2(aq)$$

and does the reaction favor the formation of reactants or products? The acid dissociation constant, K_a, of HCN is 4.0×10^{-10} and the acid dissociation constant of HNO_2 is 4.5×10^{-4}.

a) $K = 1.00$. The reaction favors neither the formation of reactants nor products.
b) $K = 8.9 \times 10^{-7}$. The reaction favors the formation of products.
c) $K = 8.9 \times 10^{-7}$. The reaction favors the formation of reactants.
d) $K = 1.1 \times 10^{6}$. The reaction favors the formation of products.
e) $K = 1.1 \times 10^{6}$. The reaction favors the formation of reactants.

Answer: c

17.6 Types of Acid-Base Reactions

36. What is the equilibrium constant for the following reaction at 25 °C.

$$HCl(aq) + KOH(aq) \rightleftharpoons H_2O(\ell) + KCl(aq)$$

a) 1.00×10^{-14} b) 1.00×10^{-7} c) 1.00
d) $1.00 \times 10^{+7}$ e) $1.00 \times 10^{+14}$

Answer: e

37. Given the following acid dissociation constants,

$$K_a\,(HCO_2H) = 1.8 \times 10^{-4}$$
$$K_a\,(NH_4^+) = 5.6 \times 10^{-10}$$

determine the equilibrium constant for the reaction below at 25 °C.

$$NH_4^+(aq) + HCO_2^-(aq) \rightleftharpoons NH_3(aq) + HCO_2H(aq)$$

a) 1.0×10^{-13} b) 3.1×10^{-6} c) 1.8×10^{-4} d) 3.2×10^{5} e) 9.9×10^{12}

Answer: b

38. Given the following equilibrium constants,

$$K_a\,(HSO_4^-) = 1.2 \times 10^{-2}$$
$$K_b\,(CH_3CO_2^-) = 5.6 \times 10^{-10}$$
$$K_w = 1.00 \times 10^{-14}$$

determine the equilibrium constant for the reaction below at 25 °C.

$$HSO_4^-(aq) + CH_3CO_2^-(aq) \rightleftharpoons SO_4^{2-}(aq) + CH_3CO_2H(aq)$$

a) 6.7×10^{-12} b) 2.1×10^{-7} c) 1.5×10^{-3} d) 6.7×10^{2} e) 2.1×10^{7}

Answer: d

CHAPTER 17: The Chemistry of Acids and Bases

<u>17.7 Calculations with Equilibrium Constants</u>

39. What is the H_3O^+ concentration in 0.66 M NH_4^+(aq)? (K_a of $NH_4^+ = 5.6 \times 10^{-10}$)

    a) 3.7×10^{-10} M   b) 8.5×10^{-10} M   c) 1.9×10^{-5} M   d) 2.4×10^{-5} M   e) 2.9×10^{-5} M

Answer: c

40. What is the OH^- concentration in 0.48 M F^-(aq)? (K_b of $F^- = 1.4 \times 10^{-11}$)

    a) 6.7×10^{-12} M   b) 1.4×10^{-11} M   c) 3.9×10^{-9} M   d) 1.7×10^{-6} M   e) 2.6×10^{-6} M

Answer: e

41. What is the pH of 0.25 M aqueous acetic acid? (K_a of $CH_3CO_2H = 1.8 \times 10^{-5}$)

    a) 0.60   b) 2.67   c) 4.74   d) 5.35   e) 9.26

Answer: b

42. What is the pH of 0.18 M aqueous formate ion? (K_b of $HCO_2^- = 5.6 \times 10^{-11}$)

    a) 3.75   b) 5.50   c) 8.50   d) 10.25   e) 13.26

Answer: c

43. The pH of aqueous 0.10 M nitrite ion is 8.17. What is the K_b of this base?

    a) 4.6×10^{-16} b) 2.2×10^{-11} c) 1.6×10^{-6} d) 1.6×10^{-5} e) 1.2

Answer: b

44. The pH of aqueous 0.50 M hypochlorous acid, HClO, is 3.88. What is the K_a of this acid?

    a) 7.6×10^{-11} b) 3.5×10^{-8} c) 2.9×10^{-7} d) 8.7×10^{-6} e) 1.3×10^{-4}

Answer: b

45. What is the pH of the solution which results from mixing 150 mL of 0.50 M CH_3CO_2H(aq) and 150 mL of 0.50 M NaOH(aq) at 25 °C? (K_a of $CH_3CO_2H = 1.8 \times 10^{-5}$)

    a) 2.67   b) 4.74   c) 4.93   d) 8.26   e) 9.07

Answer: e

46. What is the pH of the solution which results from mixing 75 mL of 0.50 M NH_3(aq) and 75
mL of 0.50 HCl(aq) at 25 °C? (K_b for $NH_3 = 1.8 \times 10^{-5}$)

a) 0.60 b) 2.67 c) 4.74 d) 4.93 e) 9.26

Answer: d

17.8 Polyprotic Acids and Bases

47. Which of the following chemical equations corresponds to K_{a2} of phosphoric acid?

a) HPO_4^{2-}(aq) + $H_2O(\ell)$ $\rightleftharpoons$ PO_4^{3-}(aq) + H_3O^+(aq)

b) PO_4^{3-}(aq) + $H_2O(\ell)$ $\rightleftharpoons$ HPO_4^{2-}(aq) + OH^-(aq)

c) H_3PO_4(aq) + $H_2O(\ell)$ $\rightleftharpoons$ $H_2PO_4^-$(aq) + H_3O^+(aq)

d) H_3PO_4(aq) + 2 $H_2O(\ell)$ $\rightleftharpoons$ HPO_4^{2-}(aq) + 2 H_3O^+(aq)

e) $H_2PO_4^-$(aq) + $H_2O(\ell)$ $\rightleftharpoons$ HPO_4^{2-}(aq) + H_3O^+(aq)

Answer: e

48. Which of the following chemical equations corresponds to K_{b2} for SO_3^{2-}?

a) HSO_3^-(aq) + $H_2O(\ell)$ $\rightleftharpoons$ H_2SO_3(aq) + OH^-(aq)

b) SO_3^{2-}(aq) + H_3O^+(aq) $\rightleftharpoons$ HSO_3^-(aq) + $H_2O(\ell)$

c) H_2SO_3(aq) + OH^-(aq) $\rightleftharpoons$ HSO_3^-(aq) + $H_2O(\ell)$

d) HSO_3^-(aq) + OH^-(aq) $\rightleftharpoons$ SO_3^{2-}(aq) + $H_2O(\ell)$

e) SO_3^{2-}(aq) + $H_2O(\ell)$ $\rightleftharpoons$ HSO_3^-(aq) + OH^-(aq)

Answer: a

49. What is the pH of 1.0 M Na_2SO_3(aq) at 25 °C? ($K_{a1} = 1.2 \times 10^{-2}$, $K_{a2} = 6.2 \times 10^{-8}$)

a) 3.40 b) 6.03 c) 7.96 d) 10.40 e) 10.60

Answer: e

17.9 The Lewis Concept of Acids and Bases

50. Identify from the following list of molecules and ions which two behave as Lewis bases:
H_2O, F^-, BH_3, Fe^{3+}.

a) H_2O an F^- b) F^- and BH_3 c) BH_3 and Fe^{3+} d) H_2O and Fe^{3+} e) F^- and Fe^{3+}

Answer: a

51. All of the following species behave as Lewis acids EXCEPT _________.

    a) BH_3 b) Al^{3+} c) SO_3 d) $Al(OH)_3$ e) NH_3

    Answer: e

17.10 Molecular Structure, Bonding, and Acid-Base Behavior

52. All of the following compounds are acids containing chlorine. Which compound is the weakest acid?

    a) HCl b) $HClO$ c) $HClO_2$ d) $HClO_3$ e) $HClO_4$

    Answer: b

53. Which of the following molecules or ions is the strongest acid?

    a) $CH_3CO_2^-$ b) CH_3CO_2H c) CFH_2CO_2H d) CF_2HCO_2H e) CF_3CO_2H

    Answer: e

Short Answer Questions

54. When a Lewis acid and a Lewis base combine, the product may be referred to as an acid-base _________.

    Answer: adduct (or complex)

55. A molecule that can behave as either a Brønsted-Lowry acid or base is termed _________.

    Answer: amphiprotic (or amphoteric)

56. When a Lewis acid combines with a Lewis base, the base supplies both electrons to the bond. This type of chemical bond is called a _________ covalent bond.

    Answer: coordinate

57. The chemical equations below show the reaction of $Al(OH)_3$ as a Lewis acid and as a Lewis base, respectively.

$$Al(OH)_3(s) + OH^-(aq) \rightleftharpoons Al(OH)_4^-(aq)$$

$$Al(OH)_3(s) + 3 H_3O^+(aq) \rightleftharpoons Al^{3+}(aq) + 3 H_2O(\ell)$$

    Substances that can behave as either Lewis acids or bases are called _________ substances.

    Answer: amphoteric

58. Write a net ionic equation for the neutralization reaction of hydrochloric acid and potassium hydroxide. Identify the spectator ion(s).

Answer: Net ionic equation: $H_3O^+(aq) + OH^-(aq) \rightleftharpoons 2\,H_2O(\ell)$; spectator ions: Cl^- and K^+.

59. Which is the stronger Brønsted-Lowry acid, $Fe(H_2O)_6^{2+}$ or $Fe(H_2O)_6^{3+}$? Explain.

Answer: $Fe(H_2O)_6^{3+}$. When a proton dissociates from the oxygen atom, the oxygen is left with an extra electron. This negative charge is stabilized by the interaction of the oxygen with the iron cation. Since the Fe^{3+} is more positively charged than the Fe^{2+}, it can more effectively stabilize the negative charge.

Chapter 18
PRINCIPLES OF REACTIVITY:
OTHER ASPECTS OF AQUEOUS EQUILIBRIA

CHAPTER OUTLINE

CHAPTER OBJECTIVES

- Understand the common ion effect (18.1)
- Understand the control of pH in aqueous solutions with buffers (18.2)
- Evaluate the pH in the course of acid-base titrations (18.3)
- Apply chemical equilibrium concepts to the solubility of ionic compounds (18.4, 18.5, 18,6)

LESSON PLAN

AP Standards

III.A.2	Precipitation reactions
III.C.2.b.(2)	Solubility product constants and their application to precipitation and the dissolution of slightly Soluble compounds
III.C.2.b.(3)	Common ion effect; buffers ; hydrolysis
IV.1	Chemical reactivity and products of chemical reaction

Suggested Time:
Three traditional classes or two blocks. This chapter stems from the topics learned in chapter 17. Depending on how well the students were able to grasp the previous chapter, you can move through 18 fairly quickly. Sections 18.1, 18.4, and 18.5 are necessary material. Topics that deal with chemical reactivity and products should at least be reviewed.

ASSESSMENT

Exercises:
18.1   Common Ion Effect (p. 814)

CHAPTER 18: Principles of Reactivity: Other Aspects of Aqueous Equilibria

Answers to Exercises p. A55-A57
Practicing Skills pp. 850-854, Answers IRM pp. 332-351
General Questions pp. 854-856, Answers IRM pp. 351-357
In the Laboratory Questions pp. 856-857, Answers IRM pp. 367-361
Summary and Conceptual Questions pp. 857-859, Answers IRM pp. 361-363

GLOSSARY

buffer A solution that resists a change in pH when hydroxide or hydronium ions are added

common ion effect The limiting of acid (or base) ionization caused by addition of its conjugate base (or conjugate acid)

equivalence point The point at which one reactant has been completely consumed by addition of another reactant

formation constant An equilibrium constant for the formation of a complex ion

Henderson-Haselbalch Equation An equation used to determine the pH of buffer solutions

indicator A substance used to signal the equivalence point of a titration by a change in some physical property such as color

ligands The molecules or anions bonded to the central metal atom in a coordination compound

saturated A stable solution in which the maximum amount of solute has been dissolved

solubility product constant (K_{sp}) An equilibrium constant relating the concentrations of the ionization products of a dissolved substance

titrant The substance being added during a titration

MEDIA MENU

The Media Menu section of this guide lists the interactive exercises, tutorials, active figures (animated, interactive figures) and additional resources available online for Chemistry Now, a unique web-based, assessment-centered personalized learning system for chemistry students. Go to www.cengagelearning.com/login.

Tutorials
- Screen 18.3   Buffer solutions
- Screen 18.4   pH of buffer solutions
- Screen 18.4   The Henderson-Hasselbalch equation
- Screen 18.5   Preparing buffer solutions
- Screen 18.6   Adding reagents to a buffer solution
- Screen 18.7   Titration curves
- Screen 18.13   Determining K_{sp} experimentally
- Screen 18.14   Estimating salt solubility using K_{sp}
- Screen 18.15   Common ion effect
- Screen 18.17   When a precipitation reaction can occur
- Screen 18.19   Complex ion formation and solubility

Active Figures
- 18.2   Buffer solutions
- 18.5   The change in pH during the titration of a weak acid with a strong base

Additional Resources
- Screen 18.2   Simulation of the common ion effect
- Screen 18.3   Simulation on buffer solutions
- Screen 18.4   Simulation of pH of buffer solutions
- Screen 18.4   Simulation on the Henderson-Hasselbalch equation
- Screen 18.7   Preparing buffer solutions
- Screen 18.7   Simulation on titration curves
- Screen 18.8   Simulation of a titration of a weak polyprotic acid
- Screen 18.9   Simulation of a titration of a weak base with a strong acid
- Screen 18.10   Acid-base indicators
- Screen 18.11   Review of precipitation reactions
- Screen 18.12   Simulation on solubility product constants
- Screen 18.15   Simulation on the common ion effect
- Screen 18.16   Self-study molecule on solubility and pH
- Screen 18.17   Simulation on when a precipitation reaction occurs

CHAPTER 18: Principles of Reactivity: Other Aspects of Aqueous Equilibria

- Screen 18.18   Information on combining equilibria
- Podcast Module 23   Controlling pH: Buffer Solutions

DIAGNOSTIC QUIZ QUESTIONS

Understand the common ion effect

1. Given a solution of 0.10 M NH_4^+(aq), what is the effect of adding NaOH(aq) to this solution?
   1. The pH will increase.
   2. The concentration of NH_3(aq) will increase.
   3. The concentration of NH_4^+(aq) will decrease.

   a) 1 only   b) 2 only   c) 3 only   d) 1 and 3   e) 1, 2, and 3

   Answer: e

Understand the control of pH in aqueous solutions with buffers

2. Which of the following statements concerning buffers are correct?
   1. Buffers are composed of a weak acid diluted with water.
   2. Buffers are resistant to pH changes when diluted with water.
   3. Buffers are used as colored indicators for acid-base reactions

   a) 1 only   b) 2 only   c) 3 only   d) 1 and 3   e) 1, 2, and 3

   Answer: b

3. Which of the following combinations would be best to buffer an aqueous solution at a pH of 9.0?

   a) H_3PO_4 and $H_2PO_4^-$, $K_{a1} = 7.5 \times 10^{-3}$ b) HNO_2 and NO_2^-, $K_a = 4.5 \times 10^{-4}$
   c) CH_3CO_2H and CH_3COO^-, $K_a = 1.8 \times 10^{-5}$ d) $H_2PO_4^-$ and HPO_4^{2-}, $K_{a2} = 6.2 \times 10^{-8}$
   e) NH_4^+ and NH_3, $K_a = 5.7 \times 10^{-10}$

   Answer: e

4. What is the pH of the buffer that results when 15.0 g of NaH_2PO_4 and 15.0 g of Na_2HPO_4 are diluted with water to a volume of 0.50 L? (K_a of $H_2PO_4^- = 6.2 \times 10^{-8}$, the molar masses of NaH_2PO_4 and Na_2HPO_4 are 120.0 g/mol and 142.0 mol, respectively)

   a) 7.13   b) 7.21   c) 7.28   d) 8.05   e) 8.39

   Answer: a

Evaluate the pH in the course of acid-base titrations

5. A 50.00 mL sample of vinegar is titrated with 0.584 M NaOH(aq). If the titration requires
    32.80 mL of NaOH(aq), what is the concentration of acetic acid in the vinegar?

    a) 0.0100 M   b) 0.0192 M   c) 0.0292 M   d) 0.383 M   e) 0.890 M

    Answer: d

6. Which one of the following conditions is always true for a titration of a weak acid with a
    strong base?

    a) A colored indicator with a pK_a less than 7 should be used.
    b) If a colored indicator is used, it must change color rapidly in the weak acid's buffer region.
    c) Equal volumes of weak acid and strong base are required to reach the equivalence point.
    d) The equivalence point occurs at a pH equal to 7.
    e) The equivalence point occurs at a pH greater than 7.

    Answer: e

7. A 25.0 mL sample of 0.10 M sodium acetate is titrated with 0.10 M HCl(aq). What is the pH
    after the addition of 35.0 mL of HCl(aq)? (K_b of $CH_3CO_2^- = 5.6 \times 10^{-10}$)

    a) 1.00   b) 1.23   c) 1.78   d) 4.75   e) 9.25

    Answer: c

Apply chemical equilibrium concepts to the solubility of ionic compounds

8. The solubility of $PbBr_2$ in water is 4.33 g in 1.0 L of water at 25 °C. What is the value of K_{sp}
    for $PbBr_2$?

    a) 1.6×10^{-6} b) 6.6×10^{-6} c) 1.3×10^{-5} d) 1.4×10^{-4} e) 19

    Answer: b

9. The K_{sp} of $PbCl_2$ is 1.7×10^{-5} at 25 °C. What is the concentration of Cl^-(aq) in a saturated
    solution of $PbCl_2$(aq)?

    a) 4.1×10^{-3} M   b) 1.6×10^{-2} M   c) 2.6×10^{-2} M   d) 3.2×10^{-2} M   e) 5.1×10^{-2} M

    Answer: d

10. What mass of NaF (molar mass = 41.99 g/mol) must be added to 1.0 L of 0.0040 M Pb^{2+}(aq) to initiate precipitation of PbF_2(s)? The K_{sp} of PbF_2 is 3.3×10^{-8}. Assume no volume change occurs upon addition of NaF.

a) 3.5×10^{-4} g   b) 0.0014 g   c) 0.0029 g   d) 0.12 g   e) 0.17 g

Answer: d

TEST BANK

<u>18.1 The Common Ion Effect</u>

1. An acid-base equilibrium system is created by dissolving 0.10 mol HF in water to a volume of 1.0 L. What is the effect of adding 0.050 mol F^-(aq) to this solution?

a) The pH of the solution will decrease.
b) Some F^-(aq) will react with H_3O^+, increasing the concentration of HF(aq) and reestablishing the solution equilibrium.
c) The addition of F^-(aq) will have no effect on the pH or the concentration of HF(aq).
d) Some HF(aq) will ionize, increasing the concentration of F^-(aq) and decreasing the pH.
e) Some HF(aq) will ionize, increasing the concentration of F^-(aq) and increasing the pH.

Answer: b

2. Given a solution of 0.10 M NH_4^+(aq), what is the effect of adding NaOH(aq) to this solution?
   1. The pH will increase.
   2. The concentration of NH_3(aq) will increase.
   3. The concentration of NH_4^+(aq) will decrease.

a) 1 only   b) 2 only   c) 3 only   d) 1 and 3   e) 1, 2, and 3

Answer: e

3. What is the pH of a solution that results from diluting 0.50 mol formic acid (HCO_2H) and 0.10 mol sodium formate ($NaHCO_2$) with water to a volume of 1.0 L? (K_a of $HCO_2H = 1.8 \times 10^{-4}$)

a) 2.22   b) 3.05   c) 3.74   d) 3.98   e) 4.44

Answer: b

4. What is the pH of a solution that results from adding 50.0 mL of 0.100 M HCl to 50.0 mL of 0.330 M NH_3? (K_b of $NH_3 = 1.8 \times 10^{-5}$)

a) 1.00   b) 1.30   c) 4.38   d) 9.62   e) 11.16

Answer: d

5. What is the pH of a solution that results from adding 25 mL of 0.50 M NaOH to 75 mL of 0.50 M CH_3CO_2H? (K_a of $CH_3CO_2H = 1.8 \times 10^{-5}$)

   a) 4.44   b) 4.74   c) 5.05   d) 9.26   e) 13.10

   Answer: a

6. What is the pH of an aqueous solution composed of 0.30 M HF and 0.10 M F^-? (K_a of HF $= 7.2 \times 10^{-4}$)

   a) 0.477   b) 1.83   c) 2.67   d) 3.62   e) 6.99

   Answer: c

18.2 Controlling pH: Buffer Solutions

7. Which of the following combinations would be best to buffer an aqueous solution at a pH of 4.5?

   a) H_3PO_4 and $H_2PO_4^-$, $K_{a1} = 7.5 \times 10^{-3}$ b) HNO_2 and NO_2^-, $K_a = 4.5 \times 10^{-4}$
   c) CH_3CO_2H and CH_3COO^-, $K_a = 1.8 \times 10^{-5}$ d) $H_2PO_4^-$ and HPO_4^{2-}, $K_{a2} = 6.2 \times 10^{-8}$
   e) NH_4^+ and NH_3, $K_a = 5.7 \times 10^{-10}$

   Answer: c

8. Which of the following combinations would be best to buffer an aqueous solution at a pH of 9.0?

   a) H_3PO_4 and $H_2PO_4^-$, $K_{a1} = 7.5 \times 10^{-3}$ b) HNO_2 and NO_2^-, $K_a = 4.5 \times 10^{-4}$
   c) CH_3CO_2H and CH_3COO^-, $K_a = 1.8 \times 10^{-5}$ d) $H_2PO_4^-$ and HPO_4^{2-}, $K_{a2} = 6.2 \times 10^{-8}$
   e) NH_4^+ and NH_3, $K_a = 5.7 \times 10^{-10}$

   Answer: e

9. Each of the following mixtures can produce an effective buffer solution EXCEPT

   a) HCl and KCl.   b) Na_2HPO_4 and Na_3PO_4.
   c) $NaHCO_3$ and Na_2CO_3.   d) NaH_2PO_4 and Na_2HPO_4.
   e) HF and NaF.

   Answer: a

10. All of the following statements concerning acid-base buffers are true EXCEPT

   a) buffers are resistant pH changes upon addition of small quantities of strong acids or bases.
   b) buffers are used as colored indicators in acid-base titrations.
   c) the pH of a buffer is close to the pK_a of the weak acid from which it is made.
   d) buffers contain appreciable quantities of a weak acid and its conjugate base.
   e) buffers are resistant to changes in pH when diluted with water.

   Answer: b

11. Which of the following mathematical expressions is the Henderson-Hasselbalch equation?

   a) $pK_a = pH + \log\left(\dfrac{[\text{conjugate base}]}{[\text{acid}]}\right)$

   b) $pH = pK_a + \log\left(\dfrac{[OH^-]}{[H_3O^+]}\right)$

   c) $pH = pK_a + \log\left(\dfrac{[\text{acid}]}{[\text{conjugate base}]}\right)$

   d) $pK_a = pH + \log\left(\dfrac{[OH^-]}{[H_3O^+]}\right)$

   e) $pH = pK_a + \log\left(\dfrac{[\text{conjugate base}]}{[\text{acid}]}\right)$

   Answer: e

12. What is the pH of a buffer composed of 0.50 M $H_2PO_4^-$(aq) and 0.30 M HPO_4^{2-}(aq)? (K_a of $H_2PO_4^-$ is 6.2×10^{-8})

   a) 4.32   b) 6.81   c) 6.99   d) 7.20   e) 7.43

   Answer: c

13. What is the pH of the buffer that results when 15.0 g of NaH_2PO_4 and 15.0 g of Na_2HPO_4 are diluted with water to a volume of 0.50 L? (K_a of $H_2PO_4^- = 6.2 \times 10^{-8}$, the molar masses of NaH_2PO_4 and Na_2HPO_4 are 120.0 g/mol and 142.0 mol, respectively)

   a) 7.13   b) 7.21   c) 7.28   d) 8.05   e) 8.39

   Answer: a

14. What is the pH of the buffer that results when 12 g sodium formate ($NaHCO_2$) is mixed with 250 mL of 0.50 M formic acid (HCO_2H) and diluted with water to 1.0 L? (K_a of $HCO_2H = 1.8 \times 10^{-4}$)

   a) 3.29   b) 3.60   c) 3.74   d) 3.89   e) 5.12

   Answer: d

15. What is the pH of a buffer that results when 0.50 mole of H_3PO_4 is mixed with 0.25 mole of NaOH and diluted with water to 1.00 L? (The acid dissociation constants of phosphoric acid are $K_{a1} = 7.5 \times 10^{-3}$, $K_{a2} = 6.2 \times 10^{-8}$, and $K_{a3} = 3.6 \times 10^{-13}$)

a) 1.82   b) 2.12   c) 6.91   d) 7.21   e) 12.44

Answer: b

16. What is the pH of a buffer that results when 0.50 mole of H_3PO_4 is mixed with 0.75 mole of NaOH and diluted with water to 1.00 L? (The acid dissociation constants of phosphoric acid are $K_{a1} = 7.5 \times 10^{-3}$, $K_{a2} = 6.2 \times 10^{-8}$, and $K_{a3} = 3.6 \times 10^{-13}$)

a) 1.82   b) 2.12   c) 6.91   d) 7.21   e) 12.44

Answer: d

17. What is the pH of a buffer that results when 0.60 mol NaF is mixed with 100.0 mL of 2.00 M HCl(aq) and diluted with water to 250 mL? (K_a of HF $= 7.2 \times 10^{-4}$)

a) 2.67   b) 2.84   c) 3.14   d) 3.44   e) 3.62

Answer: d

18. What mass of solid $NaCH_3CO_2$ (molar mass $= 82.0$ g/mol) should be added to 1.0 L of 0.50 M CH_3CO_2H to make a buffer with a pH of 7.21? (pK_a of $CH_3CO_2H = 7.21$)

a) 0.0 g   b) 1.9 g   c) 41 g   d) 71 g   e) 1.6×10^2 g

Answer: c

19. If the ratio of base to acid in a buffer increases by a factor of 10, the pH of the buffer

a) increases by 1.   b) decreases by 1.   c) increases by 10.
d) decreases by 10.   e) remains unchanged.

Answer: a

20. The K_a of bicarbonate ion, HCO_3^-, is 4.8×10^{-11}. What $[ClO_3^-]/[HClO_3]$ ratio is necessary to make a buffer with a pH of 10.00?

a) 0.48   b) 0.52   c) 1.0   d) 1.9   e) 2.1

Answer: a

21. A buffer is prepared by combining 250 mL of 0.300 M NaOH and 250 mL of a 0.800 M weak acid, HA. If the pH of the buffer is 8.44, what is the K_a of the acid?

    a) 1.4×10^{-9} b) 2.2×10^{-9} c) 3.6×10^{-9} d) 6.1×10^{-9} e) 7.3×10^{8}

    Answer: b

22. A buffer is composed of 0.250 mol $H_2PO_4^-$ and 0.250 mol HPO_4^{2-} diluted with water to a volume of 1.00 L. The pH of the buffer is 7.210. How many moles of NaOH must be added to increase the pH to 8.210?

    a) 0.045 mol   b) 0.125 mol   c) 0.205 mol   d) 0.250 mol   e) 2.50 mol

    Answer: c

23. How many moles of HCl must be added to 1.0 L of 1.0 M NH_3(aq) to make a buffer with a pH of 9.00? (pK_a of NH_4^+ = 9.25)

    a) 0.36 mol   b) 0.44 mol   c) 0.56 mol   d) 0.64 mol   e) 1.8 mol

    Answer: d

24. A buffer contains 0.50 mol CH_3CO_2H and 0.50 mol $CH_3CO_2^-$ diluted with water to 1.0 L. How many moles of NaOH are required to increase the pH of the buffer to 5.24? (pK_a of CH_3CO_2H = 4.74)

    a) 0.16 mol   b) 0.20 mol   c) 0.26 mol   d) 0.34 mol   e) 1.6 mol

    Answer: c

18.3 Acid-Base Titrations

25. Which of the following conditions is/are met at the equivalence point of the titration of a monoprotic weak base with a strong acid?
    1. The moles of acid added from the buret equals the initial moles of weak base.
    2. The volume of acid added from the buret equals the volume of base titrated.
    3. The pH of the solution is greater than 7.00.

    a) 1 only   b) 2 only   c) 3 only   d) 1 and 3   e) 2 and 3

    Answer: a

26. Which one of the following conditions is always true for a titration of a weak acid with a
strong base?

a) A colored indicator with a pK_a less than 7 should be used.
b) If a colored indicator is used, it must change color rapidly in the weak acid's buffer region.
c) Equal volumes of weak acid and strong base are required to reach the equivalence point.
d) The equivalence point occurs at a pH equal to 7.
e) The equivalence point occurs at a pH greater than 7.

Answer: e

27. A volume of 25.0 mL of 0.100 M $CH_3CO_2H(aq)$ is titrated with 0.100 M NaOH(aq). What is
the pH after the addition of 12.5 mL of NaOH? (K_a for $CH_3CO_2H = 1.8 \times 10^{-5}$)

a) 3.74   b) 4.74   c) 5.74   d) 7.00   e) 9.26

Answer: b

28. A 25.0 mL sample of 0.175 M HClO(aq) is titrated with 0.100 M NaOH(aq). What is the pH
of a solution after the addition of 25.0 mL of NaOH? (K_a of HClO $= 3.5 \times 10^{-8}$)

a) 7.00   b) 7.21   c) 7.33   d) 7.46   e) 7.58

Answer: e

29. A 25.0 mL sample of 0.300 M $NH_3(aq)$ is titrated with 0.300 M HCl(aq). What is the pH at
the equivalence point? (K_b of $NH_3 = 1.8 \times 10^{-5}$)

a) 2.78   b) 5.04   c) 7.00   d) 8.96   e) 11.22

Answer: b

30. A 25.0 mL sample of 0.200 M HF(aq) is titrated with 0.200 M NaOH(aq). What is the pH at
the equivalence point? (K_a of HF $= 7.2 \times 10^{-4}$)

a) 5.93   b) 7.00   c) 8.07   d) 10.86   e) 11.92

Answer: c

31. A 25.0 mL sample of 0.10 M sodium acetate is titrated with 0.10 M HCl(aq). What is the pH
after the addition of 35.0 mL of HCl(aq)? (K_b of $CH_3CO_2^- = 5.6 \times 10^{-10}$)

a) 1.00   b) 1.23   c) 1.78   d) 4.75   e) 9.25

Answer: c

32. Potassium hydrogen phthalate (KHP) is used to standardize sodium hydroxide. If 22.37 mL of NaOH(aq) is required to titrate 0.7414 g KHP to the equivalence point, what is the concentration of the NaOH(aq)? (The molar mass of KHP = 204.2 g/mol)

$$HC_8H_4O_4^-(aq) + OH^-(aq) \rightleftharpoons C_8H_4O_4^{2-}(aq) + H_2O(\ell)$$

a) 0.003631 M   b) 0.03314 M   c) 0.08115 M   d) 0.08122 M   e) 0.1623 M

Answer: e

33. A 50.00 mL sample of vinegar is titrated with 0.584 M NaOH(aq). If the titration requires 32.80 mL of NaOH(aq), what is the concentration of acetic acid in the vinegar?

a) 0.0100 M   b) 0.0192 M   c) 0.0292 M   d) 0.383 M   e) 0.890 M

Answer: d

34. An impure sample of sodium carbonate, Na_2CO_3, is titrated with 0.123 M HCl according to the reaction below.

$$2\ HCl(aq) + Na_2CO_3(aq) \rightleftharpoons CO_2(g) + H_2O(\ell) + 2\ NaCl(aq)$$

What is the percent of Na_2CO_3 in a 0.557 g sample if the titration requires 25.30 mL of HCl? The molar mass of Na_2CO_3 is 106.0 g/mol.

a) 0.559%   b) 29.6%   c) 55.9%   d) 59.2%   e) 118%

Answer: b

35. Which is the best colored indicator to use in the titration of 0.005 M HClO(aq) with NaOH(aq)? Why? (K_a of HClO = 3.5×10^{-8}, K_b of ClO$^-$ = 2.9×10^{-7})

Indicator	pK_a
Thymol Blue	2.0
Phenol Red	7.5
Phenolphthalein	9.2

a) Thymol Blue. The equivalence point for a weak acid titration occurs at low pH.
b) Phenol Red. The pH at the equivalence point is 7.0.
c) Phenol Red. The pK_a of HClO and the pK_a of the indicator are similar.
d) Phenolphthalein. The pK_a of ClO$^-$ and the pK_b of the indicator are similar.
e) Phenolphthalein. The pH at the equivalence point is near the pK_a of the indicator.

Answer: e

36. What color change is exhibited by phenolphthalein during a titration of aqueous acetic acid
with aqueous sodium hydroxide?

a) colorless to pink   b) pink to colorless   c) green to yellow
d) yellow to blue   e) blue to yellow

Answer: a

18.4 Solubility of Salts

37. Which of the following equations is the solubility product for lead(II) iodate, $Pb(IO_3)_2$?

a) $K_{sp} = \dfrac{[Pb(IO_3)_2]}{[Pb^{2+}][IO_3^-]^2}$ b) $K_{sp} = \dfrac{[Pb^{2+}][IO_3^-]^2}{[Pb(IO_3)_2]}$ c) $K_{sp} = [Pb^{2+}][IO_3^-]$

d) $K_{sp} = [Pb^{2+}]^2[IO_3^-]$ e) $K_{sp} = [Pb^{2+}][IO_3^-]^2$

Answer: e

38. The solubility of AgI in water is 2.16 micrograms in 1.0 L at 25 °C. What is the value of K_{sp}
for AgI?

a) 8.5×10^{-17} b) 3.4×10^{-16} c) 4.7×10^{-12} d) 9.2×10^{-9} e) 9.6×10^{-5}

Answer: a

39. The solubility of $PbBr_2$ in water is 4.33 g in 1.0 L of water at 25 °C. What is the value of K_{sp}
for $PbBr_2$?

a) 1.6×10^{-6} b) 6.6×10^{-6} c) 1.3×10^{-5} d) 1.4×10^{-4} e) 19

Answer: b

40. The K_{sp} of $PbCl_2$ is 1.7×10^{-5} at 25 °C. What is the concentration of $Cl^-(aq)$ in a saturated
solution of $PbCl_2(aq)$?

a) 4.1×10^{-3} M   b) 1.6×10^{-2} M   c) 2.6×10^{-2} M   d) 3.2×10^{-2} M   e) 5.1×10^{-2} M

Answer: d

41. The K_{sp} of $CaSO_4$ is 4.9×10^{-5} at 25 °C. What mass of $CaSO_4$ (molar mass = 136.1 g/mol)
will dissolve in 1.0 L of water at 25 °C?

a) 0.0067 g   b) 0.0070 g   c) 0.081 g   d) 0.67 g   e) 0.95 g

Answer: e

42. The K_{sp} of AgBr is 5.4×10^{-13} at 25 °C. Calculate the molar solubility of AgBr in 0.050 M $AgNO_3(aq)$ at 25 °C.

    a) 1.1×10^{-11} mol/L   b) 2.2×10^{-11} mol/L   c) 5.2×10^{-8} mol/L
    d) 7.3×10^{-7} mol/L   e) 0.050 mol/L

    Answer: a

43. What is the molar solubility of CaF_2 in 0.010 M NaF(aq) at 25 °C? The value of K_{sp} for CaF_2 is 5.3×10^{-11} at 25 °C.

    a) 5.3×10^{-9} mol/L   b) 5.3×10^{-7} mol/L   c) 7.3×10^{-6} mol/L
    d) 7.2×10^{-4} mol/L   e) 0.010 mol/L

    Answer: b

44. What is the molar solubility of $Fe(OH)_3(s)$ in a solution that is buffered at pH 3.50 at 25 °C? The K_{sp} of $Fe(OH)_3$ is 6.3×10^{-38} at 25 °C.

    a) 6.9×10^{-28} mol/L   b) 2.0×10^{-27} mol/L   c) 2.2×10^{-10} mol/L
    d) 7.4×10^{-8} mol/L   e) 2.0×10^{-6} mol/L

    Answer: e

18.5 Precipitation Reactions

45. At what pH will an aqueous solution of 0.022 M Cu^{2+} begin to precipitate as $Cu(OH)_2$ at 25 °C? The K_{sp} of $Cu(OH)_2$ is 2.2×10^{-20} at 25 °C.

    a) 5.00   b) 6.45   c) 7.55   d) 9.00   e) 18.00

    Answer: a

46. The concentration of Mg^{2+} in an aqueous solution is 1.5×10^{-3} M. What concentration of CO_3^{2-} is required to begin precipitating $MgCO_3$? The K_{sp} of $MgCO_3$ is 6.8×10^{-6}.

    a) 6.8×10^{-6} M   b) 1.3×10^{-3} M   c) 1.5×10^{-3} M   d) 2.6×10^{-3} M   e) 4.5×10^{-3} M

    Answer: e

47. What mass of NaF (molar mass = 41.99 g/mol) must be added to 1.0 L of 0.0040 M $Pb^{2+}(aq)$ to initiate precipitation of $PbF_2(s)$? The K_{sp} of PbF_2 is 3.3×10^{-8}. Assume no volume change occurs upon addition of NaF.

    a) 3.5×10^{-4} g   b) 0.0014 g   c) 0.0029 g   d) 0.12 g   e) 0.17 g

    Answer: d

48. An aqueous solution contains 0.010 M Br^- and 0.010 M I^-. If Ag^+ is added until AgBr(s) just begins to precipitate, what are the concentrations of Ag^+ and I^-? (K_{sp} of AgBr = 5.4×10^{-13}, K_{sp} of AgI = 8.5×10^{-17})

a) $[Ag^+] = 5.4 \times 10^{-11}$ M, $[I^-] = 1.0 \times 10^{-2}$ M   b) $[Ag^+] = 8.5 \times 10^{-15}$ M, $[I^-] = 1.0 \times 10^{-2}$ M
c) $[Ag^+] = 5.4 \times 10^{-11}$ M, $[I^-] = 1.6 \times 10^{-6}$ M   d) $[Ag^+] = 8.5 \times 10^{-15}$ M, $[I^-] = 6.4 \times 10^{1}$ M
e) $[Ag^+] = 8.5 \times 10^{-15}$ M, $[I^-] = 1.6 \times 10^{-6}$ M

Answer: c

49. The following anions can be separated by precipitation as silver salts: Cl^-, Br^-, I^-, CrO_4^{2-}. If Ag^+ is added to a solution containing the four anions, each at a concentration of 0.10 M, in what order will they precipitate?

Compound	K_{sp}
AgCl	1.8×10^{-10}
Ag_2CrO_4	1.1×10^{-12}
AgBr	5.4×10^{-13}
AgI	8.5×10^{-17}

a) $AgCl \rightarrow Ag_2CrO_4 \rightarrow AgBr \rightarrow AgI$ b) $AgI \rightarrow AgBr \rightarrow Ag_2CrO_4 \rightarrow AgCl$
c) $Ag_2CrO_4 \rightarrow AgCl \rightarrow AgBr \rightarrow AgI$ d) $Ag_2CrO_4 \rightarrow AgI \rightarrow AgBr \rightarrow AgCl$
e) $AgI \rightarrow AgBr \rightarrow AgCl \rightarrow Ag_2CrO_4$

Answer: e

18.7 Solubility and Complex Ions

50. Consider the reaction

$$Zn^{2+}(aq) + 4\,OH^-(aq) \rightleftharpoons Zn(OH)_4^{2-}(aq) \qquad K_f = 2.9 \times 10^{15}$$

If K_{sp} for $Zn(OH)_2$ is 3.0×10^{-17}, what is the value of the equilibrium constant, K, for the reaction below?

$$Zn(OH)_2(s) + 2\,OH^-(aq) \rightleftharpoons Zn(OH)_4^{2-}(aq)$$

a) 1.0×10^{-32} b) 1.6×10^{-9} c) 8.7×10^{-2} d) 11   e) 9.7×10^{31}

Answer: c

51. Given the following reactions,

$$AgCl(s) \rightleftharpoons Ag^+(aq) + Cl^-(aq) \qquad\qquad K_{sp} = 1.8 \times 10^{-10}$$

$$Ag^+(aq) + 2\,S_2O_3^{2-}(aq) \rightleftharpoons Ag(S_2O_3)_2^{3-}(aq) \qquad\qquad K_f = 2.0 \times 10^{13}$$

determine the equilibrium constant for the reaction below.

$$AgCl(s) + 2\,S_2O_3^{2-}(aq) \rightleftharpoons Ag(S_2O_3)_2^{3-}(aq) + Cl^-(aq)$$

a) 9.0×10^{-24} b) 2.8×10^{-4} c) 3.6×10^{3} d) 2.0×10^{13} e) 1.1×10^{23}

Answer: c

Short Answer Questions

52. Two important biological buffer systems control pH in the range between 6.9 and 7.4. These buffer systems are H_2CO_3/HCO_3^- and __________.

Answer: $H_2PO_4^-/HPO_4^{2-}$

53. For a monoprotic acid titration, the __________ point in a titration is the point where the moles of strong base added equals the moles of acid initially present.

Answer: equivalence

54. If the ratio of acid to base in a buffer is increased by a factor of 10, the buffer pH decreases by __________ .

Answer: 1

55. To make a buffer with a pH of 8.00, you should use a weak acid with a K_a close to ________.

Answer: 1.0×10^{-8}

56. Hyperventilation can cause your blood pH to rise. One way to lower your blood pH is to breath into a paper bag, thus recycling the air you exhale. Why does this procedure lower your blood pH?

Answer: By breathing recycled air, you increase your CO_2 intake. Carbon dioxide is an acidic oxide; it reacts with water in your blood to form carbonic acid, H_2CO_3, which helps neutralize the excess base in your blood stream.

57. Why do salts containing basic anions have a greater solubility in water than predicted from calculations using K_{sp} values? Examples of such salts are $CaCO_3$, PbF_2, $Ca_3(PO_4)_2$.

Answer: Basic anions react with water to form their conjugate acids, such as HCO_3^-, HF, and HPO_4^{2-}. These reactions reduce the concentration of the basic anion. Additional salt dissociates to replace the basic anions.

Chapter 19
PRINCIPLES OF REACTIVITY: ENTROPY AND FREE ENERGY

CHAPTER OUTLINE

Example 19.4 Calculating $\Delta_r G^o$ from $\Delta_r H^o$ and $\Delta_r S^o$ (p. 880)
Example 19.5 Calculating $\Delta_r G^o$ Using Free Energies of Formation (p. 881)
Free Energy and Temperature (p. 881)
Example 19.6 Effect of Temperature on $\Delta_r G^o$ (p. 883)
Case Study Thermodynamics and Living Things (p. 884)
Using the Relationship Between $\Delta_r G^o$ and K (p. 885)
Example 19.7 Calculating K_p from $\Delta_r G^o$ (p. 885)
Example 19.8 Calculating $\Delta_r G^o$ from K_{sp} for an Insoluble Solid (p. 885)

CHAPTER OBJECTIVES

- Understand the concept of entropy and its relationship to reaction spontaneity (19.2, 19.3, 19.4)
- Calculate the change in entropy for system, surroundings, and the universe to determine whether a process is spontaneous (19.4 and 19.5)
- Understand and use the Gibbs free energy (19.6 and 19.7)

LESSON PLAN

AP Standards

III.E.2	First law of thermodynamics; change in enthalpy; heat of formation; heat of reaction; Hess's law; heats of vaporization and fusion; Calorimetry
III.E.3	Second law of thermodynamics; entropy, free energy of formation; free energy of reaction; dependence of change in free energy on enthalpy and entropy changes
III.E.4	Relationship of change in free energy to equilibrium constants and electrode potentials
IV.1	Chemical reactivity and products of chemical reaction

Suggested Time:
Three traditional classes or two blocks. Sections 19.2, 19.4 and 19.7 should be covered in detail. Section 19.6 is also important, but is an introduction to 19.7 and can be shortened if time is needed. Also, if any concepts of chemical reactivity and products are in the chapter, be sure to at least review these.

ASSESSMENT

Exercises:

19.1	Entropy Comparisons (p. 870)
19.2	Calculating the Entropy Change for a Reaction $\Delta_r S^o$ (p. 871)
19.3	Determining Whether a Process is Spontaneous (p. 873)
19.4	Is a Reaction Spontaneous? (p. 876)
19.5	Effect of Temperature on Spontaneity (p. 876)
19.6	Calculating $\Delta_r G^o$ from $\Delta_r H^o$ and $\Delta_r S^o$ (p. 881)
19.7	Calculating $\Delta_r G^o$ from $\Delta_f G^o$ (p. 881)
19.8	Effect of Temperature on $\Delta_r G^o$ (p. 885)

19.9   Calculating K_p from $\Delta_r G^o$ (p. 885)
Answers to Exercises p. A57-A58
Practicing Skills pp. 887-889, Answers IRM pp. 361-370
General Questions pp. 890-892, Answers IRM pp. 370-377
In the Laboratory Questions p. 892, Answers IRM pp. 377-378
Summary and Conceptual Questions pp. 893-895, Answers IRM pp. 378-383

GLOSSARY

Boltzmann's constant Proportionality constant relating number of microstates to entropy of a system

entropy (S) A measure of the disorder of a system.

Gibbs free energy (G) A thermodynamic state function relating enthalpy, temperature, and entropy

irreversible process A process that involves nonequilibrium conditions

microstate Statistical method to indicate energy dispersal in a system

reversible process A process in which a system can go from one state to another

second law of thermodynamics The entropy of the universe increases in a spontaneous process

spontaneous A change that occurs with outside intervention, in a direction that leads to equilibrium

standard free energy (G^o) The free energy change for a reaction in which all reactants and products are in their standard states

standard molar entropy (S) The entropy of a substance in its most stable form at a pressure of 1 bar

standard molar free energy of formation ($_f G^o$) The free energy change for the formation of 1 mol of a compound from its elements, all in their standard state

third law of thermodynamics The entropy of a pure, perfectly formed crystal at 0 K is zero

MEDIA MENU

The Media Menu section of this guide lists the interactive exercises, tutorials, active figures (animated, interactive figures) and additional resources available online for Chemistry Now, a unique web-based, assessment-centered personalized learning system for chemistry students. Go to www.cengagelearning/login.

350

CHAPTER 19: Principles of Reactivity: Entropy and Free Energy

Tutorials
- Screen 19.5   Calculating ΔS for a reaction
- Screen 19.6   The second law of thermodynamics
- Screen 19.7   Gibbs free energy
- Screen 19.8   The relationship of $\Delta H°$, $\Delta S°$, and T
- Screen 19.9 $\Delta G°$ and K

Active Figures
- 19.12 The variation of $\Delta_r G°$ with temperature

Exercises
- Screen 19.7   Gibbs free energy

Additional Resources
- Screen 19.5   Simulation on calculating ΔS for a reaction
- Screen 19.6   Simulation on the second law of thermodynamics
- Screen 19.8   Simulation on the relationship of $\Delta H°$, $\Delta S°$, and T
- Screen 19.9   Simulation on $\Delta G°$ and K
- Podcast Module 24   Gibbs free energy

DIAGNOSTIC QUIZ QUESTIONS

Understand the concept of entropy and its relationship to reaction spontaneity

1. Which of the following processes leads to a decrease in entropy?

   a) Evaporating a liquid
   b) Forming mixtures from pure substances
   c) Chemical reactions that increase the number of gas particles
   d) Freezing a liquid
   e) Raising the temperature

   Answer: d

2. Which of the following linear chain alcohols is likely to have the highest standard entropy in the liquid state?

   a) CH_3Cl
   b) CH_3CH_2Cl
   c) $CH_3CH_2CH_2Cl$
   d) $CH_3CH_2CH_2CH_2Cl$
   e) $CH_3CH_2CH_2CH_2CH_2Cl$

   Answer: e

3. For which of the following reactions will the entropy of the system decrease?

a) $2 NH_3(g) \rightarrow N_2(g) + 3 H_2(g)$

b) $2 C(s) + O_2(g) \rightarrow 2 CO(g)$

c) $CaO(s) + CO_2(g) \rightarrow CaCO_3(s)$

d) $N_2O_4(g) \rightarrow 2 NO_2(g)$

e) $KOH(s) \rightarrow K^+(aq) + OH^-(aq)$

Answer: c

Calculate the change in entropy for system, surroundings, and the universe to determine whether a process is spontaneous

4. Calculate the standard entropy change for the combustion of propane at 25 °C.
$$C_3H_8(g) + 5 O_2(g) \rightarrow 3 CO_2(g) + 4 H_2O(g)$$

Species	$S°$ (J/K·mol)
$C_3H_8(g)$	270.3
$O_2(g)$	205.1
$CO_2(g)$	213.7
$H_2O(g)$	188.8

a) -72.9 J/K   b) +72.9 J/K   c) +100.5 J/K   d) +877.9 J/K   e) +2692 J/K

Answer: c

Understand and use the Gibbs free energy

5. If a chemical reaction is endothermic and spontaneous, which of the following must be true?

a) $\Delta_r G > 0$, $\Delta_r S > 0$ and $\Delta_r H > 0$

b) $\Delta_r G < 0$, $\Delta_r S > 0$ and $\Delta_r H > 0$

c) $\Delta_r G > 0$, $\Delta_r S < 0$ and $\Delta_r H > 0$

d) $\Delta_r G < 0$, $\Delta_r S < 0$ and $\Delta_r H < 0$

e) $\Delta_r G > 0$, $\Delta_r S > 0$ and $\Delta_r H < 0$

Answer: b

6. Calculate $\Delta_r G°$ for the reaction below at 25.0 °C.
$$P_4O_{10}(s) + 6 H_2O(\ell) \rightarrow 4 H_3PO_4(\ell)$$

Species	$\Delta_f H°$ (kJ/mol-rxn)	$S°$ (J/K·mol-rxn)
$P_4O_{10}(s)$	-2984.0	228.9
$H_2O(\ell)$	-285.8	69.95
$H_3PO_4(\ell)$	-1279.0	110.5

a) -6119.1 kJ   b) -632.8 kJ   c) -355.6 kJ   d) +210.6 kJ   e) +6119.1 kJ

Answer: c

7. At 298.15 K, all of the following substances have a standard free energy of formation of zero EXCEPT ____.

a) $Br_2(g)$ b) $I_2(s)$ c) $S_8(s)$ d) $Cl_2(g)$ e) $Hg(\ell)$

Answer: a

8. Calculate $\Delta_r G°$ for the reaction below at 398 K.
$$2\ CH_3OH(g)\ +\ 3\ O_2(g)\ \rightarrow\ 2\ CO_2(g)\ +\ 4\ H_2O(g)$$

Species	$\Delta_f H°$ (kJ/mol-rxn)	$S°$ (J/K·mol-rxn)
$CH_3OH(g)$	-210.1	239.7
$O_2(g)$	0	205.1
$CO_2(g)$	-393.5	213.7
$H_2O(g)$	-241.8	188.8

a) -1421.9 kJ   b) -1369.0 kJ   c) -1246.1 kJ   d) -1.9 kJ   e) +1.9 kJ

Answer: b

9. For a chemical reaction, if $\Delta_r G° = 0$, then __________.

a) $K > 1$ b) $K = 0$ c) $K = -1$ d) $K < 0$ e) $K = 1$

Answer: e

10. All of the following relationships are true EXCEPT

a) $\Delta G°_{sys} = \Delta H°_{sys} - T\Delta S°_{sys}$.   b) $\Delta G°_{sys} = -RT\ln(K)$.   c) $\Delta S°_{univ} = \Delta S°_{sys} + \Delta S°_{surr}$.

d) $\Delta H = \Delta H°_{sys} + RT\ln(K)$.   e) $\Delta G°_{sys} = -T\Delta S_{univ}$

Answer: d

TEST BANK

<u>19.1 Spontaneity and Energy Transfer as Heat</u>

1. Assume that reactant A can undergo the reaction shown below.

    $$A(aq) \rightleftharpoons B(aq)$$

    If the equilibrium constant, K, is greater than 1.0, which of the following statements are correct?
    1. The forward reaction is spontaneous.
    2. The reaction favors the formation of products.
    3. The forward rate of reaction is fast.

    a) 1 only   b) 2 only   c) 3 only   d) 1 and 2   e) 1, 2, and 3

    Answer: d

<u>19.2 Dispersal of Energy: Entropy</u>

2. The following processes occur spontaneously at 25 °C. Which of these processes is/are endothermic?
    1. NH_4NO_3 dissolving in water (which is accompanied by a cooling of the water).
    2. the expansion of a real gas into a vacuum (which is accompanied by a cooling of the gas).
    3. a pool of water evaporating on a sunny day.

    a) 1 only   b) 2 only   c) 3 only   d) 1 and 3   e) 1, 2, and 3

    Answer: e

3. Which of the following statements is/are CORRECT?
    1. Spontaneous changes only occur in the direction that leads to equilibrium.
    2. Endothermic reactions are never spontaneous.
    3. In any chemical reaction, energy must be conserved.

    a) 1 only   b) 2 only   c) 3 only   d) 1 and 3   e) 1, 2, and 3

    Answer: d

4. The second law of thermodynamics states that

    a) in a spontaneous process, the entropy of the universe increases.
    b) there is no disorder in a perfect crystal at 0 K.
    c) the total energy of the universe is always increasing.
    d) the total energy of the universe is constant.
    e) mass and energy are conserved in all chemical reactions.

    Answer: a

19.4 Entropy Measurement and Values

5. As defined by Ludwig Boltzmann, the third law of thermodynamics states that

    a) there is no disorder in a perfect crystal at 0 K.
    b) in a spontaneous process, the entropy of the universe increases.
    c) the total entropy of the universe is always increasing.
    d) the total mass of the universe is constant.
    e) mass and energy are conserved in all chemical reactions.

    Answer: a

6. Which of the following statements concerning entropy is/are CORRECT?
    1.   The entropy of a substance decreases when converted from a liquid to a gas.
    2.   The entropy of a substance decreases as its temperature increases.
    3.   Negative entropy values cannot exist.

    a) 1 only   b) 2 only   c) 3 only   d) 1 and 2   e) 1, 2, and 3

    Answer: c

7. Which of the following linear chain alcohols is likely to have the highest standard entropy in
    the liquid state?

    a) CH_3Cl b) CH_3CH_2Cl c) $CH_3CH_2CH_2Cl$
    d) $CH_3CH_2CH_2CH_2Cl$ e) $CH_3CH_2CH_2CH_2CH_2Cl$

    Answer: e

8. If a chemical reaction occurs in a direction that has a negative change in entropy then

    a) the change in enthalpy is negative.
    b) the disorder of the system decreases.
    c) heat goes from the system into the surroundings.
    d) the reaction is exothermic.
    e) the reaction is spontaneous.

    Answer: b

9. Which one of the following processes involves a decrease in entropy?

    a) the decomposition of $NH_3(g)$ to $H_2(g)$ and $N_2(g)$
    b) the condensation of steam to liquid water
    c) the sublimation of CO_2 from a solid to a gas
    d) the evaporation of ethanol
    e) the dissolution of $NH_4NO_3(s)$ in water

    Answer: b

10. All of the following statements concerning entropy are true EXCEPT

    a) entropy values for substances are greater than or equal to zero.
    b) entropy is a state function.
    c) a positive change in entropy denotes a change toward greater disorder.
    d) entropy is zero for elements under standard conditions.
    e) the entropy of a substance in the gas phase is greater than the solid phase.

Answer: d

11. All of the following processes lead to an increase in entropy EXCEPT

    a) decreasing the volume of a gas.
    b) melting a solid.
    c) chemical reactions that increase the number of moles of gas.
    d) forming mixtures from pure substances.
    e) increasing the temperature of a gas.

Answer: a

12. For which of the following reactions will the entropy of the system decrease?

    a) $2\ NH_3(g) \rightarrow N_2(g) + 3\ H_2(g)$
    b) $2\ C(s) + O_2(g) \rightarrow 2\ CO(g)$
    c) $CaO(s) + CO_2(g) \rightarrow CaCO_3(s)$
    d) $N_2O_4(g) \rightarrow 2\ NO_2(g)$
    e) $KOH(s) \rightarrow K^+(aq) + OH^-(aq)$

Answer: c

13. Calculate the standard entropy change for the combustion of propane at 25 °C.
$$C_3H_8(g) + 5\ O_2(g) \rightarrow 3\ CO_2(g) + 4\ H_2O(g)$$

Species	$S°$ (J/K·mol)
$C_3H_8(g)$	270.3
$O_2(g)$	205.1
$CO_2(g)$	213.7
$H_2O(g)$	188.8

a) -72.9 J/K b) +72.9 J/K c) +100.5 J/K d) +877.9 J/K e) +2692 J/K

Answer: c

14. Calculate the standard entropy change for the following reaction,
$$PbO(s) \rightarrow Pb(s) + 1/2\ O_2(g)$$
given $S°[PbO] = 66.5$ J/K·mol-rxn, $S°[Pb(s)] = 64.8$ J/K·mol-rxn, and $S°[O_2(g)] = 205.1$ J/K·mol-rxn.

a) -100.9 J/K   b) -28.75 J/K   c) +100.9 J/K   d) +336.4 J/K   e) +233.9 J/K

Answer: c

15. The standard entropy of formation of $SiCl_4(g)$,
$$Si(s) + 2\ Cl_2(g) \rightarrow SiCl_4(g)$$
is -134.1 J/K·mol. Calculate the standard molar entropy of $SiCl_4(g)$ given $S°[Si(s)] = 18.8$ J/K·mol-rxn and $S°[Cl_2(g)] = 223.1$ J/K·mol-rxn.

a) -599.1 J/K·mol   b) +107.8 J/K·mol   c) +293.3 J/K·mol
d) +330.9 J/K·mol   e) +599.1 J/K·mol

Answer: d

19.5 Entropy Changes and Spontaneity

16. For the following reaction at 298.2 K,
$$2\ CO(g) + O_2(g) \rightarrow 2\ CO_2(g)$$
calculate $\Delta S°$(universe) given $\Delta S°$(system) = -173.1 J/K and $\Delta H°$(system) = -566.0 kJ.

a) -2071.3 J/K   b) -739.0 J/K   c) -175.0 J/K   d) -171.3 J/K   e) +1725.0 J/K

Answer: e

17. Use the following thermodynamic data

Species	$\Delta_f H°$ (kJ/mol-rxn)	$S°$ (J/K·mol-rxn)
Fe(s)	0.0	27.8
$O_2(g)$	0.0	205.1
$Fe_2O_3(s)$	-825.5	87.4

to calculate $\Delta S°$ (universe) for the formation of $Fe_2O_3(s)$ at 298.2 K.
$$4\ Fe(s) + 3\ O_2(g) \rightarrow 2\ Fe_2O_3(s)$$

a) -2202.7 J/K   b) -546.2 J/K   c) +546.2 J/K   d) +2202.7 J/K   e) +4984.9 J/K

Answer: e

<u>19.6 Gibbs Free Energy</u>

18. When a real gas expands from high pressure to a lower pressure, its temperature decreases. Predict the signs of of ΔH and ΔS.

    a) $\Delta H < 0$ and $\Delta S < 0$ b) $\Delta H < 0$ and $\Delta S > 0$ c) $\Delta H > 0$ and $\Delta S < 0$
    d) $\Delta H > 0$ and $\Delta S > 0$ e) $\Delta H < 0$ and $\Delta S = 0$

Answer: d

19. Hydrogen gas is a non-polluting fuel. Hydrogen gas may be prepared by electrolysis of water.

$$2\,H_2O(\ell) \;\rightarrow\; 2\,H_2(g) + O_2(g)$$

Predict the signs of $\Delta_r H$ and $\Delta_r S$ for the production of hydrogen gas by electrolysis of water.

    a) $\Delta_r H > 0$ and $\Delta_r S > 0$ b) $\Delta_r H < 0$ and $\Delta_r S > 0$ c) $\Delta_r H > 0$ and $\Delta_r S < 0$
    d) $\Delta_r H < 0$ and $\Delta_r S < 0$ e) $\Delta_r H = 0$ and $\Delta_r S < 0$

Answer: a

20. If $\Delta_r G° < 0$ for a reaction at all temperatures, then $\Delta_r H°$ is __________ and $\Delta_r S°$ is __________.

    a) negative, positive   b) positive, negative   c) negative, negative
    d) positive, positive   e) positive, either positive or negative

Answer: a

21. The dissolution of ammonium nitrate occurs spontaneously in water at 25 °C. As NH_4NO_3 dissolves, the temperature of the water decreases. What are the signs of $\Delta_r H$, $\Delta_r S$, and $\Delta_r G$ for this process?

    a) $\Delta_r H > 0$, $\Delta_r S < 0$, $\Delta_r G > 0$ b) $\Delta_r H > 0$, $\Delta_r S > 0$, $\Delta_r G > 0$
    c) $\Delta_r H > 0$, $\Delta_r S > 0$, $\Delta_r G < 0$ d) $\Delta_r H < 0$, $\Delta_r S < 0$, $\Delta_r G < 0$
    e) $\Delta_r H < 0$, $\Delta_r S > 0$, $\Delta_r G > 0$

Answer: c

22. Diluting concentrated sulfuric acid with water can be dangerous. The temperature of the solution can increase rapidly. What are the signs of $\Delta_r H$, $\Delta_r S$, and $\Delta_r G$ for this process?

    a) $\Delta_r H < 0$, $\Delta_r S > 0$, $\Delta_r G < 0$ b) $\Delta_r H < 0$, $\Delta_r S < 0$, $\Delta_r G < 0$
    c) $\Delta_r H < 0$, $\Delta_r S > 0$, $\Delta_r G > 0$ d) $\Delta_r H > 0$, $\Delta_r S > 0$, $\Delta_r G < 0$
    e) $\Delta_r H > 0$, $\Delta_r S < 0$, $\Delta_r G > 0$

Answer: a

23. At what temperatures will a reaction be spontaneous if $\Delta_r H° = +45$ kJ and $\Delta_r S° = +312$ J/K?

   a) All temperatures below 144 K.
   b) All temperatures above 144 K.
   c) Temperatures between 45 K and 312 K.
   d) The reaction will be spontaneous at any temperature.
   e) The reaction will never be spontaneous.

Answer: b

24. At what temperatures will a reaction be spontaneous if $\Delta_r H° = -12.7$ kJ and $\Delta_r S° = +135$ J/K?

   a) All temperatures below 94.1 K
   b) Temperatures between 12.7 K and 135 K
   c) All temperatures above 94.1 K
   d) The reaction will be spontaneous at any temperature.
   e) The reaction will never be spontaneous.

Answer: d

25. For a reaction, $\Delta_r H° = +2112$ kJ and $\Delta_r S° = +132.9$ J/K. At what temperature will $\Delta_r G° = 0.00$ kJ?

   a) 15.89 K   b) 1979 K   c) 2244 K
   d) 1.589×10^4 K   e) ΔG is greater than 0.00 kJ at any temperature.

Answer: d

26. If a chemical reaction is endothermic and spontaneous, which of the following must be true?

   a) $\Delta_r G > 0$, $\Delta_r S > 0$ and $\Delta_r H > 0$ b) $\Delta_r G < 0$, $\Delta_r S > 0$ and $\Delta_r H > 0$
   c) $\Delta_r G > 0$, $\Delta_r S < 0$ and $\Delta_r H > 0$ d) $\Delta_r G < 0$, $\Delta_r S < 0$ and $\Delta_r H < 0$
   e) $\Delta_r G > 0$, $\Delta_r S > 0$ and $\Delta_r H < 0$

Answer: b

19.7 Calculating and Using Free Energy

27. Calculate $\Delta_r G°$ for the reaction below at 25.0 °C
$$CS_2(g) + 3\ Cl_2(g) \rightarrow S_2Cl_2(g) + CCl_4(g)$$
given $\Delta_r H° = -231.1$ kJ and $\Delta_r S° = -287.6$ J/K.

   a) -518.7 kJ   b) -316.9 kJ   c) -145.3 kJ   d) -56.5 kJ   e) +56.5 kJ

Answer: c

28. Calculate $\Delta_r G°$ for the reaction below at 25.0 °C
$$2 \, CaO(s) \rightarrow 2 \, Ca(s) + O_2(g)$$
given $\Delta_r H° = +1270.2$ kJ and $\Delta_r S° = +211.9$ J/K.

a) -1058.3 kJ   b) +5.994 kJ   c) +1058.3 kJ   d) +1207.0 kJ   e) +1482.1 kJ

Answer: d

29. Calculate $\Delta_r G°$ for the reaction below at 25.0 °C
$$CH_4(g) + H_2O(g) \rightarrow 3 \, H_2(g) + CO(g)$$
given $\Delta_f G°$ [$CH_4(g)$] = -50.8 kJ/mol-rxn, $\Delta_f G°$ [$H_2O(g)$] = -228.6 kJ/mol-rxn, $\Delta_f G°$ [$H_2(g)$] = 0.0 kJ/mol-rxn, and $\Delta_f G°$ [$CO(g)$] = -137.2 kJ/mol-rxn.

a) -416.3 kJ   b) -142.2 kJ   c) +142.2 kJ   d) +315.0 kJ   e) +416.3 kJ

Answer: c

30. The $\Delta_r G°$ for the following reaction is -140.8 kJ.
$$C_2H_2(g) + H_2(g) \rightarrow C_2H_4(g)$$
Given $\Delta_f G°$ [$C_2H_2(g)$] = +209.2 kJ/mol-rxn, calculate $\Delta_f G°$ [$C_2H_4(g)$].

a) -350.0 kJ/mol-rxn   b) -68.4 kJ/mol-rxn   c) -1.486 kJ/mol-rxn
d) +34.2 kJ/mol-rxn   e) +68.4 kJ/mol-rxn

Answer: e

31. Calculate $\Delta_r G°$ for the reaction below at 25.0 °C.
$$Fe_3O_4(s) \rightarrow 3 \, Fe(s) + 2 \, O_2(g)$$

Species	$\Delta_f H°$ (kJ/mol-rxn)	$S°$ (J/K·mol-rxn)
$Fe_3O_4(s)$	-1118.4	146.4
$Fe(s)$	0	27.8
$O_2(g)$	0	205.1

a) -1221.9 kJ   b) -1014.9 kJ   c) -771.2 kJ   d) +771.2 kJ   e) +1014.9 kJ

Answer: e

32. Calculate $\Delta_r G°$ for the reaction below at 25.0 °C.
$$P_4O_{10}(s) + 6\ H_2O(\ell) \rightarrow 4\ H_3PO_4(\ell)$$

Species	$\Delta_f H°$ (kJ/mol-rxn)	$S°$ (J/K·mol-rxn)
$P_4O_{10}(s)$	-2984.0	228.9
$H_2O(\ell)$	-285.8	69.95
$H_3PO_4(\ell)$	-1279.0	110.5

a) -6119.1 kJ   b) -632.8 kJ   c) -355.6 kJ   d) +210.6 kJ   e) +6119.1 kJ

Answer: c

33. Calculate $\Delta_r G°$ for the reaction below at 398 K.
$$2\ CH_3OH(g) + 3\ O_2(g) \rightarrow 2\ CO_2(g) + 4\ H_2O(g)$$

Species	$\Delta_f H°$ (kJ/mol-rxn)	$S°$ (J/K·mol-rxn)
$CH_3OH(g)$	-210.1	239.7
$O_2(g)$	0	205.1
$CO_2(g)$	-393.5	213.7
$H_2O(g)$	-241.8	188.8

a) -1421.9 kJ   b) -1369.0 kJ   c) -1246.1 kJ   d) -1.9 kJ   e) +1.9 kJ

Answer: b

34. At 298.15 K, all of the following substances have a standard free energy of formation of zero EXCEPT ____.

a) $Br_2(g)$ b) $I_2(s)$ c) $S_8(s)$ d) $Cl_2(g)$ e) $Hg(\ell)$

Answer: a

35. Given that
$$C(s) + O_2(g) \rightarrow CO_2(g) \qquad \Delta_r G° = -394.4\ kJ$$
$$2CO(g) + O_2(g) \rightarrow 2\ CO_2(g) \qquad \Delta_r G° = -514.4\ kJ$$
calculate $\Delta_f G°$ of carbon monoxide.
$$C(s) + 1/2\ O_2(g) \rightarrow CO(g)$$
a) -908.8 kJ   b) -651.6 kJ   c) -137.2 kJ   d) -120.0 kJ   e) +120.0 kJ

Answer: c

36. Estimate the boiling point of carbon tetrachloride given the following thermodynamic parameters.

	$CCl_4(\ell)$	$CCl_4(g)$
$\Delta_f H°$ (kJ/mol-rxn)	-128.4	-96.0
$S°$ (J/K·mol-rxn)	214.4	309.7
$\Delta_f G°$ (kJ/mol-rxn)	-57.6	-53.6

a) -272 °C   b) 25 °C   c) 67 °C   d) 69 °C   e) 109 °C

Answer: c

37. Thermodynamics can be used to determine all of the following EXCEPT

a) the rate of reaction.
b) the extent to which a reaction occurs.
c) the direction in which a reaction is spontaneous.
d) the temperature at which a reaction is spontaneous.
e) the entropy change of a reaction.

Answer: a

38. For a chemical reaction, if $\Delta_r G° = 0$, then __________.

a) $K > 1$ b) $K = 0$ c) $K = -1$ d) $K < 0$ e) $K = 1$

Answer: e

39. All of the following relationships are true EXCEPT

a) $\Delta G°_{sys} = \Delta H°_{sys} - T\Delta S°_{sys}$.   b) $\Delta G°_{sys} = -RT \ln(K)$.   c) $\Delta S°_{univ} = \Delta S°_{sys} + \Delta S°_{surr}$.

d) $\Delta H = \Delta H°_{sys} + RT\ln(K)$.   e) $\Delta G°_{sys} = -T\Delta S°_{univ}$

Answer: d

40. For a chemical system, $\Delta_r G°$ and $\Delta_r G$ are equal when

a) the system is in equilibrium.
b) the reactants and products are in standard state conditions.
c) the equilibrium constant, K, equals 0.
d) the reaction quotient, Q, is less than 1.
e) the reactants and products are in the gas phase.

Answer: b

41. The standard free energy change for a chemical reaction is +21.9 kJ/mole. What is the equilibrium constant for the reaction at 87 °C? ($R = 8.314$ J/K·mol)

a) 7.1×10^{-14} b) 6.6×10^{-4} c) 9.9×10^{-1} d) 1.0   e) 1.5×10^{3}

Answer: b

42. The standard free energy change is -33.41 kJ/mol for the formation ($\Delta_f G°$) of $AgNO_3(s)$ from elements at 25 °C. What is the equilibrium constant for the reaction? ($R = 8.314$ J/K·mol)

a) 3.3×10^{-14} b) 1.4×10^{-6} c) 13   d) 7.2×10^{5} e) 3.1×10^{13}

Answer: d

43. What is the equilibrium constant for reaction below at 25 °C? ($R = 8.314$ J/K·mol)

$$MgCO_3(s) \rightleftharpoons MgO(s) + CO_2(g)$$

given $\Delta_f G°$ [$MgCO_3(s)$] = -1028.2 kJ/mol, $\Delta_f G°$ [$MgO(s)$] = -568.8 kJ/mol, and $\Delta_f G°$ [$CO_2(g)$] = -394.4 kJ/mol.

a) 4.0×10^{-12} b) 0.97   c) 1.0   d) 1.0×10^{4} e) 2.5×10^{11}

Answer: a

44. The equilibrium constant for a reaction at 25 °C is 3.1×10^{8}. What is $\Delta_r G°$? ($R = 8.314$ J/K·mol)

a) -4.06×10^{3} J   b) -2.10×10^{4} J   c) -4.84×10^{4} J   d) -9.06×10^{4} J   e) -1.25×10^{5} J

Answer: c

45. Calculate $\Delta_r G°$ for the following reaction at 425 °C,

$$2\, HI(g) \rightleftharpoons H_2(g) + I_2(g)$$

given $K = 0.018$. ($R = 8.314$ J/K·mol)

a) 6.12×10^{3} J   b) 1.05×10^{4} J   c) 1.42×10^{4} J   d) 2.33×10^{4} J   e) 3.34×10^{5} J

Answer: d

Short Answer Questions

46. The total entropy of the universe is always increasing. This is a statement of the _________ law of thermodynamics.

Answer: second

47. A chemical reaction with an equilibrium constant greater than 1 is said to be __________-
__________.

Answer: product-favored

48. The change in entropy for any process is not dependent upon the pathway by which the
process occurs. In other words, the change in entropy for any process is a __________ function.

Answer: state

49. For any process, the change in entropy of the universe equals the sum of the entropy changes
for the system and for the __________.

Answer: surroundings

50. At what temperature (in kelvin units) is the entropy of a pure crystal 0.0 J/K.

Answer: 0 K

51. Does the formation of complex molecules such as proteins and nucleic acids from more
simple molecules contradict the second law of thermodynamics?

Answer: No. The formation of complex molecules involves a decrease in entropy locally.
The local decrease is offset by an increase in entropy of the universe.

52. Use the standard entropies for liquid water and steam to estimate the enthalpy of vaporization
of water at its normal boiling point. $\Delta S°$ [H$_2$O(ℓ)] = 69.9 J/K·mol and $\Delta S°$ [H$_2$O(g)] = 188.8
J/K·mol.

Answer: 44.4 kJ/mol

Chapter 20
PRINCIPLES OF REACTIVITY: ELECTRON TRANSFER REACTIONS

CHAPTER OUTLINE

CHAPTER OBJECTIVES

- Balance equations for oxidation –reduction reactions in acidic and basic solutions using the half-reaction method (20.1)
- Understand the principles underlying voltaic cells (20.2, 20.3)
- Understand how to use electrochemical potentials (20.4, 20.5, 20.6)
- Explore electrolysis, the use of electrical energy to produce chemical changes (20.7, 20.8)

LESSON PLAN

AP Standards

III.A.3   Oxidation-reduction reactions

III.A.3.c   Electrochemistry: electrolytic and galvanic cells; Faraday's laws, standard half-cell potentials;

         Nernst equation, prediction of the direction of redox reactions

CHAPTER 20: Principles of Reactivity: Electron Transfer Reactions

Suggested Time:
Four traditional classes or three blocks. Redox reactions are continually addressed throughout the chapter and should be covered in detail. Sections 10.1, 20.2 and 20.3 should be given extra attention.

ASSESSMENT

Exercises:

Answers to Exercises p. A58-A59
Practicing Skills pp. 940-943, Answers IRM pp. 387-396
General Questions pp. 943-946, Answers IRM pp. 396-403
In the Laboratory Questions pp. 946-947, Answers IRM pp. 403-404
Summary and Conceptual Questions p. 947, Answers IRM pp. 404-405

GLOSSARY

alkaline battery A electrochemical cell using the oxidation of zinc and the reduction of MnO_2 in NaOH or KOH

ampere (A) The unit of electric current

anode The electrode of an electrochemical cell at which oxidation occurs

battery A device consisting of two or more electrochemical cells

cathode The electrode of an electrochemical cell at which reduction occurs

Coulomb (C) The quantity of charge that passes a point in an electric circuit when a current of 1 ampere flows for 1 second

current The amount of charge per unit time (Coulombs per second or Amperes)

electrochemical cell A device that produces an electric current as a result of an electron transfer reaction

electrolysis The use of electrical energy to produce a chemical change

electromotive force (emf) Difference in potential energy per electrical charge

Faraday constant (F) The proportionality constant that relates standard free energy of reaction to standard potential; the charge carried by one mole of electrons

fuel cell A voltaic cell in which reactants are continuously added

half-cell A compartment of an electrochemical cell in which a half-reaction occurs

half-reactions The two chemical equations into which the equation for an oxidation-reduction reaction can be divided, one representing the oxidation process and the other the reduction process

half-reaction method A procedure for balancing oxidation reduction equations that involves writing separate, balanced equations, one for the oxidation and one for the reduction

lead storage battery An automobile battery, utilizing six voltaic cells, with a lead anode and a cathode covered with lead (IV) oxide

Nernst equation A mathematical expression that relates the potential of an electrochemical cell to the concentrations of the cell reactants and products

nickel-cadmium battery (Ni-cad) Lightweight, rechargeable batteries used in a variety of cordless appliances with a cadium anode and a nickel (III) oxide under basic conditions

northwest-southeast rule A product-favored reaction involves a reducing agent below and to the right of the reactant

primary battery A battery that cannot be returned to its original state by recharging

rechargeable battery(secondary battery) A battery in which the reaction can be reversed, returning it to its original condition

salt bridge A device for maintaining the balance of ion charges in the compartments of an electrochemical cell

standard conditions In an electrochemical cell, all reactants and products are pure liquids or solids, or 1.0 M aqueous solutions, or gases at a pressure of 1 bar

standard potential (E^o) The potential of an electrochemical cell measured under standard conditions

storage battery A rechargeable or secondary battery

standard hydrogen electrode (SHE) A comparison cell in which the reduction potential of hydrogen ions is assigned a value of 0 V.

volt (V) The electrical potential through which 1 coulomb of charge must pass in order to do 1 joule of work

voltaic cell (galvanic cell) A device that uses a chemical reaction to produce an electric current

work Energy transfer that occurs as a mass is moved through a distance against an opposing force

MEDIA MENU

The Media Menu section of this guide lists the interactive exercises, tutorials, active figures (animated, interactive figures) and additional resources available online for Chemistry Now, a unique web-based, assessment-centered personalized learning system for chemistry students. Go to www.cengagelearning.com/login.

Tutorials
- Screen 20.6   Standard potentials
- Screen 20.8   The Nernst equation
- Screen 20.12   Quantitative aspects of electrochemistry

Active Figures
- 20.13   A voltaic cell using $Zn|Zn^{2+}$ and $H_2|H^+$ half cells

Additional Resources
- Screen 20.2   Self-study module on various types of oxidation reduction reactions
- Screen 20.3   Self-study module on methods of balancing redox reactions in acidic solutions
- Screen 20.4   Animation of a cell based on zinc and copper
- Screen 20.5   Animation of various types of batteries
- Screen 20.5   Demonstration of the potentials of various cells
- Screen 20.6   Simulation on standard potentials
- Screen 20.11   Illustration of water electrolysis
- Podcast Module 25   Oxidation Reduction Reactions

DIAGNOSTIC QUIZ QUESTIONS

Balance equations for oxidation –reduction reactions in acidic and basic solutions using the half-reaction method (20.1)

1. Assuming the following reaction proceeds in the forward direction,
$$Cu^{2+}(aq) + Ca(s) \rightarrow Cu(s) + Ca^{2+}(aq)$$

   a) $Cu^{2+}(aq)$ is oxidized and $Ca(s)$ is reduced.
   b) $Cu^{2+}(aq)$ is oxidized and $Ca^{2+}(aq)$ is reduced.
   c) $Ca(s)$ is oxidized and $Cu^{2+}(aq)$ is reduced.
   d) $Ca(s)$ is oxidized and $Ca^{2+}(aq)$ is reduced.
   e) $Cu(s)$ is oxidized and $Ca(s)$ is reduced.

   Answer: c

2. Write a balanced chemical equation for the oxidation of $Fe^{2+}(aq)$ by concentrated nitric acid. Two products of the reaction are $NO(g)$ and $Fe^{3+}(aq)$.

   a) $HNO_3(aq) + Fe^{2+}(aq) \rightarrow Fe^{3+}(aq) + NO(g) + HO_2(\ell)$

   b) $HNO_3(aq) + 3\,Fe^{2+}(aq) \rightarrow 3\,Fe^{3+}(aq) + NO(g) + HO_2(\ell)$

   c) $2\,HNO_3(aq) + 2\,Fe^{2+}(aq) + 6\,H^+(aq) \rightarrow 3\,Fe^{3+}(aq) + 2\,NO(g) + 4\,H_2O(\ell)$

   d) $4\,HNO_3(aq) + 3\,Fe^{2+}(aq) \rightarrow 3\,Fe^{3+}(aq) + NO(g) + 2H_2O(\ell) + 3\,NO_3^-(aq)$

   e) $HNO_3(aq) + Fe^{2+}(aq) \rightarrow 3\,Fe^{3+}(aq) + NO(g)$

   Answer: d

3. Write a balanced chemical equation for the following reaction in a basic solution.
$$ClO^-(aq) + Cr(OH)_3(s) \rightarrow Cl^-(aq) + CrO_4^{2-}(aq)$$

   a) $3\,ClO^-(aq) + 2\,Cr(OH)_3(s) + 4\,OH^-(aq) \rightarrow 3\,Cl^-(aq) + 2\,CrO_4^{2-}(aq) + 5\,H_2O(\ell)$

   b) $ClO^-(aq) + Cr(OH)_3(s) + 3\,OH^-(aq) \rightarrow Cl^-(aq) + CrO_4^{2-}(aq) + 3\,H_2O(\ell)$

   c) $2\,ClO^-(aq) + 3\,Cr(OH)_3(s) + 3\,OH^-(aq) \rightarrow 2\,Cl^-(aq) + 3\,CrO_4^{2-}(aq) + 6\,H_2O(\ell)$

   d) $4\,ClO^-(aq) + Cr(OH)_3(s) + 4\,OH^-(aq) \rightarrow Cl^-(aq) + CrO_4^{2-}(aq) + 6\,H_2O(\ell)$

   e) $ClO^-(aq) + Cr(OH)_3(s) \rightarrow Cl^-(aq) + CrO_4^{2-}(aq) + 3\,H^+(aq)$

   Answer: a

CHAPTER 20: Principles of Reactivity: Electron Transfer Reactions

Understand the principles underlying voltaic cells (20.2,20.3)

4. All of the following statements concerning voltaic cells are true EXCEPT

   a) a salt bridge allows cations and anions to move between the half-cells.
   b) electrons flow from the anode to the cathode in the external circuit.
   c) oxidation occurs at the cathode.
   d) a voltaic cell can be used as a source of energy.
   e) a voltaic cell consists of two-half cells.

   Answer: c

5. What is the correct cell notation for a voltaic cell based on the reaction below?
$$Ag^+(aq) + Sn(s) \rightarrow Ag(s) + Sn^{2+}(aq)$$

   a) $Ag(s) \mid Ag^+(aq) \parallel Sn^{2+}(aq) \mid Sn(s)$ b) $Sn(s) \parallel Sn^{2+}(aq), Ag^+(aq) \mid Ag(s)$
   c) $Ag(s) \parallel Ag^+(aq), Sn^{2+}(aq) \parallel Sn(s)$ d) $Ag(s) \mid Sn^{2+}(aq) \parallel Ag^+(aq) \mid Sn(s)$
   e) $Sn(s) \mid Sn^{2+}(aq) \parallel Ag^+(aq) \mid Ag(s)$

   Answer: e

Understand how to use electrochemical potentials (20.4,20.5,20.6)

6. The unit for electromotive force, emf, is the Volt. A Volt is equal to

   a) one joule per second.   b) one coulomb per joule.
   c) one joule per coulomb.   d) one coulomb per second.
   e) one second per joule.

   Answer: c

7. Consider the following half-reactions:

$$
\begin{array}{ll}
F_2(g) + 2\,e^- \rightarrow 2\,F^-(aq) & E° = +2.87 \text{ V} \\
Cu^{2+}(aq) + 2\,e^- \rightarrow Cu(s) & E° = +0.34 \text{ V} \\
Sn^{2+}(aq) + 2\,e^- \rightarrow Sn(s) & E° = -0.14 \text{ V} \\
Al^{3+}(aq) + 3\,e^- \rightarrow Al(s) & E° = -1.66 \text{ V} \\
Na^+(aq) + e^- \rightarrow Na(s) & E° = -2.71 \text{ V}
\end{array}
$$

   Which of the above elements or ions will reduce $Sn^{2+}(aq)$?

   a) $Cu(s)$ and $Al^{3+}(aq)$ b) $F^-(aq)$ and $Cu(s)$
   c) $F_2(g)$ and $Cu^{2+}(aq)$ d) $Al(s)$ and $Na(s)$
   e) $Al^{3+}(aq)$ and $Na^+(aq)$

   Answer: d

8. Given the following two half-reactions, write the overall balanced reaction in the direction in which it is spontaneous and calculate the standard cell potential.

$$Cr^{3+}(aq) + 3 e^- \rightarrow Cr(s) \qquad E° = -0.41 \text{ V}$$
$$Sn^{2+}(aq) + 2 e^- \rightarrow Sn(s) \qquad E° = -0.14 \text{ V}$$

a) $2 Cr^{3+}(aq) + 3 Sn(s) \rightarrow 2 Cr(s) + 3 Sn^{2+}(aq)$ $E^o_{cell} = +0.55 \text{ V}$

b) $3 Cr^{3+}(aq) + 2 Sn(s) \rightarrow 3 Cr(s) + 2 Sn^{2+}(aq)$ $E^o_{cell} = -1.35 \text{ V}$

c) $2 Cr(s) + 3 Sn^{2+}(aq) \rightarrow 2 Cr^{3+}(aq) + 3 Sn(s)$ $E^o_{cell} = +0.27 \text{ V}$

d) $3 Cr(s) + 2 Sn^{2+}(aq) \rightarrow 3 Cr^{3+}(aq) + 2 Sn(s)$ $E^o_{cell} = -0.55 \text{ V}$

e) $2 Cr(s) + 3 Sn^{2+}(aq) \rightarrow 2 Cr^{3+}(aq) + 3 Sn(s)$ $E^o_{cell} = +0.40 \text{ V}$

Answer: c

9. Calculate E_{cell} for the following electrochemical cell at 25 °C

$$Pt(s) \,|\, Fe^{3+}(aq, 0.25 \text{ M}), Fe^{2+}(aq, 0.25 \text{ M}) \,||\, Cl^-(aq, 0.20 \text{ M}) \,|\, AgCl(s) \,|\, Ag(s)$$

given the following standard reduction potentials.

$$AgCl(s) + e^- \rightarrow Ag(s) + Cl^-(aq) \quad E° = +0.222 \text{ V}$$
$$Fe^{3+}(aq) + e^- \rightarrow Fe^{2+}(aq) \qquad\qquad E° = +0.771 \text{ V}$$

a) -0.590 V   b) -0.549 V   c) -0.508 V   d) -0.135 V   e) +0.549 V

Answer: c

Explore electrolysis, the use of electrical energy to produce chemical changes (20.7,20.8)

10. Cu^{2+} is reduced to Cu(s) at an electrode. If a current of 1.25 ampere is passed for 72 hours, what mass of copper is deposited at the electrode? Assume 100% current efficiency.

a) 3.0×10^{-2} g   b) 1.1×10^2 g   c) 2.1×10^2 g   d) 2.90×10^2 g   e) 4.3×10^2 g

Answer: b

11. What charge, in coulombs, is required to deposit 0.34 g Al(s) from a solution of $Al^{3+}(aq)$?

a) 4.1×10^2 C   b) 3.6×10^3 C   c) 7.2×10^3 C   d) 3.3×10^4 C   e) 9.8×10^4 C

Answer: b

TEST BANK

20.1 Oxidation-Reduction Reactions

1. Assuming the following reaction proceeds in the forward direction,
$$2 \, Ni^{2+}(aq) + Zn(s) \rightarrow 2 \, Ni(s) + Zn^{2+}(aq)$$

a) $Ni^{2+}(aq)$ is the reducing agent and $Zn(s)$ is the oxidizing agent.
b) $Zn(s)$ is the reducing agent and $Ni(s)$ is the oxidizing agent.
c) $Ni^{2+}(aq)$ is the reducing agent and $Ni(s)$ is the oxidizing agent.
d) $Zn(s)$ is the reducing agent and $Zn^{2+}(s)$ is the oxidizing agent.
e) $Zn(s)$ is the reducing agent and $Ni^{2+}(s)$ is the oxidizing agent.

Answer: e

2. Assuming the following reaction proceeds in the forward direction,
$$Cu^{2+}(aq) + Ca(s) \rightarrow Cu(s) + Ca^{2+}(aq)$$

a) $Cu^{2+}(aq)$ is oxidized and $Ca(s)$ is reduced.
b) $Cu^{2+}(aq)$ is oxidized and $Ca^{2+}(aq)$ is reduced.
c) $Ca(s)$ is oxidized and $Cu^{2+}(aq)$ is reduced.
d) $Ca(s)$ is oxidized and $Ca^{2+}(aq)$ is reduced.
e) $Cu(s)$ is oxidized and $Ca(s)$ is reduced.

Answer: c

3. The following reaction occurs spontaneously.
$$2 \, H^{+}(aq) + Sr(s) \rightarrow Sr^{2+}(aq) + H_2(g)$$
Write the balanced reduction half-reaction.

a) $2 \, H^{+}(aq) + 2 \, e^- \rightarrow H_2(g)$ b) $2 \, H^{+}(aq) \rightarrow H_2(g) + 2 \, e^-$
c) $H_2(g) \rightarrow 2 \, H^{+}(aq) + 2 \, e^-$ d) $Sr(s) + 2 \, e^- \rightarrow Sr^{2+}(aq)$
e) $Sr(s) \rightarrow Sr^{2+}(aq) + 2 \, e^-$

Answer: a

4. The following reaction occurs spontaneously.
$$2 \, Fe^{3+}(aq) + Fe(s) \rightarrow 3 \, Fe^{2+}(aq)$$
Write the balanced reduction half-reaction.

a) $Fe(s) \rightarrow Fe^{3+}(aq) + e^-$ b) $3 \, Fe(s) + 3 \, e^- \rightarrow 3 \, Fe^{3+}(aq)$
c) $2 \, Fe^{3+}(aq) + 6 \, e^- \rightarrow 2 \, Fe(s)$ d) $2 \, Fe^{3+}(aq) \rightarrow 2 \, Fe^{2+}(s) + 2 \, e^-$
e) $2 \, Fe^{3+}(aq) + 2 \, e^- \rightarrow 2 \, Fe^{2+}(aq)$

Answer: e

5. Write a balanced half-reaction for the reduction of hydrogen peroxide to water in an acidic
    solution.

a) $2\ H_2O_2(\ell)\ \rightarrow\ 2\ H_2O(\ell)\ +\ O_2(g)$

b) $2\ H_2O_2(\ell)\ +\ 2e^-\ \rightarrow\ 2\ H_2O(\ell)\ +\ O_2(g)$

c) $H_2O_2(\ell)\ +\ 2\ H^+(aq)\ +\ 2\ e^-\ \rightarrow\ 2\ H_2O(\ell)$

d) $H_2O_2(\ell)\ +\ 4\ H^+(aq)\ +\ 2\ e^-\ \rightarrow\ 2\ H_2O(\ell)\ +\ H_2(g)$

e) $H_2O_2(\ell)\ +2\ H^+(aq)\ +\ 4\ e^-\ \rightarrow\ 2\ H_2(g)\ +\ O_2(g)$

Answer: c

6. Write a balanced half-reaction for the reduction of $SO_4^{2-}(aq)$ to $SO_2(g)$ in an acidic solution.

a) $SO_4^{2-}(aq)\ +\ 4\ H^+(aq)\ \rightarrow\ SO_2(g)\ +\ 2\ H_2O(\ell)$

b) $SO_4^{2-}(aq)\ +\ 4\ H^+(aq)\ +\ 2\ e^-\ \rightarrow\ SO_2(g)\ +\ 2\ H_2O(\ell)$

c) $SO_4^{2-}(aq)\ +\ 2\ e^-\ \rightarrow\ SO_2(g)\ +\ O_2(g)$

d) $SO_4^{2-}(aq)\ +\ 4\ e^-\ \rightarrow\ SO_2(g)\ +\ O_2(g)$

e) $SO_4^{2-}(aq)\ +\ 4\ H^+(aq)\ \rightarrow\ SO_2(g)\ +\ 2\ H_2O(\ell)\ +\ 2\ e^-$

Answer: b

7. Write a balanced half-reaction for the reduction of $MnO_2(s)$ to $Mn(OH)_2(s)$ in a basic
    solution.

a) $MnO_2(s)\ +\ 2\ OH^-(aq)\ +\ 2\ e^-\ \rightarrow\ Mn(OH)_2(s)\ +\ O_2(g)$

b) $MnO_2(s)\ +\ 2\ H^+(aq)\ +\ 2\ e^-\ \rightarrow\ Mn(OH)_2(s)$

c) $MnO_2(s)\ +\ 2\ H^+(aq)\ \rightarrow\ Mn(OH)_2(s)\ +\ 2\ e^-$

d) $MnO_2(s)\ +\ 2\ H_2O(\ell)\ +\ 2\ e^-\ \rightarrow\ Mn(OH)_2(s)\ +\ 2\ OH^-(aq)$

e) $MnO_2(s)\ +\ 2\ OH^-(aq)\ \rightarrow\ Mn(OH)_2(s)\ +\ O_2(g)$

Answer: d

8. Write a balanced chemical equation for the following reaction in an acidic solution.
$$MnO_4^-(aq)\ +\ Fe(s)\ \rightarrow\ Mn^{2+}(aq)\ +\ Fe^{2+}(aq)$$

a) $2\ MnO_4^-(aq)\ +\ 5\ Fe(s)\ +\ 16\ H^+(aq)\ \rightarrow\ 2\ Mn^{2+}(aq)\ +\ 5\ Fe^{2+}(aq)\ +\ 8\ H_2O(\ell)$

b) $MnO_4^-(aq)\ +\ Fe(s)\ +\ 16\ H^+(aq)\ \rightarrow\ Mn^{2+}(aq)\ +\ Fe^{2+}(aq)\ +\ 8\ H_2O(\ell)$

c) $MnO_4^-(aq)\ +\ Fe(s)\ \rightarrow\ Mn^{2+}(aq)\ +\ Fe^{2+}(aq)\ +\ 4\ H_2O(\ell)$

d) $2\ MnO_4^-(aq)\ +\ Fe(s)\ +\ 16\ H^+(aq)\ \rightarrow\ 2\ Mn^{2+}(aq)\ +\ Fe^{2+}(aq)\ +\ 8\ H_2O(\ell)$

e) $MnO_4^-(aq)\ +\ 5\ Fe(s)\ +\ 16\ H^+(aq)\ \rightarrow\ Mn^{2+}(aq)\ +\ 5\ Fe^{2+}(aq)\ +\ 8\ H_2O(\ell)$

Answer: a

9. Write a balanced chemical equation for the oxidation of Fe^{2+}(aq) by concentrated nitric acid. Two products of the reaction are NO(g) and Fe^{3+}(aq).

a) HNO_3(aq) + Fe^{2+}(aq) $\rightarrow$ Fe^{3+}(aq) + NO(g) + HO_2(ℓ)

b) HNO_3(aq) + 3 Fe^{2+}(aq) $\rightarrow$ 3 Fe^{3+}(aq) + NO(g) + HO_2(ℓ)

c) 2 HNO_3(aq) + 2 Fe^{2+}(aq) + 6 H^+(aq) $\rightarrow$ 3 Fe^{3+}(aq) + 2 NO(g) + 4 H_2O(ℓ)

d) 4 HNO_3(aq) + 3 Fe^{2+}(aq) $\rightarrow$ 3 Fe^{3+}(aq) + NO(g) + 2H_2O(ℓ) + 3 NO_3^-(aq)

e) HNO_3(aq) + Fe^{2+}(aq) $\rightarrow$ 3 Fe^{3+}(aq) + NO(g)

Answer: d

10. Write a balanced chemical equation for the following reaction in a basic solution.
$$ClO^-(aq) + Cr(OH)_3(s) \rightarrow Cl^-(aq) + CrO_4^{2-}(aq)$$

a) 3 ClO^-(aq) + 2 $Cr(OH)_3$(s) + 4 OH^-(aq) $\rightarrow$ 3 Cl^-(aq) + 2 CrO_4^{2-}(aq) + 5 H_2O(ℓ)

b) ClO^-(aq) + $Cr(OH)_3$(s) + 3 OH^-(aq) $\rightarrow$ Cl^-(aq) + CrO_4^{2-}(aq) + 3 H_2O(ℓ)

c) 2 ClO^-(aq) + 3 $Cr(OH)_3$(s) + 3 OH^-(aq) $\rightarrow$ 2 Cl^-(aq) + 3 CrO_4^{2-}(aq) + 6 H_2O(ℓ)

d) 4 ClO^-(aq) + $Cr(OH)_3$(s) + 4 OH^-(aq) $\rightarrow$ Cl^-(aq) + CrO_4^{2-}(aq) + 6 H_2O(ℓ)

e) ClO^-(aq) + $Cr(OH)_3$(s) $\rightarrow$ Cl^-(aq) + CrO_4^{2-}(aq) + 3 H^+(aq)

Answer: a

20.2 Simple Voltaic Cells

11. All of the following statements concerning voltaic cells are true EXCEPT

a) a salt bridge allows cations and anions to move between the half-cells.
b) electrons flow from the anode to the cathode in the external circuit.
c) oxidation occurs at the cathode.
d) a voltaic cell can be used as a source of energy.
e) a voltaic cell consists of two-half cells.

Answer: c

12. What is the correct cell notation for a voltaic cell based on the reaction below?
$$Ag^+(aq) + Sn(s) \rightarrow Ag(s) + Sn^{2+}(aq)$$

a) Ag(s) | Ag^+(aq) || Sn^{2+}(aq) | Sn(s)
c) Ag(s) || Ag^+(aq), Sn^{2+}(aq) || Sn(s)
e) Sn(s) | Sn^{2+}(aq) || Ag^+(aq) | Ag(s)

b) Sn(s) || Sn^{2+}(aq), Ag^+(aq) | Ag(s)
d) Ag(s) | Sn^{2+}(aq) || Ag^+(aq) | Sn(s)

Answer: e

13. What is a correct cell notation for a voltaic cell based on the reaction below?
$$PbBr_2(s) + 2 Zn(s) \rightarrow Pb(s) + Zn^{2+}(aq) + 2 Br^-(aq)$$

a) $Pb(s) | Pb^{2+}(aq) || Br^-(aq), Zn^{2+}(aq) || Zn(s)$
b) $Zn(s) | Zn^{2+}(aq) || Br^-(aq) | PbBr_2(s) | Pb(s)$
c) $Zn(s) | Zn^{2+}(aq), Br^-(aq) | PbBr_2(s) | Pb(s)$
d) $Zn(s) | Zn^{2+}(aq), Br^-(aq) || PbBr_2(s) || Pb(s)$
e) $Pb(s), PbBr_2(s) | Br^-(aq) || Zn^{2+}(aq) || Zn(s)$

Answer: b

14. Write a balanced chemical equation for the overall reaction represented by the cell notation below.
$$Pt | H_2(g) | H^+(aq) || SO_4^{2-}(aq) | PbSO_4(s) | Pb(s)$$

a) $Pb(s) + 2 H^+(aq) \rightarrow Pb^{2+}(s) + H_2(g)$
b) $Pb(s) + 2 H^+(aq) + SO_4^{2-}(aq) \rightarrow PbSO_4(s) + H_2(g)$
c) $PbSO_4(s) + H_2(g) \rightarrow Pb(s) + 2 H^+(aq) + SO_4^{2-}(aq)$
d) $PbSO_4(s) + 2 H^+(aq) \rightarrow Pb(s) + H_2(g) + SO_4^{2-}(aq)$
e) $Pb^{2+}(s) + H_2(g) \rightarrow Pb(s) + 2 H^+(aq)$

Answer: c

15. Write a balanced net ionic equation for the overall reaction represented by the cell notation below.
$$Pt | Fe^{2+}(aq), Fe^{3+}(aq) || I^-(aq) | AgI(s) | Ag(s)$$

a) $Fe^{3+}(aq) + Ag(s) + I^-(aq) \rightarrow Fe^{2+}(aq) + AgI(s)$
b) $Fe^{2+}(aq) + Ag(s) + I^-(aq) \rightarrow Fe^{3+}(aq) + AgI(s)$
c) $Fe^{3+}(aq) + AgI(s) \rightarrow Fe^{2+}(aq) + Ag(s) + I^-(aq)$
d) $Fe^{2+}(aq) + AgI(s) \rightarrow Fe^{3+}(aq) + Ag(s) + I^-(aq)$
e) $Fe^{3+}(aq) + 2 I^-(aq) \rightarrow Fe^{2+}(aq) + I_2(s)$

Answer: d

20.3 Commercial Voltaic Cells

16. Which of the following statements is/are CORRECT?
   1. A nickel-cadmium battery is an example of a secondary or rechargable battery, often used in rechargable cordless appliances.
   2. Hydrogen-oxygen fuel cells use the heat of combustion of hydrogen to recharge lead storage batteries.
   3. LeClanché cells are the most efficient rechargable batteries, but they are rarely used due their high cost of production.

a) 1 only   b) 2 only   c) 3 only   d) 1 and 3   e) 1, 2, and 3

Answer: a

17. The fuel cells used aboard NASA's Space Shuttles are electrochemical cells that generate electricity from which overall chemical reaction?

a) $Fe_2O_3(s) + Al(s) \rightarrow Fe(s) + Al_2O_3(s)$
b) $Pb(s) + PbO_2(s) + 2\ H_2SO_4(aq) \rightarrow 2\ PbSO_4(s) + 2\ H_2O(\ell)$
c) $2\ NiO(OH)(s) + Cd(s) + H_2O(\ell) \rightarrow 2\ Ni(OH)_2(s) + Cd(OH)_2(s)$
d) $2\ H_2(g) + O_2(g) \rightarrow 2\ H_2O(\ell)$
e) $N_2H_4(\ell) + O_2(g) \rightarrow N_2(g) + 2\ H_2O(\ell)$

Answer: d

20.4 Standard Electrochemical Potentials

18. The unit for electromotive force, emf, is the Volt. A Volt is equal to

a) one joule per second.　　　b) one coulomb per joule.　　c) one joule per coulomb.
d) one coulomb per second.　　e) one second per joule.

Answer: c

19. An SHE electrode has been assigned a standard reduction potential, $E°$, of 0.00 Volts. Which reaction occurs at this electrode?

a) $2\ H_2O(\ell) + 2\ e^- \rightarrow H_2(g) + 2\ OH^-(aq)$　　　　b) $O_2(g) + 4\ e^- \rightarrow 2\ O^{2-}(aq)$
c) $Hg_2Cl_2(s) + 2\ e^- \rightarrow 2\ Hg(\ell) + 2\ Cl^-(aq)$　　　　d) $Li^+(aq) + e^- \rightarrow Li(s)$
e) $2\ H^+(aq) + 2\ e^- \rightarrow H_2(g)$

Answer: e

20. Use the standard reduction potentials below to determine which element or ion is the best oxidizing agent.

$$Hg_2^{2+}(aq) + 2\ e^- \rightarrow 2\ Hg(\ell) \quad E° = +0.789\ V$$
$$I_2(s) + 2\ e^- \rightarrow 2\ I^-(aq) \quad E° = +0.535\ V$$
$$Ni^{2+}(aq) + 2\ e^- \rightarrow Ni(s) \quad E° = -0.25\ V$$

a) $I_2(s)$　　　　b) $Hg_2^{2+}(aq)$　　　c) $I^-(aq)$　　　　d) $Ni^{2+}(aq)$　　　e) $Hg(\ell)$

Answer: b

21. Use the standard reduction potentials below to determine which element or ion is the best reducing agent.

$$Pd^{2+}(aq) + 2 e^- \rightarrow Pd(s) \quad E° = +0.90 \text{ V}$$
$$2 H^+(aq) + 2 e^- \rightarrow H_2(g) \quad E° = 0.00 \text{ V}$$
$$Mn^{2+}(aq) + 2 e^- \rightarrow Mn(s) \quad E° = -1.18 \text{ V}$$

a) $Pd^{2+}(aq)$ b) $Pd(s)$ c) $H^+(aq)$ d) $Mn^{2+}(aq)$ e) $Mn(s)$

Answer: e

22. Which of the following species are likely to behave as oxidizing agents: $K(s)$, $H_2(g)$, $Cr_2O_7^{2-}$ (aq), and $I_2(s)$?

a) $K(s)$ only

b) $Cr_2O_7^{2-}(aq)$ and $I_2(s)$

c) $H_2(g)$ and $I_2(s)$

d) $K(s)$ and $Cr_2O_7^{2-}(aq)$

e) $I_2(s)$ only

Answer: b

23. Consider the following half-reactions:

$$Cu^{2+}(aq) + 2 e^- \rightarrow Cu(s) \quad E° = +0.34 \text{ V}$$
$$Sn^{2+}(aq) + 2 e^- \rightarrow Sn(s) \quad E° = -0.14 \text{ V}$$
$$Fe^{2+}(aq) + 2 e^- \rightarrow Fe(s) \quad E° = -0.44 \text{ V}$$
$$Al^{3+}(aq) + 3 e^- \rightarrow Al(s) \quad E° = -1.66 \text{ V}$$
$$Mg^{2+}(aq) + 2 e^- \rightarrow Mg(s) \quad E° = -2.37 \text{ V}$$

Which of the above metals or metal ions will oxidize $Fe(s)$?

a) $Cu^{2+}(aq)$ and $Sn^{2+}(aq)$

b) $Cu(s)$ and $Sn(s)$

c) $Al^{3+}(aq)$ and $Mg^{2+}(aq)$

d) $Al(s)$ and $Mg(s)$

e) $Sn(s)$ and $Al^{3+}(aq)$

Answer: a

24. Consider the following half-reactions:

$$F_2(g) + 2 e^- \rightarrow 2 F^-(aq) \quad E° = +2.87 \text{ V}$$
$$Cu^{2+}(aq) + 2 e^- \rightarrow Cu(s) \quad E° = +0.34 \text{ V}$$
$$Sn^{2+}(aq) + 2 e^- \rightarrow Sn(s) \quad E° = -0.14 \text{ V}$$
$$Al^{3+}(aq) + 3 e^- \rightarrow Al(s) \quad E° = -1.66 \text{ V}$$
$$Na^+(aq) + e^- \rightarrow Na(s) \quad E° = -2.71 \text{ V}$$

Which of the above elements or ions will reduce $Sn^{2+}(aq)$?

a) $Cu(s)$ and $Al^{3+}(aq)$

b) $F^-(aq)$ and $Cu(s)$

c) $F_2(g)$ and $Cu^{2+}(aq)$

d) $Al(s)$ and $Na(s)$

e) $Al^{3+}(aq)$ and $Na^+(aq)$

Answer: d

25. Given the following two half-reactions, write the overall balanced reaction in the direction in which it is spontaneous and calculate the standard cell potential.

$$Cr^{3+}(aq) + 3\ e^- \rightarrow Cr(s) \qquad E° = -0.41\ V$$
$$Sn^{2+}(aq) + 2\ e^- \rightarrow Sn(s) \qquad E° = -0.14\ V$$

a) $2\ Cr^{3+}(aq) + 3\ Sn(s) \rightarrow 2\ Cr(s) + 3\ Sn^{2+}(aq)$ $\qquad E°_{cell} = +0.55\ V$

b) $3\ Cr^{3+}(aq) + 2\ Sn(s) \rightarrow 3\ Cr(s) + 2\ Sn^{2+}(aq)$ $\qquad E°_{cell} = -1.35\ V$

c) $2\ Cr(s) + 3\ Sn^{2+}(aq) \rightarrow 2\ Cr^{3+}(aq) + 3\ Sn(s)$ $\qquad E°_{cell} = +0.27\ V$

d) $3\ Cr(s) + 2\ Sn^{2+}(aq) \rightarrow 3\ Cr^{3+}(aq) + 2\ Sn(s)$ $\qquad E°_{cell} = -0.55\ V$

e) $2\ Cr(s) + 3\ Sn^{2+}(aq) \rightarrow 2\ Cr^{3+}(aq) + 3\ Sn(s)$ $\qquad E°_{cell} = +0.40\ V$

Answer: c

26. Given the following two half-reactions, write the overall reaction in the direction in which it is spontaneous and calculate the standard cell potential.

$$Pb^{2+}(aq) + 2\ e^- \rightarrow Pb(s) \qquad E° = -0.126\ V$$
$$Cu^{2+}(aq) + 2\ e^- \rightarrow Cu(s) \qquad E° = +0.337\ V$$

a) $Pb^{2+}(aq) + Cu(s) \rightarrow Pb(s) + Cu^{2+}(aq)$ $\qquad E°_{cell} = +0.463\ V$

b) $Pb^{2+}(aq) + Cu(s) \rightarrow Pb(s) + Cu^{2+}(aq)$ $\qquad E°_{cell} = +0.211\ V$

c) $Pb(s) + Cu^{2+}(aq) \rightarrow Pb^{2+}(aq) + Cu(s)$ $\qquad E°_{cell} = -0.211\ V$

d) $Pb(s) + Cu^{2+}(aq) \rightarrow Pb^{2+}(aq) + Cu(s)$ $\qquad E°_{cell} = +0.463\ V$

e) $Pb(s) + Cu^{2+}(aq) \rightarrow Pb^{2+}(aq) + Cu(s)$ $\qquad E°_{cell} = +0.926\ V$

Answer: d

27. Calculate $E°_{cell}$ for the reaction below,

$$2\ Ag^+(aq) + Pb(s) + SO_4^{2-}(aq) \rightarrow 2\ Ag(s) + PbSO_4(s)$$

given the following standard reduction potentials.

$$PbSO_4(s) + 2\ e^- \rightarrow Pb(s) + SO_4^{2-}(aq) \qquad E° = -0.356\ V$$
$$Ag^+(aq) + e^- \rightarrow Ag(s) \qquad E° = +0.799\ V$$

a) -1.155 V $\qquad$ b) -0.443 V $\qquad$ c) +0.443 V $\qquad$ d) +1.155 V $\qquad$ e) +1.954 V

Answer: d

28. Calculate E°_{cell} for the reaction below,

$$5\ Cd^{2+}(aq) + 2\ Mn^{2+}(aq) + 8\ H_2O(\ell) \rightarrow 5\ Cd(s) + 2\ MnO_4^{-}(aq) + 16\ H^{+}(aq)$$

given the following standard reduction potentials.

$$MnO_4^{-}(aq) + 8\ H^{+}(aq) + 5\ e^{-} \rightarrow Mn^{2+}(aq) + 4\ H_2O(\ell) \qquad E^{\circ} = +1.52\ V$$
$$Cd^{2+}(aq) + 2\ e^{-} \rightarrow Cd(s) \qquad E^{\circ} = -0.40\ V$$

a) -1.92 V   b) +1.04 V   c) +1.12V   d) +5.04 V   e) +8.40 V

Answer: a

29. Calculate E°_{cell} for the electrochemical cell below,

$$Pb(s)\ |PbCl_2(s)\ |\ Cl^{-}(aq,\ 1.0\ M)\ ||\ Fe^{3+}(aq,\ 1.0\ M),\ Fe^{2+}(aq,\ 1.0\ M)\ |\ Pt(s)$$

given the following reduction half-reactions.

$$Pb^{2+}(aq) + 2\ e^{-} \rightarrow Pb(s) \qquad E^{\circ} = -0.126\ V$$
$$PbCl_2(s) + 2\ e^{-} \rightarrow Pb(s) + 2\ Cl^{-}(aq) \qquad E^{\circ} = -0.267\ V$$
$$Fe^{3+}(aq) + e^{-} \rightarrow Fe^{2+}(aq) \qquad E^{\circ} = +0.771\ V$$
$$Fe^{2+}(aq) + e^{-} \rightarrow Fe(s) \qquad E^{\circ} = -0.44\ V$$

a) -0.504 V   b) -0.062 V   c) +0.504 V   d) +1.038 V   e) +1.604 V

Answer: d

30. Calculate E°_{cell} for the electrochemical cell below,

$$Pb(s)\ |\ PbSO_4(s)\ |\ Pb^{2+}(aq)\ ||\ Cu^{2+}(aq)\ |\ Cu(s)$$

given the following standard reduction potentials.

$$Cu^{2+}(aq) + 2\ e^{-} \rightarrow Cu(s) \qquad E^{\circ} = +0.337\ V$$
$$PbSO_4(s) + 2\ e^{-} \rightarrow Pb(s) + SO_4^{2-}(aq) \qquad E^{\circ} = -0.356\ V$$

a) -0.693 V   b) -0.019 V   c) +0.019 V   d) +0.693 V   e) +1.386 V

Answer: d

20.5 Electrochemical Cells under Nonstandard Conditions

31. Which of the following equations is a correct form of the Nernst equation?

a) $E = E^{\circ} - \left(\dfrac{RT}{nF}\right)\ln Q$ b) $E = E^{\circ} + \log\left(\dfrac{RT}{nF}\right)$ c) $E = E^{\circ} - \left(\dfrac{nF}{RT}\right)\log Q$

d) $E^{\circ} = E - \left(\dfrac{RT}{nF}\right)\ln Q$ e) $E = E^{\circ} - \left(\dfrac{nF}{RT}\right)\ln Q$

Answer: a

32. A Faraday, F, is defined as

   a) the charge on a single electron.
   b) the charge, in coulombs, carried by one mole of electrons.
   c) the voltage required to reduce one mole of reactant.
   d) the moles of electrons required to reduce one mole of reactant.
   e) the charge passed by one ampere of current in one second.

Answer: b

33. Calculate E_{cell} for the following electrochemical cell at 25 °C
$$Pt(s) \mid H_2(g, 1.00 \text{ atm}) \mid H^+(aq, 1.00 \text{ M}) \parallel Pb^{2+}(aq, 0.150 \text{ M}) \mid Pb(s)$$
given the following standard reduction potentials.
$$Pb^{2+}(aq) + 2 e^- \rightarrow Pb(s) \quad E° = -0.126 \text{ V}$$
$$2 H^+(aq) + 2 e^- \rightarrow H_2(g) \quad E° = 0.000 \text{ V}$$

   a) -0.150 V   b) -0.126 V   c) -0.102 V   d) +0.102 V   e) +0.150 V

Answer: a

34. Calculate E_{cell} for the following electrochemical cell at 25 °C
$$Pt(s) \mid Fe^{3+}(aq, 0.25 \text{ M}), Fe^{2+}(aq, 0.25 \text{ M}) \parallel Cl^-(aq, 0.20 \text{ M}) \mid AgCl(s) \mid Ag(s)$$
given the following standard reduction potentials.
$$AgCl(s) + e^- \rightarrow Ag(s) + Cl^-(aq) \quad E° = +0.222 \text{ V}$$
$$Fe^{3+}(aq) + e^- \rightarrow Fe^{2+}(aq) \quad E° = +0.771 \text{ V}$$

   a) -0.590 V   b) -0.549 V   c) -0.508 V   d) -0.135 V   e) +0.549 V

Answer: c

35. Calculate the cell potential, at 25 °C, based upon the overall reaction
$$3 Cr^{2+}(aq) + 2 Al(s) \rightarrow 3 Cr(s) + 2 Al^{3+}(aq)$$
if $[Cr^{2+}] = 0.15$ M and $[Al^{3+}] = 0.0040$ M. The standard reduction potentials are as follows:
$$Cr^{2+}(aq) + 2 e^- \rightarrow Cr(s) \quad E° = -0.91 \text{ V}$$
$$Al^{3+}(aq) + 3 e^- \rightarrow Al(s) \quad E° = -1.66 \text{ V}$$

   a) -2.64 V   b) -0.75 V   c) +0.44 V   d) +0.73 V   e) +0.77 V

Answer: e

36. The following electrochemical cell has a potential of +0.217 V at 25 °C.
$$Pt \mid H_2(g, 1.00 \text{ atm}) \mid H^+(aq, 1.00 \text{ M}) \parallel Cu^{2+}(aq) \mid Cu$$
The standard reduction potential, $E°$, of $Cu^{2+} = +0.337$ V. What is the $Cu^{2+}(aq)$ concentration?

   a) 1.9×10^{-19} M   b) 4.6×10^{-10} M   c) 8.8×10^{-5} M   d) 0.12 M   e) 0.65 M

Answer: c

37. What is the pH of the solution at the cathode if E°_{cell} = -0.094 V for the following
electrochemical cell at 25 °C?
$$Pt \mid H_2(g, 1.0 \text{ atm}) \mid H^+(aq, 1.00 \text{ M}) \parallel H^+(aq) \mid H_2(g, 1.0 \text{ atm}) \mid Pt$$

a) 0.79   b) 1.03   c) 1.59   d) 3.18   e) 4.78

Answer: c

38. Which factor will increase the measured cell potential of the following galvanic cell?
$$Pt \mid Sn^{4+}(aq, 1.0 \text{ M}), Sn^{2+}(aq, 1.0 \text{ M}) \parallel Cu^{2+}(aq, 0.200 \text{ M}) \mid Cu$$

a) switching from a platinum to a graphite anode
b) increasing the size of the anode
c) decreasing the concentration of Cu^{2+}
d) increasing the concentration of Sn^{4+}
e) decreasing the temperature of the cell

Answer: e

20.6 Electrochemistry and Thermodynamics

39. E°_{cell} for the following galvanic cell is -0.39 V.
$$Sn^{4+}(aq) + 2 I^-(aq) \rightarrow Sn^{2+}(aq) + I_2(s)$$
What is $\Delta_r G^\circ$ for this reaction?

a) -75 kJ   b) -38 kJ   c) 2.0 kJ   d) 38 kJ   e) 75 kJ

Answer: e

40. Calculate $\Delta_r G^\circ$ for the disproportionation reaction of Cu^+ at 25 °C,
$$2 Cu^+(aq) \rightarrow Cu^{2+}(aq) + Cu(s)$$
given the following thermodynamic information.
$$Cu^+(aq) + e^- \rightarrow Cu(s) \qquad E^\circ = +0.518 \text{ V}$$
$$Cu^{2+}(aq) + 2 e^- \rightarrow Cu(s) \qquad E^\circ = +0.337 \text{ V}$$

a) -165 kJ   b) -135 kJ   c) -34.9 kJ   d) +17.5 kJ   e) +135 kJ

Answer: c

41. If $\Delta_r G^\circ$ for the following reaction is -1880 kJ, calculate E°_{cell}.
$$2 MnO_4^-(aq) + 5 Fe(s) + 16 H^+(aq) \rightarrow 2 Mn^{2+}(aq) + 5 Fe^{2+}(aq) + 8 H_2O(\ell)$$

a) -1.95 V   b) -0.558 V   c) +0.02 V   d) +0.513 V   e) +1.95 V

Answer: e

42. Calculate the equilibrium constant for the following reaction at 25 °C,
$$Co(s) + 2\,Cr^{3+}(aq) \rightarrow Co^{2+}(aq) + 2\,Cr^{2+}(aq)$$
given the following thermodynamic information.
$$Co^{2+}(aq) + 2\,e^- \rightarrow Co(s) \quad E° = -0.28\ V$$
$$Cr^{3+}(aq) + e^- \rightarrow Cr^{2+}(aq) \quad E° = -0.41\ V$$

a) 5×10^{-24} b) 7×10^{-11} c) 4×10^{-5} d) 2×10^{4} e) 1×10^{10}

Answer: c

43. Given the following standard reduction potentials,
$$Hg_2^{2+}(aq) + 2\,e^- \rightarrow 2\,Hg(\ell) \qquad E° = +0.789\ V$$
$$Hg_2Cl_2(s) + 2\,e^- \rightarrow 2\,Hg(\ell) + 2\,Cl^-(aq) \quad E° = +0.271\ V$$
determine K_{sp} for $Hg_2Cl_2(s)$ at 25 °C.

a) 4.5×10^{-41} b) 9.0×10^{-36} c) 1.4×10^{-36} d) 3.0×10^{-18} e) 7.2×10^{35}

Answer: d

44. Calculate the standard reduction potential for the following reaction at 25 °C,
$$AuCl_4^-(aq) + 3\,e^- \rightarrow Au(s) + 4\,Cl^-(aq)$$
given the following thermodynamic information.
$$Au^{3+}(aq) + 3\,e^- \rightarrow Au(s) \qquad E° = +1.50\ V$$
$$Au^{3+}(aq) + 4\,Cl^-(aq) \rightarrow AuCl_4^-(aq) \qquad K_f = 2.3 \times 10^{25}$$

a) -1.28 V b) -0.50 V c) $+1.00$ V d) $+1.28$ V e) $+3.85$ V

Answer: c

20.7 Electrolysis: Chemical Change using Electrical Energy

45. What charge, in coulombs, is required to deposit 0.34 g Al(s) from a solution of $Al^{3+}(aq)$?

a) 4.1×10^{2} C b) 3.6×10^{3} C c) 7.2×10^{3} C d) 3.3×10^{4} C e) 9.8×10^{4} C

Answer: b

46. Cu^{2+} is reduced to Cu(s) at an electrode. If a current of 1.25 ampere is passed for 72 hours, what mass of copper is deposited at the electrode? Assume 100% current efficiency.

a) 3.0×10^{-2} g b) 1.1×10^{2} g c) 2.1×10^{2} g d) 2.90×10^{2} g e) 4.3×10^{2} g

Answer: b

47. One kind of battery used in watches contains mercury(II) oxide. As current flows, the mercury oxide is reduced to mercury.

$$HgO(s) + H_2O(\ell) + 2\ e^- \rightarrow Hg(\ell) + 2\ OH^-(aq)$$

If 1.8×10^{-5} amperes flows continuously for 1461 days, what mass of $Hg(\ell)$ is produced?

a) 2.4 g   b) 4.7 g   c) 9.4 g   d) 5.3 g   e) 1.1×10^3 g

Answer: a

Short Answer Questions

48. In an electrolytic cell, reduction occurs at the _________ and oxidation occurs at the _________.

Answer: cathode, anode

49. Gold and platinum are commonly used as inert electrodes in laboratory experiments. In commercial applications, such as batteries, _________ is more commonly used for inert electrodes because it is far less expensive.

Answer: graphite

50. The use of electrical energy to produce chemical change is known as _________. An example of this process is the reduction of $NaCl(\ell)$ to produce $Na(s)$.

Answer: electrolysis

51. If electric current is passed through a solution of molten NaBr, the product at the anode is _________.

Answer: $Br_2(g)$

52. When a secondary battery provides electrical energy, it is acting as a(n) _________ cell. When the battery is recharging, it is operating as an electrolytic cell.

Answer: voltaic

53. Write balanced reduction and oxidation half-reactions for the processes that occur at the cathode and anode of a fuel cell used aboard NASA's space shuttles.

Answer: Cathode: $O_2(g) + 2\ H_2O(\ell) + 4\ e^- \rightarrow 4\ OH^-(aq)$

           Anode: $H_2(g) \rightarrow 2\ H^+(aq) + 2\ e^-$

54. Explain the function of a salt bridge in a voltaic cell.

Answer: The salt bridge physically connects the redox couples in an electrochemical cell, thereby enabling electrical conduction through the cell. In addition, the salt bridge helps maintain electroneutrality as reduction and oxidation occur at the cathode and anode, respectively. As oxidation occurs, anions flow from the salt bridge toward the anode compartment and cations flow from the anode compartment into the salt bridge. Likewise, cations flow from the salt bridge toward the cathode compartment and anions flow from the cathode compartment into the salt bridge.

Chapter 21
THE CHEMISTRY OF THE MAIN GROUP ELEMENTS

CHAPTER OUTLINE

CHAPTER 21: The Chemistry of the Main Group Elements

CHAPTER OBJECTIVES

- Relate the formulas and properties of compounds to the periodic table (21.2)
- Describe the chemistry of the main group of A-Group elements, particularly H; Na and K; Mg and Ca; B and Al; Si; N and P; O and S; and F and Cl (21.3-21.10)
- Apply the principles of Stoichiometry, thermodynamics, and electrochemistry to the chemistry of the main group elements

LESSON PLAN

AP Standards
IV.2   Relationships in the periodic table: horizontal, vertical, and diagonal with examples from the alkali metals, alkaline earth metals, halogens, and the first series of transition elements

Suggested Time:
Two traditional classes or one block. The only section crucial to the AP material is 21.2 which concerns relationships in the periodic table. If time permits, the remainder of the chapter would be beneficial for students to learn because of its application to important economical elements and chemistry, along with environmental issues.

ASSESSMENT

Exercises:

Answers to Exercises p. A59-A60
Practicing Skills pp. 1011-1014, Answers IRM pp. 408-417
General Questions pp. 1014-1016, Answers IRM pp. 417-422
In the Laboratory Questions p. 1016, Answers IRM pp. 422-423
Summary and Conceptual Questions p. 1017, Answers IRM pp. 424-426

GLOSSARY

disproportionation reaction A reaction in which an element or compound is simultaneously oxidized and reduced

MEDIA MENU

The Media Menu section of this guide lists the interactive exercises, tutorials, active figures (animated, interactive figures) and additional resources available online for Chemistry Now, a unique web-based, assessment-centered personalized learning system for chemistry students. Go to www.cengagelearning.com/login.

Active Figures

- 21.17 Industrial production of aluminum
- 21.23 Compounds and oxidation numbers for nitrogen
- 21.33 A membrane cell for the production of NaOH and Cl_2 gas from a saturated, aqueous solution of NaCl (brine)
- Screen 20.12 Quantitative aspects of el

Exercises

- Screen 21.4   The structure of the family of boron-hydrogen compounds
- Screen 21.5   The chemistry of aluminum compounds
- Screen 21.6   The structural chemistry of silicon-oxygen compounds
- Screen 21.8   Sulfur chemistry

CHAPTER 21: The Chemistry of the Main Group Elements

- Screen 21.9   The structural chemistry of sulfur compounds

DIAGNOSTIC QUIZ QUESTIONS

Relate the formulas and properties of compounds to the periodic table (21.2)

1 Predict the chemical formula for the bromate ion.

   a) Br^- b) BrO_4^- c) BrO_3^- d) BrO_2^- e) BrO^-

Answer: c

2. Use your knowledge of group properties to predict the chemical formula for selenous acid.

   a) $HSeO_2$ b) H_2SeO_3 c) H_2SeO_4 d) H_3SeO_3 e) H_3SeO_4

Answer: b

3. Which of the following compound formulas is incorrect?

   a) $Ca(IO_3)_2$ b) Al_2S_3 c) $Al_2(SO_4)_3$ d) $NaHCO_3$ e) Sr_3PO_4

Answer: e

Describe the chemistry of the main group of A-Group elements, particularly H; Na and K; Mg and Ca; B and Al; Si; N and P; O and S; and F and Cl (21.3-21.10)

4. Write a balanced chemical equation for the reaction of potassium and water.

   a) $K(s) + H_2O(\ell) \rightarrow KO(s) + H_2(g)$

   b) $2\ K(s) + 2\ H_2O(\ell) \rightarrow 2\ KOH(aq) + H_2(g)$

   c) $2\ K(s) + 2\ H_2O(\ell) \rightarrow K_2O_2(s) + 2\ H_2(g)$

   d) $2\ K(s) + H_2O(\ell) \rightarrow K_2O(s) + H_2(g)$

   e) $4\ K(s) + 2\ H_2O(\ell) \rightarrow 4\ KH(s) + O_2(g)$

Answer: b

5. Which of the following statements is INCORRECT?

   a) Carbon is classified as a nonmetal.   b) Silicon is classified as a metalloid.
   c) Germanium is classified as a metal.   d) Tin is classified as a metal.
   e) Lead is classified as a metal.

Answer: c

6. In which one of the following compounds does the oxidation number of the nitrogen atom(s) equal -2?

a) NH_3 b) NH_4Cl c) N_2H_4 d) N_2O e) NO

Answer: c

7. Pure oxygen can be produced by the catalyzed decomposition of potassium chlorate. Write a balanced chemical equation for this reaction.

a) $KClO_3(s) \rightarrow KClO(s) + O_2(g)$
b) $2\ KClO_3(s) \rightarrow 2\ K(s) + Cl_2(g) + 3\ O_2(g)$
c) $2\ KClO_3(s) \rightarrow 2\ KCl(s) + 3\ O_2(g)$
d) $2\ KClO_3(s) \rightarrow 2\ KClO_2(s) + O_2(g)$
e) $KClO_3(s) \rightarrow K(s) + ClO(g) + O_2(g)$

Answer: c

Apply the principles of Stoichiometry, thermodynamics, and electrochemistry to the chemistry of the main group elements

8. Ammonium perchlorate can decompose violently according to the chemical equation below.
$$2\ NH_4ClO_4(s) \rightarrow N_2(g) + Cl_2(g) + 2\ O_2(g) + 4\ H_2O(g)$$
What volume of chlorine gas is produced by the decomposition of 125 g NH_4ClO_4 at 575 K and 1.0 atm? ($R = 0.08206$ L·atm/K·mol)

a) 0.532 L   b) 25.1 L   c) 37.7 L   d) 50.2 L   e) 101 L

Answer: b

9. Molten Al_2O_3 is reduced by electrolysis at low potentials and high currents. If 1.5×10^4 amperes of current is passed through molten Al_2O_3 for 8.0 hours, what mass of Al is produced? Assume 100% current efficiency.

a) 1.5×10^3 g   b) 4.0×10^4 g   c) 1.2×10^5 g   d) 1.3×10^5 g   e) 3.6×10^6 g

Answer: b

10. The only acid that reacts with silicon dioxide is __________.

a) CF_3CO_2H b) HF c) HNO_3 d) H_2SO_4 e) CH_3CO_2H

Answer: b

TEST BANK

21.1 Element Abundances

1. Which list contains the three most abundant elements in the Earth's crust?

    a) hydrogen, helium, and lithium
    b) sodium, calcium, and iron
    c) oxygen, nitrogen, and iron
    d) carbon, nitrogen, and oxygen
    e) oxygen, aluminum, and silicon

    Answer: e

2. Which two elements are most abundant in our solar system?

    a) hydrogen and helium
    b) hydrogen and oxygen
    c) helium and iron
    d) hydrogen and iron
    e) carbon and oxygen

    Answer: a

21.2 The Periodic Table: A Guide to the Elements

3. What is the charge on a nonmetallic, monatomic ion from Group 6A?

    a) 6+   b) 2+   c) 2-   d) 4-   e) 6-

    Answer: c

4. What is the highest oxidation state of selenium?

    a) -2   b) 0   c) +4   d) +6   e) +8

    Answer: d

5. Predict the chemical formula for the bromate ion.

    a) Br^- b) BrO_4^- c) BrO_3^- d) BrO_2^- e) BrO^-

    Answer: c

6. Use your knowledge of group properties to predict the chemical formula for selenous acid.

    a) $HSeO_2$ b) H_2SeO_3 c) H_2SeO_4 d) H_3SeO_3 e) H_3SeO_4

    Answer: b

7. Which of the following compound formulas is incorrect?

a) $Ca(IO_3)_2$ b) Al_2S_3 c) $Al_2(SO_4)_3$ d) $NaHCO_3$ e) Sr_3PO_4

Answer: e

21.3 Hydrogen

8. Approximately 300 billion liters of hydrogen is produced annually. The majority of the hydrogen is used in the production of __________.

a) ammonia   b) hydrogen chloride   c) methane
d) sulfuric acid   e) water

Answer: a

9. Which of the following chemical equations shows the formation of synthesis gas?

a) $2\ H_2O(g)\ \rightarrow\ 2\ H_2(g)\ +\ O_2(g)$
b) $16\ CH_4(g)\ +\ 7\ O_2(g)\ \rightarrow\ 2\ C_8H_{18}(g)\ +\ 14\ H_2O(g)$
c) $C(s)\ +\ 2\ H_2(g)\ \rightarrow\ CH_4(g)$
d) $CO(g)\ +\ H_2O(g)\ \rightarrow\ CO_2(g)\ +\ H_2(g)$
e) $C(s)\ +\ H_2O(g)\ \rightarrow\ H_2(g)\ +\ CO(g)$

Answer: e

10. Metal hydrides are used to remove traces of water from nonpolar solvents. Write a balanced chemical equation for the reaction of calcium hydride and water.

a) $CaH_2(s)\ +\ 2\ H_2O(\ell)\ \rightarrow\ Ca(s)\ +\ 2\ H_3O^+(aq)$

b) $CaH_2(s)\ +\ 2\ H_2O(\ell)\ \rightarrow\ Ca(OH)_2(s)\ +\ 2\ H_2(g)$

c) $CaH_2(s)\ +\ H_2O(\ell)\ \rightarrow\ CaO(s)\ +\ 2\ H_2(g)$

d) $CaH_2(s)\ +\ H_2O(\ell)\ \rightarrow\ CaO(s)\ +\ 4\ H^+(aq)$

e) $CaH_2(s)\ +\ 4\ H_2O(\ell)\ \rightarrow\ 2\ CaO_2(s)\ +\ 5\ H_2(g)$

Answer: b

11. All of the following elements or compounds react with water to produce hydrogen gas EXCEPT __________.

a) $K(s)$ b) $MgH_2(s)$ c) $Ca(s)$ d) $LiH(s)$ e) $Cl_2(g)$

Answer: e

21.4 The Alkali Metals, Group 1A

12. All of the following statements concerning sodium are correct EXCEPT

    a) most sodium compounds are water soluble.
    b) sodium is soft and easily cut with a knife.
    c) sodium is usually found in nature as the reduced metal or as a sulfide salt.
    d) sodium ions help regulate osmotic pressure in the human body.
    e) sodium is commercially prepared by electrolysis of molten NaCl.

    Answer: c

13. Which of the following statements concerning potassium is/are CORRECT?
    1. The potassium concentration in seawater is over three times greater than the concentration of sodium.
    2. Potassium is an important factor in plant growth.
    3. Roman soldiers were paid in potassium chloride.

    a) 1 only   b) 2 only   c) 3 only   d) 1 and 3   e) 1, 2, and 3

    Answer: b

14. The principal product of the reaction of sodium and oxygen is not Na_2O. The product is named _________ and has the formula _________.

    a) disodium trioxide, Na_2O_3 b) sodium monoxide, NaO
    c) sodium superoxide, NaO_2 d) sodium peroxide, Na_2O_2
    e) trisodium dioxide, Na_3O_2

    Answer: d

15. Elemental sodium is commercially produced by electrolysis of _________. A valuable by-product, _________, is simultaneously produced during the electrolysis.

    a) $NaCl(\ell)$, $Cl_2(g)$ b) $NaH(g)$, $H_2(g)$ c) $NaCl(aq)$, $NaOH(aq)$
    d) $Na_2O_2(\ell)$, $O_2(g)$ e) $Na_2S(s)$, $S_8(s)$

    Answer: a

16. Which of the following statements is/are CORRECT?
    1. Sodium carbonate (Na_2CO_3) is used in making glass.
    2. Sodium bicarbonate ($NaHCO_3$) is used in cooking.
    3. Lithium carbonate (Li_2CO_3) is used in treating bipolar disorder.

    a) 1 only   b) 2 only   c) 3 only   d) 2 and 3   e) 1, 2, and 3

    Answer: e

17. Write a balanced chemical equation for the reaction of potassium and water.

   a) $K(s) + H_2O(\ell) \rightarrow KO(s) + H_2(g)$

   b) $2\,K(s) + 2\,H_2O(\ell) \rightarrow 2\,KOH(aq) + H_2(g)$

   c) $2\,K(s) + 2\,H_2O(\ell) \rightarrow K_2O_2(s) + 2\,H_2(g)$

   d) $2\,K(s) + H_2O(\ell) \rightarrow K_2O(s) + H_2(g)$

   e) $4\,K(s) + 2\,H_2O(\ell) \rightarrow 4\,KH(s) + O_2(g)$

Answer: b

18. Which of the following statements are CORRECT?
    1.   Chili saltpeter is a common name for $NaNO_3$.
    2.   Trona is a common name for $Na_2CO_3 \cdot NaHCO_3 \cdot 2\,H_2O$.
    3.   Soda ash is a common name for Na_2CO_3.

   a) 1 only   b) 2 only   c) 3 only   d) 1 and 3   e) 1, 2, and 3

Answer: e

21.5 The Alkaline Earth Elements, Group 2A

19. Write a balanced chemical equation for the reaction of lime and water to form slaked lime.

   a) $CaO(s) + 2\,H_2O(\ell) \rightarrow CaCO_3(s) + 2\,H_2(g)$

   b) $CaO(s) + H_2O(\ell) \rightarrow Ca(OH)_2(s)$

   c) $Ca_3(PO_4)_2(s) + 2\,H_2O(\ell) \rightarrow 2\,CaHPO_4(aq) + Ca(OH)_2(aq)$

   d) $CaCO_3(s) + H_2O(\ell) \rightarrow Ca(s) + H_3O^+(aq) + CO_2(g)$

   e) $CaCO_3(s) + H_2O(\ell) \rightarrow CaO(s) + H_2CO_3(aq)$

Answer: b

20. Which of the following statements is INCORRECT?

   a) Diagnostic imaging of the human digestive tract often requires a patient to consume a slurry of $BaSO_4(s)$, a compound that contains the toxic Ba^{2+} ion.
   b) Magnesium oxide is commonly known as lime. It is a primary component of mortar.
   c) Beryllium is toxic. Exposure (by breathing) to beryllium can cause berylliosis.
   d) Most magnesium is obtained from seawater, where it is present at a roughly 0.05 M concentration.
   e) Magnesium is the central element in chlorophyll.

Answer: b

21. At high temperatures limestone decomposes according to which balanced chemical equation?

    a) $Ca(OH)_2(s) \rightarrow CaO(s) + H_2O(g)$
    b) $2\ CaO(s) \rightarrow 2\ Ca(s) + O_2(g)$
    c) $CaCO_3(s) \rightarrow CaO(s) + CO_2(g)$
    d) $Ca(OH)_2(s) \rightarrow CaH_2(s) + O_2(g)$
    e) $2\ CaCO_3(s) \rightarrow 2\ Ca(s) + O_2(g) + CO_2(g)$

Answer: c

22. Oxides of the alkaline earth family form _________ upon reaction with water.

    a) basic solutions   b) acidic solutions   c) alkaline earth hydrides
    d) oxygen gas   e) hydrogen gas

Answer: a

21.6 Boron, Aluminum, and the Group 3A Elements

23. Aluminum is resistant to corrosion because

    a) the surface of aluminum is coated with unreactive AlN.
    b) the surface of aluminum is coated with unreactive Al_2O_3.
    c) aluminum is a noble metal.
    d) aluminum has the greatest density of all naturally occurring metals.
    e) aluminum has a large positive standard reduction potential.

Answer: b

24. Molten Al_2O_3 is reduced by electrolysis at low potentials and high currents. If 1.5×10^4 amperes of current is passed through molten Al_2O_3 for 8.0 hours, what mass of Al is produced? Assume 100% current efficiency.

    a) 1.5×10^3 g   b) 4.0×10^4 g   c) 1.2×10^5 g   d) 1.3×10^5 g   e) 3.6×10^6 g

Answer: b

25. Who developed the process for obtaining aluminum from bauxite ore?

    a) Antoine Lavoisier   b) Fritz Haber   c) Michael Faraday
    d) Herman Frasch   e) Charles Hall

Answer: e

26. In what way is cryolite (Na_3AlF_6) used in the production of aluminum?

a) Cryolite is a catalyst, it increases the rate of aluminum reduction.
b) Cryolite-bauxite mixtures melt at lower temperatures than pure bauxite.
c) Cryolite makes aluminum more corrosion resistant.
d) Cryolite is reduced to aluminum at the cathode.
e) Cryolite is oxidized to $F_2(g)$ at the anode.

Answer: b

27. Which of the following statements concerning boron compounds is INCORRECT?

a) More than 20 neutral boron hydrides (having formulas B_xH_y) are known to exist.
b) Boric acid, $B(OH)_3$, is used as a flame retardant for cellulose insulation.
c) Borax, $Na_2B_4O_7 \cdot 10\ H_2O$, is used as a low-melting flux in metallurgy.
d) One allotrope of boron is an icosohedron of boron atoms.
e) Boron is used as an inert electrode in rechargeable batteries.

Answer: e

21.7 Silicon and the Group 4A Elements

28. The building block of silicate minerals is the

a) linear SiO unit.
c) trigonal-bipyramidal SiO_5 unit.
e) the linear SiO_2 unit.

b) the bent SiO_2 unit.
d) tetrahedral SiO_4 unit.

Answer: d

29. Which of the following statements is INCORRECT?

a) Carbon is classified as a nonmetal.
c) Germanium is classified as a metal.
e) Lead is classified as a metal.

b) Silicon is classified as a metalloid.
d) Tin is classified as a metal.

Answer: c

30. Quartz is the pure crystalline form of

a) silica, SiO_2.
c) calcium orthosilicate, Ca_2SiO_4.
e) silicon carbide, SiC.

b) sodium silicate, $NaSiO_4$.
d) silicon tetrachloride, $SiCl_4$.

Answer: a

CHAPTER 21: The Chemistry of the Main Group Elements

31. The only acid that reacts with silicon dioxide is __________.

    a) CF_3CO_2H b) HF   c) HNO_3 d) H_2SO_4 e) CH_3CO_2H

    Answer: b

21.8 Nitrogen, Phosphorus, and the Group 5A Elements

32. All of the following statements concerning phosphorus are correct EXCEPT

    a) P_4 is an allotrope of phosphorus named white phosphorus.
    b) red phosphorus is a polymer of P_4 units.
    c) phosphorus was first purified from human wastes.
    d) stable $P_4(g)$ provides a nonoxidizing atmosphere for prepackaged foods.
    e) Over 200 compounds containing phosphate ions (PO_4^{3-}) are known to exist.

    Answer: d

33. The Ostwald process is an industrial method of producing __________ from the oxidation of __________.

    a) nitric acid, ammonia   b) ammonia, nitric acid   c) ammonia, nitrogen
    d) nitric acid, nitrogen   e) dinitrogen monoxide, nitrogen

    Answer: a

34. In which one of the following compounds does the oxidation number of the nitrogen atom(s) equal -2?

    a) NH_3 b) NH_4Cl c) N_2H_4 d) N_2O e) NO

    Answer: c

35. Dinitrogen monoxide, or laughing gas, can be made by decomposing ammonium nitrate at 250 °C. Write a balanced equation for this reaction.

    a) $NH_4NO_3(s) \rightarrow N_2O(g) + 2\,H_2(g) + O_2(g)$
    b) $NH_4NO_3(s) + 2\,N_2(g) \rightarrow 3\,N_2O(g) + 2\,H_2(g)$
    c) $NH_4NO_3(s) \rightarrow N_2O(g) + 2\,H_2O(g)$
    d) $2\,NH_4NO_3(s) \rightarrow N_2O(g) + 2\,HNO_2(g) + H_2O(g) + 2\,H_2(g)$
    e) $3\,NH_4NO_3(s) \rightarrow N_2O(g) + 2\,N_2(g) + 8\,H_2O(g)$

    Answer: c

36. Hydrazine, N_2H_4, is produced by the Raschig process –the oxidation of ammonia with alkaline _________.

a) $HNO_3(aq)$ b) $NaOCl(aq)$ c) $N_2(g)$ d) $HPO_4^{2-}(aq)$ e) $N_2O(g)$

Answer: b

37. Which of the following nitrogen oxides are free radicals: NO, N_2O, NO_2, and/or N_2O_3?

a) NO and N_2O b) N_2O and NO_2 c) NO_2 and N_2O_3
d) NO and NO_2 3) N_2O and N_2O_3

Answer: d

38. The active ingredients in "strike anywhere" matches are _________ and the powerful oxidizing agent $KClO_3(s)$.

a) $P_4O_{10}(s)$ b) $P_2O_5(s)$ c) $Na_3PO_4(s)$ d) $H_3PO_4(\ell)$ e) $P_4S_3(s)$

Answer: e

39. Which one of the following phosphorus compounds is a diprotic acid?

a) phosphoric acid; H_3PO_4 b) phosphorous acid; H_3PO_3
c) metaphosphoric acid; $(HPO_3)_3$ d) hypophosphorous acid; H_3PO_2
e) pyrophosphoric acid; $H_4P_2O_7$

Answer: b

40. Which of the following statements is INCORRECT?

a) Phosphoric acid gives a tart taste to carbonated beverages such as colas.
b) Sodium phosphate, Na_3PO_4, is a relatively strong base used in paint strippers.
c) $Ca(H_2PO_4)_2 \cdot H_2O$ is used as an acid leavening agent in baking powder.
d) $Ca_3(PO_4)_2$ is a dark green compound used in latex paints.
e) Calcium monohydrogen phosphate is an abrasive and polishing agent in toothpaste.

Answer: d

21.9 Oxygen, Sulfur and the Group 6A Elements

41. In the United States, most sulfur is obtained from

a) the reduction of cinnabar, HgS.   b) S_8 deposits along the Gulf of Mexico.
c) the reduction of iron pyrite, FeS_2.   d) the reduction of copper sulfide, CuS.
e) the oxidation of H_2S gas.

Answer: b

42. Pure oxygen can be produced by the catalyzed decomposition of potassium chlorate. Write a balanced chemical equation for this reaction.

a) $KClO_3(s) \rightarrow KClO(s) + O_2(g)$
b) $2\ KClO_3(s) \rightarrow 2\ K(s) + Cl_2(g) + 3\ O_2(g)$
c) $2\ KClO_3(s) \rightarrow 2\ KCl(s) + 3\ O_2(g)$
d) $2\ KClO_3(s) \rightarrow 2\ KClO_2(s) + O_2(g)$
e) $KClO_3(s) \rightarrow K(s) + ClO(g) + O_2(g)$

Answer: c

43. A commercially important reaction of sulfur dioxide is its oxidation to _________, which then reacts with water to form sulfuric acid.

a) SO_3 b) H_2S c) S_8 d) H_2SO_3 e) SO_2^+

Answer: a

44. Write a balanced chemical equation for the roasting of CuS.

a) $CuS(s) + 2\ HCl(aq) \rightarrow CuCl_2(aq) + H_2(g) + S(s)$
b) $CuS(s) + O_2(g) \rightarrow CuO_2(s) + S(s)$
c) $CuS(s) + O_2(g) \rightarrow CuO(s) + SO(g)$
d) $CuS(s) + 2\ HCl(aq) \rightarrow CuCl_2(aq) + H_2S(g)$
e) $2\ CuS(s) + 2\ O_2(g) \rightarrow 2\ CuO(s) + SO_2(g)$

Answer: e

21.10 The Halogens, Group 7A

45. Ammonium perchlorate can decompose violently according to the chemical equation below.
$$2\ NH_4ClO_4(s) \rightarrow N_2(g) + Cl_2(g) + 2\ O_2(g) + 4\ H_2O(g)$$
What volume of chlorine gas is produced by the decomposition of 125 g NH_4ClO_4 at 575 K and 1.0 atm? ($R = 0.08206$ L·atm/K·mol)

a) 0.532 L   b) 25.1 L   c) 37.7 L   d) 50.2 L   e) 101 L

Answer: b

46. What is the molecular formula of chloric acid?

a) $HClO$ b) $HClO_2$ c) $HClO_3$ d) $HClO_4$ e) HCl

Answer: c

47. The most reactive nonmetallic element is _________. This element has been shown to form
compounds with every other element except He and Ne.

a) oxygen   b) fluorine   c) lithium   d) phosphorus   e) carbon

Answer: b

48. Anhydrous hydrogen fluoride is used in all of the following applications EXCEPT

a) etching fluorescent light bulbs.   b) the production of refrigerants.
c) the reduction of alkali and alkaline earth metals.   d) the production of plastics.
e) the production of herbicides.

Answer: c

Short Answer Questions

49. A _________ reaction is one where an element or compound is simultaneously oxidized and
reduced.

Answer: disproportionation

50. Orthorhombic sulfur (S_8) and plastic sulfur (S) are examples of two different forms of the
same element. Two or more different forms of an element are called _________.

Answer: allotropes

51. Silicon is purified in a process called _________, where a narrow segment of a silicon rod is
melted and allowed to slowly recrystallize.

Answer: zone refining

52. The primary product of the reaction of potassium and oxygen is _________. This compound is
used in oxygen generators.

Answer: potassium superoxide or KO_2

53. A type of aluminosilicate, called a _________, contains regularly spaced holes and cavities.
This type of aluminosilicate is often used as a catalyst.

Answer: zeolites

54. Concentrated nitric acid reacts with copper to form $NO_2(g)$. Dilute nitric acid reacts with copper to form $NO(g)$. Why does the reaction involving the dilute acid give a different nitrogen oxide than the concentrated acid?

Answer: The oxidation states of the nitrogen in each compound are listed below.

Compound	Oxidation State
HNO_3	+5
NO_2	+4
NO	+2

The more dilute acid contains fewer nitrate ions that can react with the copper. To fully oxidize the copper, each nitrogen needs to gain more electrons. Likewise, because of the large number of nitrate ions present in the concentrated acid, each nitrogen only gains a single electron.

55. One commercial method of producing Cl_2 is by electrolysis of aqueous NaCl.

$$NaCl(aq) + 2\,H_2O(\ell) \rightarrow Cl_2(g) + 2\,NaOH(aq) + H_2(g)$$

Owing to environmental concerns, some manufacturers are shifting to the soda-lime method for NaOH production. Explain why this method is environmentally preferred. Write a balanced chemical equation for the production of NaOH from soda ash and lime.

Answer: The electrolysis method of manufacturing sodium hydroxide also produces chlorine. Chlorine and its by-products are considered environmentally unfriendly substances. The balanced equation for the soda-lime process is below.

$$Na_2CO_3(s) + CaO(s) + H_2O(\ell) \rightarrow 2\,NaOH(aq) \text{ and } CaCO_3(s)$$

Chapter 22
THE CHEMISTRY OF THE TRANSITION ELEMENTS

CHAPTER OUTLINE

CHAPTER 22: The Chemistry of the Transition Elements

CHAPTER OBJECTIVES

- Identify and explain the chemical and physical properties of the transition elements (22.1, 22.2)
- Understand the composition, structure, and bonding in coordination compounds (22.3, 22.4, 22.5)
- Relate ligand field theory to the magnetic and spectroscopic properties of complexes (22.6)
- Apply the effective atomic number rule to simple organometallic compounds of the transition elements (22.7)

LESSON PLAN

AP Standards
I.B.3   Geometry of molecules and ions, structural isomerism, of simple organic molecules and coordination complexes; dipole moments of molecules; relation of properties to structure

Suggested Time:
Two traditional classes or one block. The concepts of isomerism and coordination complexes are the only necessary topics to be covered in this chapter. The remainder can be done, if time permits, as a post-AP lesson for advanced students.

ASSESSMENT

Exercises:
22.1   Formulas of Coordination Compounds (p. 1034)

GLOSSARY

bidentate ligand A ligand which coordinate to a metal ion via two donor atoms with Lewis base attractions

chelating ligands A ligand that forms more than one coordinate covalent bond with the central metal ion in a complex

coordination complex (complex ion) Ions that compose a single structural unit, containing a metal ion and one or more molecules or ions bonded in a coordinate covalent bond

coordination compound A compound in which a metal ion or atom if bonded to one or more molecules or anions to define a structural unit

coordination geometry The geometrical arrangement maintained by the ligands in a coordination compound.

coordination isomers Two or more complexes in which a coordinated ligand and a non-coordinated ligand are exchanged

coordination number The number of ligands attached to the central metal atom in a coordination compound

coordination chemistry The study of coordination compounds

_d_-block elements Transition elements whose occurrence in the periodic table coincides with the filling of the _d_ orbitals

_d_-to-_d_ transition The change that occurs when an electron moves between two orbitals having different eneries in a complex

effective atomic number rule (18-electron rule) Organometallic compounds in which the number of metal valence electrons plus the number of electrons donated by the ligand groups total 18 are likely to be stable.

CHAPTER 22: The Chemistry of the Transition Elements

f-**block elements** Transition elements whose occurrence in the periodic table coincides with the filling of the f orbitals

gangue A mixture of sand and clay in which a desired mineral is usually found

geometric isomers Isomers in which the atoms of the molecule are arranged in different geometric relationships

high spin The electron configuration for a coordination complex with the maximum number of unpaired electrons

hydrometallurgy Recovery of metals from their ores by reactions in aqueous solution

lanthanide contraction The decrease in ionic radius that results from filling of the $4f$ orbitals

ligand The molecules or anions bonded to the central metal atom in a coordination compound

ligand field model A model of metal-ligand bonding in coordination compounds

ligand field splitting The difference in the potential energy between sets of d orbitals in a metal atom or ion surrounded by ligands

linkage isomers Two or more complexes in which a ligand is attached to the metal atom through different atoms

low spin The electron configuration for a coordination complex with the minimum number of unpaired electrons

metallurgy The process of obtaining metals from their ores

monodentate ligand A ligand that coordinates to the metal via a single Lewis base atom

optical isomers Isomers that are nonsuperimposable mirror images of each other

ore A sample of matter containing a desired mineiral or element, usually with large quantities of impurities

pairing energy The additional potential energy due to the electrostatic repulsion between two electrons in the same orbital

pyrometallurgy Recovery of metals from their ores by high-temperature processes

spectrochemical series An ordering of ligands by the magnitudes of the splitting energies they cause

structural isomers Two or more compounds with the same molecular formula but with different atoms bonded to each other

MEDIA MENU

The Media Menu section of this guide lists the interactive exercises, tutorials, active figures (animated, interactive figures) and additional resources available online for Chemistry Now, a unique web-based, assessment-centered personalized learning system for chemistry students. Go to www.cengagelearning.com/login.

Active Figures
- Figure 22.8   A blast furnace

Exercises
- Screen 22.2   Transition metal compounds
- Screen 22.5   Compound geometries
- Screen 22.6   Isomerism in coordination chemistry

Additional Resources
- Chapter 22 Contents   Simulation of bonding and electronic structure in transition metal complexes
- Screen 22.8   Simulation on the absorption and transmission of light by transition metal complexes

DIAGNOSTIC QUIZ QUESTIONS

Identify and explain the chemical and physical properties of the transition elements

1. What is the highest oxidation state for manganese?

    a) +3   b) +6   c) +7   d) +8   e) +10

    Answer: c

2. Which element from the first transition series does NOT react with hydrochloric acid?

    a) titanium   b) chromium   c) cobalt   d) nickel   e) copper

    Answer: e

3. All of the following metals may be found in nature as free elements EXCEPT __________.

    a) Ir   b) Au   c) Pd   d) Zn   e) Cu

    Answer: d

CHAPTER 22: The Chemistry of the Transition Elements

Understand the composition, structure, and bonding in coordination compounds

4. Ions such as $[Co(H_2O)_6]^{3+}$ and $[Ag(CN)_2]^-$ are called __________.

    a) ligands
    d) alloys
    b) Lewis bases
    e) coordination complexes
    c) chelates

    Answer: e

5. What is the name of the compound having the formula $K_3[CoCl_6]$?

    a) potassium chlorocobaltate(III)
    b) potassium hexachlorocobaltate(III)
    c) potassium cobaltohexachlorate(II)
    d) potassium cobaltohexachlorate(III)
    e) tripotassium hexachlorochromate(II)

    Answer: b

6. What are the possible geometries of a metal complex with a coordination number of 6?
    1. square planar
    2. tetrahedral
    3. octahedral

    a) 1 only b) 2 only c) 3 only d) 1 and 2 e) 1, 2, and 3

    Answer: c

Relate ligand field theory to the magnetic and spectroscopic properties of complexes

7. All of the following statements concerning crystal field theory are true EXCEPT

    a) in low-spin complexes, electrons are concentrated in the d_{xy}, d_{yz}, and d_{xz} orbitals.

    b) in an isolated atom or ion, the five d orbitals have identical energy.

    c) low-spin complexes contain the maximum number of unpaired electrons.

    d) the crystal field splitting is larger in low-spin complexes than high-spin complexes.

    e) the energy difference between d orbitals often corresponds to an energy of visible light.

    Answer: c

8. What is the number of unpaired electrons in an octahedral, high-spin Mn(II) complex?

    a) 1 b) 2 c) 3 d) 4 e) 5

    Answer: e

9. The spectrochemical series

a) lists ligands in order from monodentate to hexadentate.
b) lists ligands in order of their tendency to split *d* orbitals.
c) lists coordination compounds in order of their color.
d) lists complex ions in order of their color.
e) lists ligands in the order of their electronegativity.

Answer: b

Apply the effective atomic number rule to simple organometallic compounds of the transition elements

10. According to the effective atomic number rule, Ru(0) should coordinate to ______ carbonyl ligands.

a) 4   b) 5   c) 6   d) 7   e) 8

Answer: b

TEST BANK

22.1 Properties of the Transition Elements

1. Which element has the highest melting point?

a) Zn   b) Ni   c) W   d) Fe   e) Hg

Answer: c

2. ________ is a crucial component of vitamin B_{12}.

a) Fe   b) Co   c) Zn   d) Mg   e) Cd

Answer: b

3. Which of the following transition metals is a constituent of Prussian Blue, a pigment used in engineering blueprints?

a) Ti   b) Cu   c) Cr   d) Co   e) Fe

Answer: e

4. Which ion produces the red color in rubies?

a) Mn^{2+} b) Zr^{4+} c) Pd^{2+} d) Cr^{3+} e) Ti^{4+}

Answer: d

5. Which of the following ions have an $[Ar]3d^6$ electron configuration: Co^{2+}, Fe^{2+}, and Ni^{2+}?

   a) Co^{2+} only   b) Fe^{2+} only   c) Ni^{2+} only   d) Co^{2+} and Fe^{2+} e) Co^{2+} and Ni^{2+}

   Answer: b

6. Identify the element with the ground state electron configuration $[Xe]4f^{14}5d^76s^2$.

   a) Ir   b) Ta   c) Os   d) Rh   e) Au

   Answer: a

7. What is the highest oxidation state for manganese?

   a) +3   b) +6   c) +7   d) +8   e) +10

   Answer: c

8. Which element from the first transition series does NOT react with hydrochloric acid?

   a) titanium   b) chromium   c) cobalt   d) nickel   e) copper

   Answer: e

9. Which of the following statements is/are CORRECT?
   1. The atomic radii of transition metals increase initially, then decrease as one moves from left to right across a period of the periodic table.
   2. The atomic radii of the *d*-block elements in the 5th and 6th periods in each group are almost identical.
   3. The densities of the period 5 transition metals are greater than those from either the period 4 or period 6 transition metals.

   a) 1 only   b) 2 only   c) 3 only   d) 1 and 2   e) 1, 2, and 3

   Answer: b

10. The product of the oxidation of iron in most soil in the absence of excess oxygen is black magnetite. The chemical formula of this compound is __________.

   a) FeO   b) FeO_2 c) Fe_2O_3 d) Fe_3O_2 e) Fe_3O_4

   Answer: e

11. If an element can be obtained profitably from a mineral, the mineral is called a(n) _________.

   a) gangue   b) ore   c) slag   d) roast   e) alloy

   Answer: b

12. Ores are often found mixed with impurities such as sand and clay. In metallurgy, these impurities are called _________.

   a) gangue   b) alloys   c) slags   d) roasts   e) precipitates

   Answer: a

13. All of the following metals may be found in nature as free elements EXCEPT _________.

   a) Ir   b) Au   c) Pd   d) Zn   e) Cu

   Answer: d

14. All of the following metals are generally found in nature as either oxides or sulfides, EXCEPT _________.

   a) Fe   b) Al   c) Cr   d) Pt   e) Pb

   Answer: d

15. What element or compound is used to reduce Fe_2O_3 to Fe in a blast furnace?

   a) $H_2(g)$ b) $Cl_2(g)$ c) $C(s)$ d) $CaSiO_3(s)$ e) $CO_2(g)$

   Answer: c

16. The flexibility, strength, hardness, and malleability of carbon steel is affected by a heating and cooling process called _________.

   a) tempering   b) flotation   c) reduction   d) oxidation   e) recrystallization

   Answer: a

17. Copper ores are enriched to increase the percentage of copper by a process called _________.

   a) electrolysis b) flotation   c) roasting   d) tempering   e) filtering

   Answer: b

CHAPTER 22: The Chemistry of the Transition Elements

22.3 Coordination Compounds

18. Ions such as $[Co(H_2O)_6]^{3+}$ and $[Ag(CN)_2]^-$ are called __________.

   a) ligands   b) Lewis bases   c) chelates
   d) alloys   e) coordination complexes

   Answer: e

19. Which molecule or ion does NOT act as a ligand?

   a) H_2O b) NH_4^+ c) Br^- d) OH^- e) CO

   Answer: b

20. All of the following molecules or ions can act as polydentate ligands EXCEPT __________.

   a) phenanthroline; $C_{12}H_8N_2$ b) acetylacetate ion; $CH_3COCHCOCH_3^-$
   c) oxalate ion; $C_2O_4^{2-}$ d) ethylenediamine; $H_2NCH_2CH_2NH2$
   e) dimethylamine; $(CH_3)_2NH_2$

   Answer: e

21. What is the coordination number of the central metal ion in $[Fe(H_2O)_4(CN)_2]Cl$?

   a) 2   b) 3   c) 4   d) 6   e) 7

   Answer: d

22. What is the coordination number of the central metal ion in $[Co(C_2O_4)_2(OH)_2]^{3-}$?

   a) 2   b) 4   c) 6   d) 8   e) 10

   Answer: c

23. What is the oxidation state of iron in $K[Fe(NH_3)_2(CN)_4]$?

   a) -1   b) +1   c) +2   d) +3   e) +6

   Answer: d

24. What is the oxidation state of cobalt in $[Co(NH_3)_3(H_2O)_2Cl]Cl_2$?

   a) +1   b) +2   c) +3   d) +4   e) +6

   Answer: c

25. What is the name of the compound having the formula $K_3[CoCl_6]$?

    a) potassium chlorocobaltate(III)   b) potassium hexachlorocobaltate(III)
    c) potassium cobaltohexachlorate(II)   d) potassium cobaltohexachlorate(III)
    e) tripotassium hexachlorochromate(II)

Answer: b

26. What is the name of the compound having the formula $[Cr(en)_2(NH_3)_2]Cl_2$?

    a) dihydroxodiethlyenediamminechromate(II) chloride
    b) diamminebis(ethylenediamine)chlorochromate(II)
    c) bis(ethylenediamine)diamminechromium(II) dichloride
    d) diamminebis(ethylenediamine)dichlorochromium(II)
    e) diamminebis(ethylenediamine)chromium(II) chloride

Answer: e

27. What is the name of the compound having the formula $(NH_4)_4[Ni(CN)_6]$?

    a) ammonium hexacyanonickelate(II)   b) tetraammonium hexacyanonickelate(II)
    c) tetraamminehexacyanonickelate(II)   d) hexacyanotetraamminenickel(II)
    e) hexacyanonickelate(II) ammonia

Answer: a

28. What is the name of the compound having the formula $[Fe(CO)_6](NO_3)_3$?

    a) hexacarbonmonoxideiron(III) nitrate
    b) hexacarbonmonoxideiron(II) nitrogen trioxide
    c) trinitratehexacarbonylironate(III)
    d) hexacarbonyliron(II) nitrite
    e) hexacarbonyliron(III) nitrate

Answer: e

29. What is the formula for dibromobis(ethylenediamine)titanium(IV) bromide?

    a) $[Ti(en)_2]Br_4$ b) $[TiBr_4(en)_2]$ c) $Br_2[TiBr_2(en)_2]$
    d) $[TiBr_2(en)_2]Br_2$ e) $[CrBr(en)_2]Br_3$

Answer: d

30. What is the formula for potassium diamminetetrachlorovanadate(III)?

a) $[KV(NH_3)_2Cl_4]$

b) $K_2[V(NH_3)_2Cl_4]$

c) $K[V(NH_3)_2Cl_4]$

d) $[KV(NH_3)_2]Cl_4$

e) $K[V(NH_3)_2]Cl_4$

Answer: c

22.4 Structures of Coordination Compounds

31. What are the possible geometries of a metal complex with a coordination number of 6?
    1. square planar
    2. tetrahedral
    3. octahedral

a) 1 only   b) 2 only   c) 3 only   d) 1 and 2   e) 1, 2, and 3

Answer: c

32. Which of the following pairs of coordination compounds are coordination isomers?

a) $[Ni(H_2O)_6]SO_4$ and $[Ni(H_2O)_6]SO_3$
b) $[Fe(NH_3)_5Br]SO_4$ and $[Fe(NH_3)_5SO_4]Br$
c) $[Co(NH_3)_4(H_2O)_2]Cl_2$ and $[Co(NH_3)_2(H_2O)_4]Cl_2$
d) $[Fe(NH_3)_6]Br_2$ and $[Fe(NH_3)_6]Br_3$
e) $[Fe(NH_3)_6]Cl_2$ and $[Co(NH_3)_6]Cl_2$

Answer: b

33. Which of the following species have geometric isomers: $[Fe(CN)_6]^{3-}$, $[Fe(CN)_5H_2O]^{2-}$, $[Fe(CN)_4(H_2O)_2]$, and $[Fe(CN)_3(H_2O)_3]$?

a) $[Fe(CN)_6]^{3-}$ only
b) $[Fe(CN)_5H_2O]^{2-}$ only
c) $[Fe(CN)_3(H_2O)_3]$ only
d) $[Fe(CN)_4(H_2O)_2]^-$ and $[Fe(CN)_3(H_2O)_3]$
e) $[Fe(CN)_5H_2O]^{2-}$, $[Fe(CN)_4(H_2O)_2]^-$, and $[Fe(CN)_3(H_2O)_3]$

Answer: d

34. Which of the following compounds may have linkage isomers?

a) $[Cr(H_2O)_4Cl_2]Cl$

b) $[Cr(H_2O)_4Cl_2]NO_2$

c) $[Cr(H_2O)_5(NO_2)]SO_4$

d) $[Cr(H_2O)_5Cl](SCN)_2$

e) $[Cr(NH_3)_4Cl_2]Cl$

Answer: c

35. Which one of the following complex ions has an optical isomer?

a) $[Cu(CN)_4]^{2-}$ b) $[Zn(phen)_2]^{2+}$ c) $[Zn(NH_3)_2en]^{2+}$
d) $[Co(H_2O)_4en]^{2+}$ e) $[Ni(en)_3]^{2+}$

Answer: e

36. How many geometric isomers may exist for the square-planar complex ion $[Pt(CN)_2Cl_2]^{2-}$?

a) 2 b) 3 c) 4 d) 5 e) 6

Answer: a

37. Which of the following statements is/are CORRECT?
    1. Complex ions with the structure $[ML_2]^{n\pm}$ are always linear.
    2. Complex ions with the structure $[ML_4]^{n\pm}$ are always tetrahedral.
    3. Complex ions with the structure $[ML_6]^{n\pm}$ are always octahedral.

a) 1 only b) 2 only c) 3 only d) 1 and 3 e) 2 and 3

Answer: d

22.5 Bonding in Coordination Compounds

38. In an octahedral complex, electrons in d_{z^2} and $d_{x^2-y^2}$ orbitals experience a greater repulsion from the lone pairs of electrons on ligands because these orbitals

a) contain more electrons than the other three d orbitals.
b) each contain one unpaired electron.
c) are oriented directly toward the incoming ligand electron pairs.
d) are located a greater distance from the metal ion nucleus.
e) are roughly spherical in shape, much like an s orbital.

Answer: c

39. All of the following statements concerning crystal field theory are true EXCEPT

a) in low-spin complexes, electrons are concentrated in the d_{xy}, d_{yz}, and d_{xz} orbitals.

b) in an isolated atom or ion, the five d orbitals have identical energy.
c) low-spin complexes contain the maximum number of unpaired electrons.
d) the crystal field splitting is larger in low-spin complexes than high-spin complexes.
e) the energy difference between d orbitals often corresponds to an energy of visible light.

Answer: c

40. What is the number of unpaired electrons in an octahedral, high-spin Mn(II) complex?

    a) 1 b) 2 c) 3 d) 4 e) 5

    Answer: e

41. Determine the number of unpaired electrons in an octahedral, low-spin Cr(II) complex.

    a) 1 b) 2 c) 3 d) 4 e) 5

    Answer: b

42. If an octahedral cobalt(III) complex is paramagnetic, which of the following sets of conditions best describes the complex?

    a) low-spin, Δ_0 small b) low-spin, Δ_0 large c) high-spin, Δ_0 small
    d) high-spin, Δ_0 large e) none of the above

    Answer: c

43. As bound ligands, which of the following causes the largest splitting of d-orbitals?

    a) phen b) H_2O c) I^- d) Cl^- e) $C_2O_4^{2-}$

    Answer: a

44. Determine the number of unpaired electrons in the tetrahedral complex ion $[Co(NH_3)_4]^{2+}$.

    a) 0 b) 1 c) 2 d) 3 e) 4

    Answer: d

45. Determine the number of unpaired electrons in a Ni(II) tetrahedral complex.

    a) 0 b) 1 c) 2 d) 3 e) 4

    Answer: c

46. All of the following complexes will be colored (in the visible) EXCEPT _________.

    a) $[Zn(en)_3]^{2+}$ b) $[Fe(CN)_6]^{4-}$ c) $[Ni(H_2O)_6]^{2+}$ d) $[Cr(NH_3)_6]^{3+}$ e) $[Cu(NH_3)_4]^{2+}$

    Answer: a

47. The spectrochemical series

    a) lists ligands in order from monodentate to hexadentate.
    b) lists ligands in order of their tendency to split d orbitals.
    c) lists coordination compounds in order of their color.
    d) lists complex ions in order of their color.
    e) lists ligands in the order of their electronegativity.

Answer: b

22.7 Organometallic Chemistry: The Chemistry of Low-Valent Metal-Organic Complexes

48. The effective atomic number (EAN) rule states that organometallic coordination compounds are likely to be stable if

    a) the oxidation state of the metal is zero.
    b) the sum of the metal valence electrons plus the electrons donated by the ligand equals 18.
    c) the ligand field splitting is small.
    d) the metal has an even atomic number.
    e) the metal has an odd atomic number.

Answer: b

49. According to the effective atomic number rule, Mo(0) should coordinate to ______ carbonyl ligands.

    a) 4   b) 5   c) 6   d) 7   e) 8

Answer: c

50. According to the effective atomic number rule, Ru(0) should coordinate to ______ carbonyl ligands.

    a) 4   b) 5   c) 6   d) 7   e) 8

Answer: b

Short Answers Questions

51. The period 6 transition metals have greater densities than either the period 4 or 5 transition metals. An explanation concerns the $4f$ orbitals, which are filled with electrons just prior to the $5d$ orbitals. As the $4f$ orbitals are filled, the radii of the elements decrease. The decrease in size is referred to as the ________ contraction.

Answer: lanthanide

52. One method of recovering metals from their ores involves aqueous solutions; a second method involves heat. The method involving heat is called __________.

    Answer: pyrometallurgy

53. Phenanthroline ($C_{12}H_8N_2$) can bind to metal ions using both its ammine groups. It is referred to as a __________ ligand, meaning it is "two-toothed."

    Answer: bidentate

54. $EDTA^{4-}$, a hexadentate ligand, is most effective at chelating metal ions when the pH is greater than 12. Explain why low pH decreases the chelating ability of the ligand.

    Answer: At low pH, $EDTA^{4-}$ is protonated (in successive fashion) forming $HEDTA^{3-}$, H_2EDTA^{2-}, H_3EDTA^-, H_4EDTA (and even H_5EDTA^+ and H_6EDTA^{2+}). Once protonated, a chelating ligand site loses it ability to form a coordinate-covalent bond with a metal ion.

55. The complex ion $[NiCl_4]^{2-}$ is paramagnetic, but the complex ion $[Ni(NH_3)_4]^{2+}$ is diamagnetic. Why do the magnetic properties of these two nickel complexes differ?

    Answer: $[NiCl_4]^{2-}$ must have a tetrahedral geometry. In this geometry, the *d*-orbitals are split into two energies, with 3 orbitals (d_{xy}, d_{yz}, and d_{xz}) higher than the other two orbitals (d_{z^2} and $d_{x^2-y^2}$). Two of nickel's 8 *d* electrons are unpaired. The tetraammine complex must be square planar. In this geometry, the *d* orbitals split into four energy levels. The highest energy orbital is unoccupied. The lower orbitals are fi

Chapter 23
NUCLEAR CHEMISTRY

CHAPTER OUTLINE

CHAPTER OBJECTIVES

- Identify radioactive elements, and describe natural and artificial nuclear reactions (23.1, 23.2, 23.3)
- Calculate the binding energy and binding energy per nucleon for a particular isotope (23.3)
- Understand the rates of radioactive decay (23.4)
- Understand artificial nuclear reactions (23.6 and 23.7)
- Understand issues of health and safety with respect to radioactivity (Sections 23.8 and 23.9)
- Be aware of some radioactive isotopes in science and medicine (23.9)

LESSON PLAN

AP Standards
I.C.1   Nuclear equations
I.C.3   Radioactivity; chemical applications

Suggested Time:
Two traditional classes or one block. Section 23.2 and the concept of half-life need to be covered. There are some interesting applications to medicine and the nuclear industry that can be taught post-AP or for those interested in these fields.

ASSESSMENT

Exercises:

Answers to Exercises p. A61
Practicing Skills pp. 1091-1093, Answers IRM pp. 442-448
General Questions pp. 1093-1094, Answers IRM pp. 448-449
In the Laboratory Questions p. 1094, Answers IRM pp. 449-450
Summary and Conceptual Questions p. 1095, Answers IRM pp. 450-452

GLOSSARY

activity(A) A measure of the rate of nuclear decay, the number of disintegrations observed in a sample per unit time

alpha particle A helium nucleus ejected at high speed from certain radioactive substances

alpha radiation Radiation that is readily absorbed

antineutrinos Massless, chargeless particles that are emitted accompanying beta decay

band of stability Narrow range of isotope stability demonstrated by a graph of the number of neutrons versus the number of protons

becquerel The SI unit of radioactivity, 1 decomposition per second

beta particle An electron ejected at high speed from certain radioactive substances

beta radiation Radiation of a penetrative character

binding energy per mol nucleon The energy required to separate a nucleus into individual protons and neutrons, divided by the number of nucleons

chain reaction A reaction in which each step generates a reactant to continue the reaction

curies A unit or radioactivity

electron capture A nuclear process in which an innershell electron is captured

gamma radiation High energyelectromagnetic radiation

gray (Gy) The SI unit of radiation dosage

mass defect The "missing mass" equivalent to the energy that holds nuclear particles together

neutron activation analysis An analytical technique which is nondestructive where a sample is irradiated with neutrons, causing it to emit gamma rays

neutrinos Massless, chargeless particles which are concurrently emitted by a nucleus undergoing positron emission

nuclear binding energy The energy required to separate the nucleus the nucleus of an atom into protons and
neutrons

nuclear fission The highly exothermic process by which very heavy nuclei split to form lighter nuclei

nuclear fusion A reaction in which several small nuclei react to form a larger nucleus

nuclear reactions A reaction involving one or more atomic nuclei, resulting in ta change in the identities of the isotopes

nuclear reactor A container in which a controlled nuclear reaction occurs

nucleon A nuclear particle, either a neutron or a proton

positron emission Emission of a nuclear particle having the same mass as an electron but a positive charge

rad (radiation absorbed dose) A unit of radiation dosage; 1 rad = 0.01 J of energy absorbed per kilogram of tissue

radioactive decay series A series of nuclear reactions by which a radioactive isotope decays to form a stable isotope

rem (roentgen equivalent man) A unit of radiation dosage to biological tissue

roentgen A unit of radiation dosage

sievert (Sv) The SI unit of radiation dosage to biological tissue; 1 Sv = 1 rem

transuranium elements Elements with atomic numbers greater than 92

MEDIA MENU

The Media Menu section of this guide lists the interactive exercises, tutorials, active figures (animated, interactive figures) and additional resources available online for Chemistry Now, a unique web-based, assessment-centered personalized learning system for chemistry students. Go to www.cengagelearning.com/login.

Tutorials
- Screen 23.2   Balancing equations for nuclear reactions
- Screen 23.3   Modes of radioactive decay
- Screen 23.3   Predicting modes of radioactive decay
- Screen 23.5   Calculating binding energy
- Screen 23.6   Half-life and radiochemical dating

Exercises
- Screen 23.3   The Geiger counter

Additional Resources
- Screen 23.4   Simulation on isotope stability

DIAGNOSTIC QUIZ QUESTIONS

Identify radioactive elements, and describe natural and artificial nuclear reactions

1. Place alpha (α), beta (β), and gamma (γ) radiation in order from least to most penetrating.

   a) $\alpha < \beta < \gamma$ b) $\alpha < \gamma < \beta$ c) $\beta < \alpha < \gamma$
   d) $\beta < \gamma < \alpha$ e) $\gamma < \beta < \alpha$

   Answer: a

2. By what (single step) process does erbium-160 change to holmium-160?

   a) α particle emission   b) β particle emission   c) gamma ray emission
   d) electron capture   e) neutron capture

   Answer: d

3. If a nucleus emits an alpha particle

    a) its atomic number decreases by four and its mass number decreases by two.
    b) its atomic number decreases by two and its mass number decreases by four.
    c) its atomic number decreases by two and its mass number decreases by two.
    d) its atomic number decreases by four and its mass number is unchanged.
    e) its atomic number is unchanged and its mass number decreases by four.

Answer: b

Calculate the binding energy and binding energy per nucleon for a particular isotope

4. The point of maximum stability in the binding energy curve occurs in the vicinity of which one of the following isotopes?

    a) $^{12}_{6}C$　　　　b) $^{56}_{26}Fe$　　　　c) $^{4}_{2}He$　　　　d) $^{209}_{83}Bi$　　　　e) $^{244}_{94}Pu$

Answer: b

5. The molar nuclear mass of boron-10 is 10.012937 g/mol. The molar mass of a proton is 1.007825 g/mol. The molar mass of a neutron is 1.008665 g/mol. Calculate the binding energy (in J/mol) of B-10. ($c = 2.998 \times 10^8$ m/s)

    a) 2.084×10^7 J/mol　　　　b) 6.248×10^{12} J/mol　　　　c) 9.000×10^{13} J/mol
    d) 9.000×10^{14} J/mol　　　　e) 6.248×10^{15} J/mol

Answer: b

Understand the rates of radioactive decay

6. The half-life of carbon-14 is 5730 years. What percentage of carbon-14 remains in a sample after 1.50×10^4 years?

    a) 0.0%　　　　b) 0.13%　　　　c) 6.1%　　　　d) 16%　　　　e) 38%

Answer: d

7. Uranium-235 has a half-life of 7.04×10^8 years. How many years will it take for 99% of a U-235 sample to decay?

    a) 6.5×10^{-9} yr　　　b) 7.1×10^8 yr　　　c) 3.2×10^9 yr　　　d) 4.7×10^9 yr　　　e) 7.0×10^{10} yr

Answer: d

Understand artificial nuclear reactions

8. What particle(s) are produced in the following reaction?
$$^{238}_{92}U + ^{16}_{8}O \rightarrow ^{250}_{100}Fm + \underline{}$$

a) $^{1}_{+1}p$ b) $2\ ^{0}_{-1}\beta$ c) $4\ ^{0}_{+1}\beta$ d) $^{4}_{2}He$ e) $4\ ^{1}_{0}n$

Answer: e

9. Which of the following statements is/are CORRECT?
    1. Naturally occurring isotopes account for only a small fraction of known radioactive isotopes.
    2. Nuclear reactions involving neutron capture followed by gamma ray emission are referred to as (n, γ) reactions.
    3. Bombarding nuclei with high energy beta particles is an efficient means of creating heavier isotopes.

a) 1 only   b) 2 only   c) 3 only   d) 1 and 2   e) 1, 2, and 3

Answer: d

10. Complete the following fission reaction.
$$^{1}_{0}n + ^{235}_{92}U \rightarrow ^{236}_{92}U \rightarrow ^{141}_{56}Ba + \underline{} + 3\ ^{1}_{0}n$$

a) $^{92}_{36}Kr$ b) $^{95}_{36}Kr$ c) $^{92}_{36}U$ d) $^{95}_{95}Am$ e) $^{95}_{35}Br$

Answer: a

Understand issues of health and safety with respect to radioactivity. Be aware of some radioactive isotopes in science and medicine

11. In boron neutron capture therapy (BNCT), boron-10 is injected into a tumor. When the tumor is irradiated with neutrons, boron nuclei capture the neutrons and disintegrate into two particles. These particles are lithium-7 and a(n) __________.

a) alpha particle   b) beta particle   c) positron
d) deuterium nuclei   e) helium nuclei

Answer: a

TEST BANK

<u>23.1 Natural Radioactivity</u>

1. Place alpha (α), beta (β), and gamma (γ) radiation in order from least to most penetrating.

a) $\alpha < \beta < \gamma$ b) $\alpha < \gamma < \beta$ c) $\beta < \alpha < \gamma$
d) $\beta < \gamma < \alpha$ e) $\gamma < \beta < \alpha$

Answer: a

2. Which symbol represents an alpha particle?

a) $^{0}_{+1}e$ b) $^{0}_{-1}e$ c) $^{1}_{0}n$ d) $^{0}_{0}\gamma$ e) $^{4}_{2}He$

Answer: e

3. Which symbol represents a positron?

a) $^{4}_{2}He$ b) $^{0}_{-1}\beta$ c) $^{1}_{0}n$ d) $^{0}_{0}\gamma$ e) $^{0}_{+1}\beta$

Answer: e

<u>23.2 Nuclear Reactions and Radioactive Decay</u>

4. Which of the following reactions is an example of positron emission?

a) $^{232}_{90}Th \rightarrow {}^{228}_{88}Ra + {}^{4}_{2}He$ b) $^{7}_{4}Be + {}^{0}_{-1}e \rightarrow {}^{7}_{3}Li$

c) $^{14}_{6}C \rightarrow {}^{14}_{7}N + {}^{0}_{-1}\beta$ d) $^{139}_{62}Sm \rightarrow {}^{139}_{61}Pm + {}^{0}_{+1}\beta$

e) $^{58}_{26}Fe + {}^{1}_{0}n \rightarrow {}^{59}_{26}Fe$

Answer: d

5. Which of the following reactions is an example of electron capture?

a) $^{82}_{35}Br \rightarrow {}^{82}_{36}Kr + {}^{0}_{-1}\beta$ b) $^{135}_{57}La + {}^{0}_{-1}e \rightarrow {}^{135}_{56}Ba$

c) $^{58}_{26}Fe + {}^{1}_{0}n \rightarrow {}^{59}_{26}Fe$ d) $^{227}_{89}Ac \rightarrow {}^{223}_{87}Fr + {}^{4}_{2}He$

e) $^{7}_{3}Li + {}^{1}_{1}p \rightarrow 2\,{}^{4}_{2}He$

Answer: b

6. If a nucleus undergoes beta particle emission

   a) its atomic number decreases by one and its mass number is unchanged.
   b) its atomic number is unchanged and its mass number increases by one.
   c) its atomic number increases by one and its mass number is unchanged.
   d) both its atomic number and its mass number are unchanged.
   e) its atomic number is unchanged and its mass number decreases by one.

   Answer: c

7. If a nucleus emits an alpha particle

   a) its atomic number decreases by four and its mass number decreases by two.
   b) its atomic number decreases by two and its mass number decreases by four.
   c) its atomic number decreases by two and its mass number decreases by two.
   d) its atomic number decreases by four and its mass number is unchanged.
   e) its atomic number is unchanged and its mass number decreases by four.

   Answer: b

8. If a nucleus gains a neutron and then undergoes electron capture,

   a) its atomic number decreases by one and its mass number increases by one.
   b) its atomic number is unchanged and its mass number increases by one.
   c) both its atomic number and its mass number are unchanged.
   d) its atomic number increases by one and its mass number is unchanged.
   e) its atomic number increases by one and its mass number increases by one.

   Answer: a

9. If a nucleus decays by successive α, β, α particle emissions, its atomic number will

   a) decrease by eight. b) decrease by five. c) decrease by four.
   d) decrease by three. e) decrease by two.

   Answer: d

10. What nucleus decays by successive α, β, β, α emissions to produce uranium-236?

    a) $^{243}_{95}\text{Am}$ b) $^{242}_{96}\text{Cu}$ c) $^{246}_{96}\text{Cu}$ d) $^{240}_{90}\text{Th}$ e) $^{244}_{94}\text{Pu}$

    Answer: e

11. If californium-253 decays by successive β, α, β emissions, what nucleus is produced?

    a) $^{249}_{98}\text{Cf}$ b) $^{249}_{94}\text{Pu}$ c) $^{247}_{97}\text{Bk}$ d) $^{253}_{92}\text{U}$ e) $^{249}_{100}\text{Fm}$

    Answer: a

12. What nucleus is produced if palladium-106 decays by electron capture?

    a) $^{105}_{46}Pd$ b) $^{106}_{45}Rh$ c) $^{105}_{47}Ag$ d) $^{106}_{47}Ag$ e) $^{107}_{45}Rh$

    Answer: b

13. What nucleus decays by beta emission to produce tin-127?

    a) $^{128}_{50}Sn$ b) $^{126}_{50}Sn$ c) $^{127}_{49}In$ d) $^{126}_{48}Cd$ e) $^{127}_{48}Cd$

    Answer: c

14. Complete the following nuclear reaction.
    $$^{100}_{45}Rh + ^{0}_{-1}\beta \rightarrow \underline{\quad}$$

    a) $^{101}_{45}Rh$ b) $^{100}_{46}Pd$ c) $^{100}_{44}Ru$ d) $^{99}_{44}Ru$ e) $^{99}_{45}Rh$

    Answer: c

15. By what (single step) process does radon-222 decay to polonium-218?

    a) α particle emission   b) β particle emission   c) positron emission
    d) electron capture   e) neutron capture

    Answer: a

16. By what (single step) process does erbium-160 change to holmium-160?

    a) α particle emission   b) β particle emission   c) gamma ray emission
    d) electron capture   e) neutron capture

    Answer: d

23.3 Stability of Atomic Nuclei

17. All of the following statements concerning nuclei are true EXCEPT

    a) only hydrogen-1 and helium-3 have more protons than neutrons.
    b) from He to Ca, stable nuclei have roughly equal numbers of protons and neutrons.
    c) isotopes with a low neutron to proton ratio always decay by alpha particle emission.
    d) the neutron to proton ratio in stable nuclei increases as mass increases.
    e) beyond calcium, the neutron to proton ratio is always greater than one for stable nuclei.

    Answer: c

18. A plot of the number of neutrons versus the number of protons in nuclei shows a narrow
band of stable isotopes. _________ occurs for isotopes that have a high neutron to proton
ratio.

a) beta emission.
d) gamma ray emission.
b) positron emission.
e) neutron capture.
c) electron capture.

Answer: a

19. The point of maximum stability in the binding energy curve occurs in the vicinity of which
one of the following isotopes?

a) $^{12}_{6}C$
b) $^{56}_{26}Fe$
c) $^{4}_{2}He$
d) $^{209}_{83}Bi$
e) $^{244}_{94}Pu$

Answer: b

20. The molar nuclear mass of boron-10 is 10.012937 g/mol. The molar mass of a proton is
1.007825 g/mol. The molar mass of a neutron is 1.008665 g/mol. Calculate the binding
energy (in J/mol) of B-10. ($c = 2.998 \times 10^8$ m/s)

a) 2.084×10^7 J/mol
d) 9.000×10^{14} J/mol
b) 6.248×10^{12} J/mol
e) 6.248×10^{15} J/mol
c) 9.000×10^{13} J/mol

Answer: b

21. Calculate the energy released (per mole of tritium consumed) for the following fusion
reaction,

$$^{4}_{2}He + ^{3}_{1}H \rightarrow ^{6}_{3}Li + ^{1}_{0}n$$

given the following molar masses of nucleons and nuclei. ($c = 2.998 \times 10^8$ m/s)

particle	mass (g/mol)
proton	1.007825
neutron	1.008665
tritium	3.01605
helium-4	4.00260
lithium-6	6.01512

a) 1.54×10^3 J/mol
d) 4.62×10^{11} J/mol
b) 3.57×10^{11} J/mol
e) 3.86×10^{14} J/mol
c) 3.86×10^{11} J/mol

Answer: d

23.4 Rates of Nuclear Decay

22. Cobalt-56 has a half-life of 78 days. Starting with 1.42 mg of this isotope, how much would remain after 312 days?

  a) 0.0 mg   b) 0.044 mg   c) 0.089 mg   d) 0.18 mg   e) 0.36 mg

  Answer: c

23. The half-life for the spontaneous decay of iodine-131 is 8.04 days. How much of a 0.50 g sample of this isotope remains after 3.0 days?

  a) 0.11 g   b) 0.13 g   c) 0.19 g   d) 0.28 g   e) 0.39 g

  Answer: e

24. The half-life of thorium-226 is 31 minutes. What percentage of thorium-226 remains in a sample after 6.0 hours?

  a) 0.032%   b) 0.086%   c) 0.87%   d) 8.6%   e) 87%

  Answer: a

25. The half-life of carbon-14 is 5730 years. What percentage of carbon-14 remains in a sample after 1.50×10^4 years?

  a) 0.0%   b) 0.13%   c) 6.1%   d) 16%   e) 38%

  Answer: d

26. Uranium-235 has a half-life of 7.04×10^8 years. How many years will it take for 99% of a U-235 sample to decay?

  a) 6.5×10^{-9} yr   b) 7.1×10^8 yr   c) 3.2×10^9 yr   d) 4.7×10^9 yr   e) 7.0×10^{10} yr

  Answer: d

27. What is the half-life of an isotope if the decay constant is 1.6×10^{-2} yr^{-1}?

  a) 4.1 yr   b) 4.8 yr   c) 19 yr   d) 43 yr   e) 63 yr

  Answer: d

28. $^{241}_{95}$Am is used in many home smoke alarms. If 33% of the americium in a smoke detector decays in 690 years, what is the half-life of this isotope?

a) 1.0×10^{-3} yr   b) 1.4×10^2 yr   c) 2.3×10^2 yr   d) 3.6×10^2 yr   e) 4.3×10^2 yr

Answer: e

29. The number of disintegrations per minute (dsm) of a sample of iodine-133 is measured at 5.5 $\times 10^2$ dsm. When the sample is measured 605 minutes later, the number of disintegrations has decreased to 3.9×10^2 dsm. What is the half-life of iodine-133?

a) 5.7×10^{-4} m.   b) 2.6 m.   c) 1.2×10^3 m.   d) 3.9×10^2 m.   e) 9.7×10^4 m.

Answer: c

30. A sample of magnesium-28 is found to have an activity of 750 disintegrations per hour (dph). After 6.0 hours the activity has decreased to 620 dph. What is the rate constant for the decay of magnesium-28?

a) 0.014 hr^{-1} b) 0.032 hr^{-1} c) 0.14 hr^{-1} d) 0.81 hr^{-1} e) 5.0 hr^{-1}

Answer: b

23.5 Artificial Nuclear Reactions

31. Which of the following statements is/are CORRECT?
    1. Naturally occurring isotopes account for only a small fraction of known radioactive isotopes.
    2. Nuclear reactions involving neutron capture followed by gamma ray emission are referred to as (n, γ) reactions.
    3. Bombarding nuclei with high energy beta particles is an efficient means of creating heavier isotopes.

a) 1 only   b) 2 only   c) 3 only   d) 1 and 2   e) 1, 2, and 3

Answer: d

23.6 Nuclear Fission

32. Complete the following fission reaction.
$$^{1}_{0}n + {}^{235}_{92}U \rightarrow {}^{236}_{92}U \rightarrow {}^{141}_{56}Ba + \underline{\quad} + 3\,^{1}_{0}n$$

a) $^{92}_{36}$Kr   b) $^{95}_{36}$Kr   c) $^{92}_{36}$U   d) $^{95}_{95}$Am   e) $^{95}_{35}$Br

Answer: a

33. Enriched uranium is uranium that has a greater proportion of _________.

    a) cadmium-112   b) uranium-235   c) uranium-238   d) radium-226   e) plutonium-248

    Answer: b

34. Which two nuclei are the two fissionable isotopes most commonly used in nuclear reactors?

    a) $^{235}_{92}$U and $^{239}_{94}$Pu   b) $^{238}_{92}$U and $^{232}_{90}$Th   c) $^{235}_{92}$U and $^{238}_{93}$U

    d) $^{235}_{92}$U and $^{226}_{88}$Ra   e) $^{232}_{90}$Th and $^{239}_{94}$Pu

    Answer: a

35. What role do the cadmium control rods play in a fission reactor?

    a) The rods control the rate of fission by absorbing neutrons.
    b) The cadmium combines with spent uranium fuel to produce a non-radioactive product.
    c) The rods focus the neutrons toward the center of the reactor.
    d) The cadmium acts as a catalyst, enabling fission to occur at lower temperatures.
    e) The rods move forward and backward, driving the pistons that turn the turbines.

    Answer: a

36. What do scientists call the sequence of rapidly occurring reactions that results when a nuclear fission reaction produces enough neutrons to produce more fission reactions?

    a) nucleation   b) binding energy   c) disintegration
    d) (n, γ) reactions   e) chain reaction

    Answer: e

23.7 Nuclear Fusion

37. Which of the following elements undergoes nuclear fusion to provide the primary source of energy from the sun?

    a) helium   b) uranium   c) hydrogen   d) carbon   e) boron

    Answer: c

38. Complete the following fusion reaction.
    $$^{3}_{1}\text{H} + ^{9}_{4}\text{Be} \rightarrow ^{1}_{0}\text{n} + \underline{\hspace{1cm}}$$

    a) $^{11}_{4}$Be   b) $^{12}_{6}$C   c) $^{12}_{7}$N   d) $^{12}_{5}$B   e) $^{11}_{5}$B

    Answer: e

39. What particle(s) are produced in the following reaction?
$$^{238}_{92}U + {}^{16}_{8}O \rightarrow {}^{250}_{100}Fm + \underline{\quad}$$

a) $^{1}_{+1}p$ b) $2\ ^{0}_{-1}\beta$ c) $4\ ^{0}_{+1}\beta$ d) $^{4}_{2}He$ e) $4\ ^{1}_{0}n$

Answer: e

40. What percentage of the world's electricity is supplied by nuclear fusion reactors?

a) 0%   b) 3%   c) 8%   d) 17%   e) 39%

Answer: a

41. At the high temperatures necessary for fusion, matter exists as a plasma, which is

a) a liquid form of a radioactive element.
b) a gaseous cloud of electrons and positrons.
c) a mixture of unbound nuclei and electrons.
d) an equal mixture of matter and anti-matter.
e) a dense solid composed of neutrons.

Answer: c

23.8 Radiation and Human Health

42. For an average person, which of the following is likely to be the **most** significant source of exposure to radiation?

a) cosmic radiation
b) transuranium elements found naturally in human tissue
c) americium in smoke detectors
d) emissions from nuclear power plants
e) fallout from nuclear testing

Answer: a

43. The becquerel is an SI unit for the measurement of radiation. One Bq represents

a) 1 J energy absorbed per kg of tissue.
b) 3.7×10^{10} disintegrations per second.
c) 1 disintegration per second.
d) 0.01 J energy absorbed per kg of tissue.
e) 1 calorie energy absorbed per kg of tissue.

Answer: c

23.9 Applications of Nuclear Chemistry

44. Which of the following isotopes is used in the treatment of thyroid disorders?

a) $^{67}_{31}$Ga

b) $^{201}_{81}$Tl

c) $^{10}_{5}$B

d) $^{18}_{9}$F

e) $^{131}_{53}$I

Answer: e

45. In positron emission tomography (PET), a positron emitted from an unstable isotope travels a short distance before it is annihilated by a(n) __________, creating two gamma rays that travel in opposite directions.

a) proton
d) alpha particles

b) neutron
e) tachyon

c) electron

Answer: c

46. In boron neutron capture therapy (BNCT), boron-10 is injected into a tumor. When the tumor is irradiated with neutrons, boron nuclei capture the neutrons and disintegrate into two particles. These particles are lithium-7 and a(n) __________.

a) alpha particle
d) deuterium nuclei

b) beta particle
e) helium nuclei

c) positron

Answer: a

Short Answer Questions

47. All isotopes of atomic number greater than __________ are unstable and radioactive.

Answer: 83

48. Technetium-99*m* is routinely used in medical imaging. The italics m means the nucleus is __________.

Answer: metastable

49. __________ emission tomography uses gamma rays produced by matter-antimatter annihilations to map regions of interest in the human body.

Answer: Positron

50. Neutron __________ analysis is a non-destructive process in which a sample is irradiated with neutrons. The neutrons react with nuclei to form isotopes with masses one unit higher than the original nuclei. The nuclei are formed in excited states and they emit gamma radiation that can be used to both identify the presence of an element and quantify how much is present.

Answer: activation

51. Explain the difference between 1 rad and 1 rem of radiation.

Answer: One radiation absorbed dose (rad) equals 0.01 J of energy absorbed per kilogram of tissue. One roentgen equivalent man (rem) equals one ram multiplied by the quality factor for the type of radiation involved. The quality factor accounts for variations in tissue damaging power of different types of radiation.

52. Explain the treatment named boron neutron capture therapy (BNCT).

Answer: Boron-10 is injected into tumors and then bombarded with neutrons. Boron-10 is effective at capturing neutrons. The resulting boron-11 nuclei disintegrate to produce alpha particles and lithium-7 atoms. The alpha particles destroy tissue in the localized area where the boron was injected.